Atlas of terrestrial mammals of the Ionian and Aegean islands

Edited by Marco Masseti

Atlas of terrestrial mammals of the Ionian and Aegean islands

Edited by Marco Masseti

DE GRUYTER

Editor

Marco Masseti
Laboratories of Anthropology and Ethnology
Department of Evolutionistic Biology "Leo Pardi" of the University of Florence
Via del Proconsolo, 12
50122 Florence
Italy

This book has 167 figures and 6 tables.

Front cover image
Cretan wild goats or *agrimia*, on the islet of Dhia, off the northern shores of Crete. September 1979 (photo by Marco Masseti)

ISBN 978-3-11-025457-0
e-ISBN 978-3-11-025458-7

Library of Congress Cataloging-in-Publication Data
A CIP catalogue record for this book is available from the Library of Congress

Bibliografic information published by the Deutsche Nationalbibliothek

The Deutsche Nationalbibliothek lists this publication in the Deutsche Nationalbibliografie; detailed bibliographic data are available in the Internet at http://dnb.d-nb.de.

Printing and Binding: Hubert & Co. GmbH & Co. KG, Göttingen
Printed in Germany

www.degruyter.com

To my wife, Silvia

Preface

By Jean-Denis Vigne, CNRS-MNHN, Paris

The *"Atlas of terrestrial mammals of the Ionian and Aegean islands"* fills a gap in the literature. Because of the physical and political fragmentation of the Aegean and Ionian areas and because of the diversity of the 44 species which live on these hundreds of islands, the zoological information in this area was scattered over more than 1,200 books, dissertations and journal articles. Marco Masseti diligently collected and studied this dispersed literature over a period of more than 20 years, and finally presents it here in a remarkably synthetic, comprehensive and very attractive book. This is an impressive source of information, for the most part presented species by species (chap. 6), with a map and extensive information and references for each of the 44 species. This main chapter is completed by a chapter on dubiously or erroneously recorded species (chap. 7), another on several exotic species of monkey, antelope or large carnivores which have been recorded on these islands in the archaeo(zoo)logical documentation (chap. 8), and appendixes on some Aegean domestic breeds (appendix A), and on non-terrestrial mammals (bats and marine mammals; appendix B–D).

Thus, this book is a reference and a splendid tool for scientists, not only for those who work on the North Central Mediterranean area, but also for those like me who are mainly interested in other areas of the Mediterranean, South Europe, the Near East or North Africa.

However, this Atlas also has a lot of other qualities. It is not one of those ever more frequent compilations of literature where all data are equally presented with no distinction between primary and secondary sources, or without any critical approach or general vision. Marco Masseti checked or even produced a lot of these data by himself, since he is one of the leading experts of Aegean zoology and zoological literature. He travelled throughout the entire studied area, trapping mammals and dissecting thousands of owl pellets on more than 40 islands or islets. He spent countless days working on the collections of all the natural History Museums throughout Europe. He knows most of the living authors working in this area and has enjoyed endless discussions with them. He himself is the author of nearly 200 scientific publications, in which he has elucidated numerous biogeographical enigmas such as, to cite a recent example, the mystery surrounding the presence of the porcupine on the Northern shore of the Mediterranean. The more substantial original and attractive contributions that he makes for some of the species, such as rabbits, small carnivores, fallow deer or wild goats, stem directly from his own research, in such a way that the reader gains access in the same book to most of the personal scientific results of Marco Masseti.

What is particularly original and also contributes to making this book easy to read (even for the general public, I presume), is that Marco Masseti presents us with a genuine work of an Encyclopaedist, bringing together in simple words and with often impressive figures or photos (e.g. the photo of the Cretan wild cat on page 131), all kinds of information, beyond the scope of the different disciplines. Taking advantage of his extensive culture and of his tireless curiosity, he blends classical morphological or biological observations, genetics, systematics, historical reports or iconography, ethnozoology, as well as archaeological and paleontological data. This approach gives the Atlas both marked historical and anthropological dimensions, which are the two main components of modern biogeography, and the two main conditions for reaching the goal that Marco Masseti was aiming towards, i.e., a strong contribution to the ongoing debates about biological invasions and biodiversity conservation, especially in areas endangered by the development of tourism and economic pressures. This allows the author to develop, in chap. 9, an interesting reflection about the biogeographical pattern and processes in these distinctive Mediterranean areas, where the islands are so numerous that the life of the different civilisations (namely the Classic Greek civilisation) and of the animal communities was organized within a sea rather than within a continent. This also allows Marco Masseti to raise a very interesting open discussion about the decision to eradicate the modern invasive species on some islands, or to preserve them as both a biological and a cultural legacy, a kind of inevitable result of the natural and cultural history of the Mediterranean.

One of the main products of this Atlas is the high number of new biogeographical scenarios which are proposed by the author. All of them are exciting pieces of anthropozoology and some of them will contribute to bridging the history between the Near East and Europe. This is especially striking for species such as rats and mice, or the European fallow deer.

The quality and innovative nature of such works can be measured by the number of unsolved questions that they raise. The Atlas of Marco Masseti highlights *a lot* of gaps in our knowledge, e.g. about the origins of Pleistocene faunas and about the early occupation of the "true" islands by Humans. For me, however, the two most conspicuous gaps for the Aegean and Ionian area are our poor empirical knowledge concerning the evolution of the shorelines during the Holocene, and the excessively small number of large well dated archaeological bone collections with reference to the multiplicity and diversity of island contexts.

The lack of information concerning the shorelines makes it difficult to resolve the question of the biogeographical history of numerous species or islands. Marco Masseti rightfully points out this difficulty, first in the general introductive presentation of the current issues of island biogeography, then concerning several species such as the moles, the voles, the jackal or the leopard. Were these species trapped on islands which were separated from the mainland because of the post-glacial transgression, and are they thus relicts of ancient populations or communities? Or did they invade the island/islet overseas at some stage after the isolation of the island? The answer is crucial not only for understanding the historical processes and for constructing a general image of the biogeography of the studied areas, but also for making decisions related to biology conservation. We would not interpret the significance of the presence on Euboea of species such as the Eurasian pygmy shrew, Millers' water shrew, the grey hamster, and the common, the Thomas' and the snow voles which are only attested on this island in the same way if we had clear data concerning the time at which this large island was definitively separated from the continent. On a more local scale, this also applies to the Mesolithic red deer on the Sporades island of Youra. And in the present state of palaeogeographical data, it is difficult to put forward pertinent global interpretations of the biogeographical evolution of terrestrial mammals in the Cyclades islands, which probably made up a large central Aegean island at the maximum marine regression, but for which we have no accurate information concerning the time(s) of their separations. As emphasized by Marc Masseti in the natural setting (chap. 3), the strong neotectonic and sedimentological dynamics in these areas are too fast to simply refer to modern bathymetric maps as a guide for shoreline reconstructions.

It is not surprising that the second main gap of knowledge concerns the archaeozoological data, in such a high number of islands. I myself (not alone!) spent years and years trying to establish the succession of the introductions and extinctions of the terrestrial mammal species on just two large islands (Corsica and Cyprus), and two islets (Zembra and Lavezzi), for a very imperfect result. Undertaking the same research for the hundreds of islands of the Ionian and Aegean areas would take decades, and will often yield disappointing results, because of strong soil erosion and archaeological destruction. However, it is surprising that in such a rich and active archaeological area as the Aegean Sea, there is so little and sometimes such badly contextualized archaeozoological information. This is certainly due to the scant attention that most archaeologists paid to the animal (and plant) remains of the Classical and historical sites for such a long time. A good deal of this information has doubtlessly been irremediably lost, but there is still time to make a strong concerted effort for systematically collecting with the appropriate techniques (including water sieving for small remains), and for accurately dating and studying the animal bone series of all the sites in these areas, whatever the time period, including the most recent ones. This is the key for answering the pendant questions and testing the hypothesis that Marco Masseti formulates.

Hopefully, in addition to the fantastic leap forward that it represents for the knowledge of the historical biogeography of the north central Mediterranean, this book will stimulate such new developments in Greek and Turkish geomorphological and archaeological research.

August 2012

Jean-Denis Vigne,

Senior Researcher of the CNRS
Director of the laboratory "Archaeozoology, Archaeobotany"
of the National Museum of Natural History, Paris
Director of the Excellence Laboratory
"Biological and Cultural Diversity"
Silver Medal of the CNRS

Acknowledgments

This study has been partially made possible by the financial support of the EEC Human Capital and Mobility Project CHRX-CT94-0597 *"Adaptations of the hunter-gatherer groups to environmental modifications in the circum-Mediterranean area"*, in cooperation with The Hellenic Zoological Society (Athens), The Hellenic Ornithological Society (Athens), and also thanks to the facilities provided by the former Environmental Organisation of the Municipality of the city of Rhodes. It is also based on contacts with the Ministry of Agriculture of the Hellenic Republic, the Forest Department of Rhodes, the Public Forest Service of Crete, the Forest Station of Lemnos, the Forest Department of Lesbos, the Forest Department of Lefkada, the Goulandris Natural History Museum (Athens), the Department of Biology of the University of Athens, the Department of Biology of the University of Crete, the Department of Environmental Studies of the University of the Aegean, the Department of Agriculture and Rural Environment, University of Thessaly, the Game Fund, Ministry of Interior of Cyprus, the Turkish General Direction of National Parks, Hunting Activities and Wildlife (Ankara), the Forest Department of Antalya, the Direction of the Termessos National Park (Antalya); the Akdeniz University of Antalya; and the Istituto Agronomico per l'Oltremare of the Italian Ministry for Foreign Affairs (Florence, Italy).

While preparing this volume I have been fortunate in having had the help and the assistance of many friends and colleagues. I am particularly grateful to Awatef Abiadh, Laboratoire d'écologie animale, faculté des sciences de Tunis; Triandophilos Akriotis, University of the Aegean (Lesbos, Greece); Tamer Albayrak, Department of Biology, Mehmet Akif Ersoy University, Burdur (Turkey); Haralambos Alivizatos, Athens (Greece); Cilia Antoniou, Department of Genetics and Molecular Biotechnology, Hellenic Centre for Marine Research. Heraklion, Crete (Greece); Fausto Barbagli, Zoological Musuem "La Specola" of the University of Florence; Andrea Bonetti, Pylos and Lugano; Vassilis P. Chondropoulos, Section of Animal Biology, University of Patras (Greece); Anna M. De Marinis, Istituto Superiore per la Protezione e la Ricerca Ambientale ISPRA, Ozzano dell'Emilia (Bologna), Italy; Simon J.M. Davis, Instituto Português de Arqueologia, Lisbon (Portugal); Anastasia Devetzi, Society for the Promotion of Studies on Prehistoric Thera, Athens; Maria Dimaki, The Goulandris Natural History Museum, Kifissia (Athens, Greece); Achilleas Dimitropoulos, The Goulandris Natural History Museum, Kifissia (Athens, Greece); Florian Fladerer, Prähistorische Kommission, Österreichische Akademie der Wissenschaften, Wien, Austria; Luigi Forte, Museo-Orto Botanico, University of Bari; Jakob Fric, Hellenic Ornithological Society (HOS), Athens (Greece); Stella E. Fraguedakis-Tsolis, Section of Animal Biology, University of Patras (Greece); Martin Gaetlich, Athens (Greece); Giorgos Giannatos, Department of Zoology of the University of Athens; Xavier Grémillet, Sizun (France); S. Güsar, Samsundag National Park (Izmir, Turkey); George Hadrinos, Department of National Park and Wildlife Management, Ministry of Agriculture of the Hellenic Republic, Athens; Yannis Ioannidis, The Goulandris Natural History Museum, Kifissia (Athens, Greece); Niki Kardakari, Hellenic Ornithological Society, Athens (Greece); Max Kasparek, Kasparek Verlag, Heidelberg (Germany); M. Süleyman Kaçar, Southwest Anatolia Forest Reasearch Institute, Antalya

(Turkey); Boris Kryštufekk, Slovenian Museum of Natural History, Ljubljana (Slovenia); Adrian Lister and Victoria Herridge, Department of Palaeontology, Natural History Museum (London); Pietro Lo Cascio, Associazione Nesos, Lipari (Messina, Italy); Petros Lymberakis, Myoses Milonas, Kalloust Paragamian, and Apostolos Trichas, Natural History Museum of Crete, Herakleion (Greece); Manolis Mandragos, Rhodes (Greece); Toula Marketou, 22nd Ephorate of Prehistoric and Classical Antiquity, Rhodes; Frieder Mayer, Juliane Schaer, and Saskia Jancke, Museum für Naturkunde Leibniz-Institut für Evolutions- und Biodiversitätsforschung an der Humboldt-Universität zu Berlin; Lyvia Morgan, Brigthon (UK); Joannis Chr. Ondrias, Section of Animal Biology, University of Patras (Greece); Beytullah Özkan, Faculty of Arts and Sciences, Departement of Biology of the Trakya University, Edirne (Turkey); Aliki Panou, Archipelagos, Environment and development. Argostoli (Chefallonia); Costas Papaconstantinou, Hellenic Ornithological Society, Athens (Greece); Carlo Peretto, Dipartimento di Scienze Geologiche e Paleontologiche of the University of Ferrara (Italy); Harald Pieper, Zoologisches Museum Kiel (Germany); Kostas Poirazidis, Technological Educational Institute of Ionian Islands. Department of Environmental Technology and Ecology, Zakynthos; Thomas Rathgeber, National History Museum of Stuttgart; David S. Reese, Guilford, CT; Dietrich Ristow, Neubiberg (Germany); Adamantios Sampson, University of the Aegean, Rhodes; Stefania Sani, Istituto Agronomico per l'Oltremare of the Italian Ministry for Foreign Affairs, Florence; Maurizio Sarà, Department of Animal Biology of the University of Palermo (Italy); Sevket Sen, Paléoenvironnements, Muséum National d'Histoire Naturelle, Département Histoire de la Terre, Paris; Spyros Sfenthourakis, Department of Biology of the University of Patras (Greece); Athanasios Sfougaris, Department of Agriculture, University of Thessaly (Volos, Greece); Franz Suchentrunk, Research Institute of Wildlife Ecology. University of Veterinary Medicine, Vienna; Andreas Sioulas, National Center for Marine Research, Hydrobiologiocal Station of Rhodes; Efstratios D. Valakos, Department of Biology of the University of Athens; Costantinos Vallianos, Museum of Cretan Ethnology (Vory, Crete); Zaphiris Vaos, Folk and Historic Museum of Milos; Jean-Denis Vigne, Muséum National d'Histoire Naturelle, Paris; and Barbara Wilkens, Dipartimento di Storia dell'Università di Sassari (Italy).

Special thanks are due to Paul Mazza, Department of Earth Sciences, University of Florence, for the critical reading of the manuscript and for her suggestions, to Anastasios Aliferis, Municipality of Tilos; Costas Papaconstantinou, Hellenic Ornithological Society, Athens; Thomas Roussos, Public Forest Service of Crete; the rangers of the Forest Stations of Lesbos and Lemnos, Michali Kaikis, Municipality of Rhodes; Philippos Kontizis and Katerina Kaikis, Forest Direction of the Dodecanese; Nikos Theodoridis, Decentrated Administration of Aegean, Directorate General of Forest and Agriculture, Piraeus (Greece) for the invaluable logistic assistance and in the field; to K. Trantalidou, for the enjoyable and stimulating discussions on the mammal remains provided by the archaeological exploration of the Aegean islands, to Anastasios Legakis, Zoological Museum, Department of Biology, University of Athens (Greece), and Christiane Denys, Department of Systématics and Evolution, Muséum National d'Histoire Naturelle, Paris, without whose help this study would never have been completed and probably not even begun.

Finally, special thanks are due to Maria Corsini-Foka, Hellenic Center for Marine Research, Hydrobiological Station of Rhodes, to her husband Sotiris Fokas, Municipality of Rhodes, and to my wife Silvia, for their patience in having put up with me and my research over the years.

Contents

1 Introduction

Εὕδουσι δ' ὀρέων κορυφαί τε καὶ φάραγγες
πρώονές τε καὶ χαράδραι
φῦλά τ' ἑρπέτ' ὅσα τρέφει μέλαινα
γαῖα
θῆρές τ' ὀρεσκῷοι καὶ γένος μελισσᾶν καὶ
κνώδαλ' ἐν βένθεσσι πορφυρέας ἁλός
εὕδουσι δ' οἰωνῶν φῦλα τανυπτερύγων.

(Alcman, 7th century BC)

And the mountain-peaks are asleep and
the ravines,
the headlands and the torrent-beds,
all the creeping tribes that the black earth
nourishes,
the wild animals of the mountains,
the race of bees and the monsters in the depths
of the surging sea;
and the tribes of long-winged birds are asleep

(Alcman, 7th century BC)

The Mediterranean may perhaps be considered the place that has been most intensively and at length affected by the activity of man. The succession of different civilisations over a period of 10 000–12 000 years has entirely modified the landscapes, disrupted or even destroyed the native biocenoses and introduced new animals and plants, to the extent that practically no ecosystem has been left intact. Since prehistory, human settlers have brought about a radical turnover between ancient and modern mammalian faunas, introducing a variety of allochthonous continental species. In particular, the data available for the islands point to endemic mammalian extinction being largely the result of human activities of land clearance and the introduction of allochthonous animals. Today this fauna is no longer characterised by the majority of the endemic mammals previously reported. It displays a practically identical species composition to the continent, consisting almost exclusively of continental mammals whose appearance on the islands has essentially been influenced by man [1–8], and dominated by generalist species (compare with [9]).

The invasion of ecosystems by exotic taxa is currently viewed as one of the most important causes of the loss of biodiversity, and today the terrestrial non-flying mammalian fauna of most of the Mediterranean islands displays an undoubtedly homogeneous composition. It consists mainly of species that are more or less common to the current fauna of all the islands, and reveals a generic continental origin [1, 3–6] influenced by the faunistic composition of the nearest mainland (compare with [8–9]).

In the light of current knowledge, less than about one-fourth of the mammalian species found in the continental Mediterranean region have been described as endemic to the area (compare with [10]). The number of endemics decreases drastically, however, if we consider the composition of the extant fauna on the islands. In fact, in comparison to the Pleistocene, it can be observed that the extant insular mammalian fauna is richer in species but with a greatly reduced level of endemics (compare with [6, 11]). Apart from certain native taxa identified from a few islands such as the Siculo-Maltese insular complex (Italy and Malta), Crete (Greece) and perhaps Cyprus, the almost total absence of endemic species among the current insular mammals is quite surprising. The extant endemics of the Mediterranean islands are restricted to just a very few taxa, comprising two species of shrew, the Sicilian shrew, *Crocidura sicula* Miller, 1900 [12–14] and the Cretan white-toothed shrew,

Crocidura zimmermanni Wettstein, 1953 [15–17], and possibly the recently discovered Cypriot mouse, *Mus cypriacus* [18–19] (see ▶Fig. 1). Recent morphometric [20] and genetic [21] studies have shown in fact that the previously presumed *Dipodillus zakariai* Cockrum, Vaughan and Vaughan, 1976 [22], endemic of the Kerkennah islands can be regarded as a synonym of the continental North-African lesser short-tailed gerbil, *Gerbillus simoni* Lataste, 1881. All these elements appear as the relics, the last remnants of the much richer Pleistocene endemic mammalian faunas that characterised the lost world of the Mediterranean islands.

Several years ago, it was decided to carry out a study designed to integrate existing knowledge on the non-flying terrestrial mammals of the Ionian and Aegean islands, including not only those comprised within the Greek territory, but also those belonging to Albania and Turkey. The aim of this work is to outline the occurrence and the distribution, and provide a mapping of the non-volant mammals of these territories, in order to offer a starting point for future studies. The study also considers the role played by man in the exploitation of the insular environment, and more specifically, in the alteration of the original mammalian faunas of the islands. It is primarily for this reason that only the insular non-flying terrestrial mammals, that is, those species unable to overcome physical barriers such as even short stretches of sea without human help, have been taken into consideration. It is above all on the islands that the impact of extraneous elements on the unspoilt ecological system can be identified and its chronology specified with sufficient precision, due to the evidence left and the relative rapidity of the consequences produced (compare with [23]).

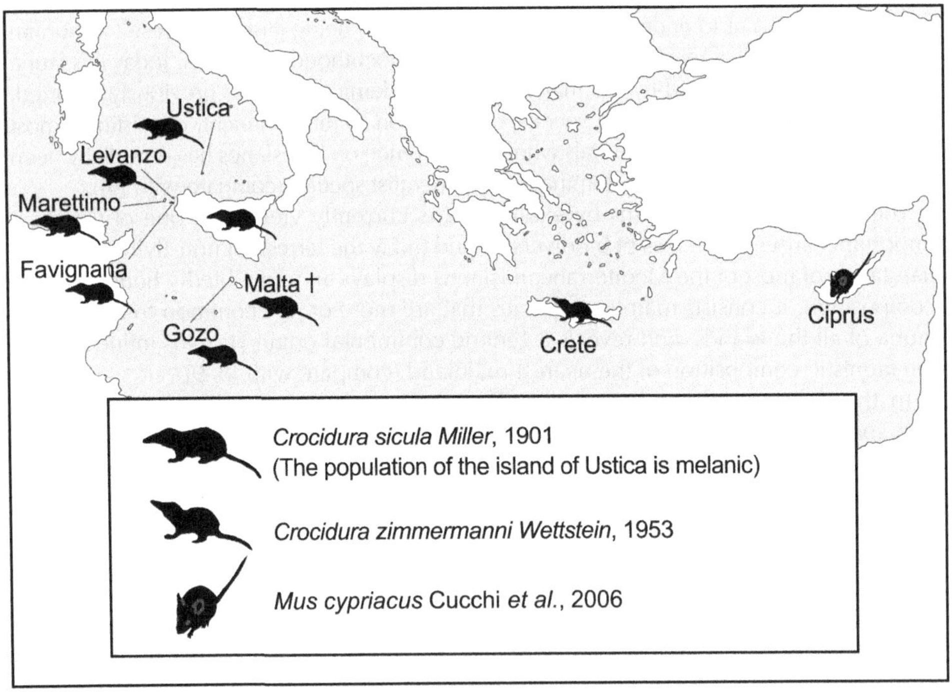

Fig. 1: The extant endemic mammals of the Mediterranean islands are restricted to just a very few species, including two shrew and one mouse.

The non-volant mammalian fauna of the Ionian and Aegean islands is to date still imperfectly known and scientific publications are relatively few. Up to the beginning of the 1970s, most of the faunistic studies were carried out by foreign scholars, although in the last 40 years these have been taken over by Greek zoologists [24]. According to Legakis [25], the visits of foreign scientists and their publications increased largely from the end of the 19th century on. During this period the unknown Ionian and Aegean islands attracted the interest of European researches; one of the most significant developments occurred with the "invasion" of Italian scientists after Dodecanese archipelago came under Italian rule in 1912. Nevertheless, although on the increase, biogeographical studies on the Ionian and Aegean islands are still few, and are unsatisfactory in terms of the extent of current knowledge and the need to identify and preserve island biodiversity hotspots [24]. Even in the early '90s, for example, in the compilation of the global list of islands inhabited by rabbits, Flux and Fullagar [26] demonstrated that they were aware of the distribution of the species on only a very few islands of the Aegean Sea, namely Delos, Makria, Pachia, Elasa, Chios, and possibly Anaphi, whereas the occurrence of these lagomorphs in the remaining Aegean islands was generically referred to the, not better identified, Cyclades. The Albanian and Greek islands of the Ionian Sea were not even taken into consideration, given the evident total lack of information in this respect. However, the last decades of the 20th century and the years immediately following witnessed an interesting production of studies and scientific publications. Extant knowledge of these appears heterogeneous, the studies carried out up to now being referable to episodic research mostly in specific areas, or to specialist investigations on taxonomy and systematics, morphometrics, genetics, ecology and distribution. These are, however, almost all studies that consider the insular mammals within the broader context of continental mammalian fauna. Objectively speaking, apart from some papers on the Rhodian fallow deer, *Dama dama dama* Linnaeus, 1758 [27–32] (see ▶Fig. 2), the wild goat, *Capra aegagrus*

Fig. 2: The Rhodian fallow deer, *Dama dama dama* Linnaeus, 1758, has been the subject of a number of scientific studies in the course of the last decades. Photo by Marco Masseti.

Erxleben, 1777 [33–45] (see ►Fig. 3), the Cretan spiny mouse, *Acomys cahirinus* Desmarest, 1819 [46–47] (see ►Fig. 4), and a very small number of other taxa [48], as far as is presently known, very few specialist studies on the insular mammals appear to be available.

Complete data on the non-flying terrestrial mammalian compositions are, instead, available for Corfu [49], Zakynthos [50], Milos [51], Kythera [52–53], Crete [54], Karpathos [55–56], Rhodes [54–55, 57–60], Astypalaia [61–62], Tilos [63], Kos [55, 59, 64], Samos [65–66], Chios [67–69], Lesbos [64, 70–73], Gökçeada (Imbros) [74–76], Bozcaada (Tenedos) [74–76], and Samothrace [77]. There is, however, no information on the distribution of mammals on the Albanian islands (see [78]). In this regard, the scientific literature is available only for the continental distribution – albeit coastal, in some cases – of a few mammalian species, such as the Balkan pine vole, *Microtus felteni* Malec and Storch, 1963 [79], the underground vole, *Microtus thomasi* Barrett-Hamilton, 1903 [80], the otter, *Lutra lutra* Linnaeus, 1758 ([81], compare with [82]), the red deer, *Cervus elaphus* Linnaeus, 1758 [83], and a few others. According to Bego et al. [84], more comprehensive studies, since they were published in Albanian [85–87], went largely unnoticed by mammalogists outside the country. For the same resason, by the end of the 20th century, the mammalian fauna of Albania was the least known in Europe [78, 84, 88]. For Turkey, instead, there is already an interesting scientific

Fig. 3: The first scientific description of the Cretan wild goat, or *agrimi*, *Capra aegagrus cretica* Schinz, 1838, already figures in the *Tavole di animali* (Tome 1, Volume V, Carta 24) of Ulisse Aldrobandi (1522–1605), preserved in the university library of Bologna (Italy).

Fig. 4: The Cretan spiny mouse, *Acomys cahirinus minous* Bate, 1906. Photo by Apostolos Trichas.

literature available on the subject (compare with [89–90] and [91] and references therein), even if it is not specific to the islands. Within this, I would mention in particular the contribution that appears in the aforementioned Özkan [74–75], and Kryštufek et al. [92], as well as Kryštufek and Vohralík [93–95].

2 Materials and methods

The compilation of an atlas of the non-volant mammals of the Ionian and Aegean islands has come a very long way. The checklist is indebted to many European travellers, traders, colonial officers, museum collectors, curators and biologists who described the species that are known to us today. The goal of this book is to foster better knowledge of the mammalian fauna of the Mediterranean islands. The atlas presents the current state of knowledge of the past and present distribution of the non-flying terrestrial mammals of the Ionian and Aegean islands. It also focuses on the importance of the role of human beings in the redefinition of the insular ecological equilibrium, as well as on the environmental impact produced as a consequence of biological invasions. The protection and study of this fauna can provide an opportunity for testing a range of different evolutionary theories.

This research is the result of a series of studies carried out by the author in the course of the last 20 years. Data were collected through a review of all previous knowledge of the insolar mammals and their history; original sightings and direct observations of tracks and other field signs, excrement and food remains; search for the roosts of owls (Aves: Strigiformes) and pellet analysis; live-trapping stations on consecutive nights in various insular habitat types, employing plastic traps, and trip-traps (very similar to longworths). Surveys were carried out directly by the author in the islands of Aegina, Aghii Pantes (Crete), Alonissos, Amorgos, Antimilos, Antiparos, Chios, Crete, Delos, Dhia (Crete), Euboea, Hydra, Ikaria, Ios, Kalymnos, Kea, Kira Panagia, Kos, Lefkada, Lemnos, Leros, Lesbos, Lipsoi, Milos, Mykonos, Naxos, Nissiros, Paros, Patmos, Poros, Rhodes, Samos, Sapientza, Simi, Skyros, Thassos, Theodorou (Crete), Thera, Tilos, Youra, and several Turkish islets off the south-western Anatolian coast between the Datça peninsula and the gulf of Fethiye.

Additional data were also obtained through the evaluation and examination of the materials conserved in the following museums:

- Goulandris Natural History Museum, Athens (GNHM)
- Museo di Zoologia dell'Università di Palermo (MZUPA)
- Muséum national d'Histoire naturelle, Paris (MNHN)
- Natural History Museum, London (BMNH)
- Natural History Museum of the Aegean, Mytelenii (Samos) (NHMAM)
- Natural History Museum of Crete (NHMC)
- Natural History Museum of Genoa (MSNG)
- Naturhistorisches Museum, Wien (NMW)
- Senckenberg Research Institute and Natural History Museum, Frankfurt am Main (SMF)
- Zoological Museum "La Specola" of the University of Florence (MZUF)
- Zoological Museum of the University of Athens (ZMUA)
- Zoological Museum of the University of Patras (ZMUP)
- Zoological Research Museum Alexander Koenig, Bonn (ZFMK)
- Zoologische Staatssammlung, München (ZSM).

The catalogue numbers and the location of the materials conserved in these museums are given in "Chapter 6. Atlas of the non-volant mammals of the Ionian and the Aegean islands. The species account."

This work is accompanied by maps on the current distribution of the species and the geographical position of the islands that yielded archaeozoological data on the first appearance and the former diffusion of the various non-volant mammals in the area under study. Shaded areas in maps do not necessarily reflect continuous distributions. The hatching of the areal distribution of the individual species refers only to the insular areas and not those of the continent.

An additional aim of this work is the concentration of the greatest possible number of bibliographical references dealing with the land mammals of the Ionian and Aegean archipelagos.

The volume also includes appendices on livestock, bats and marine mammals of the Ionian and Aegean islands. However, despite the fact that I have managed to collect data from more than 140 islands, there is still much work to be done.

3 The natural setting

"The Aegean Sea is a little universe, a microcosm that perfectly displays its characteristics of the limited infinite [...] Here island always follows island, and little archipelagos develop within the larger, in a continual play of concentric expansions and fragmentations"

Mauro Pieroni [96] – Il fuoco imperfetto (= *The imperfect focus*)

Apart from a few islands that belong to Albania and Turkey, most of the territories surrounded by the Ionian and the Aegean seas are today comprised within the political borders of the Hellenic Republic (see ▶Fig. 5). According to Handrinos and Akriotis [97], Greece has 9 838 islands and islets, 8 110 of them in the Aegean Sea, and 1 315 in the Ionian. Only 217 are inhabited. They range in size from islets of less than 1 km^2

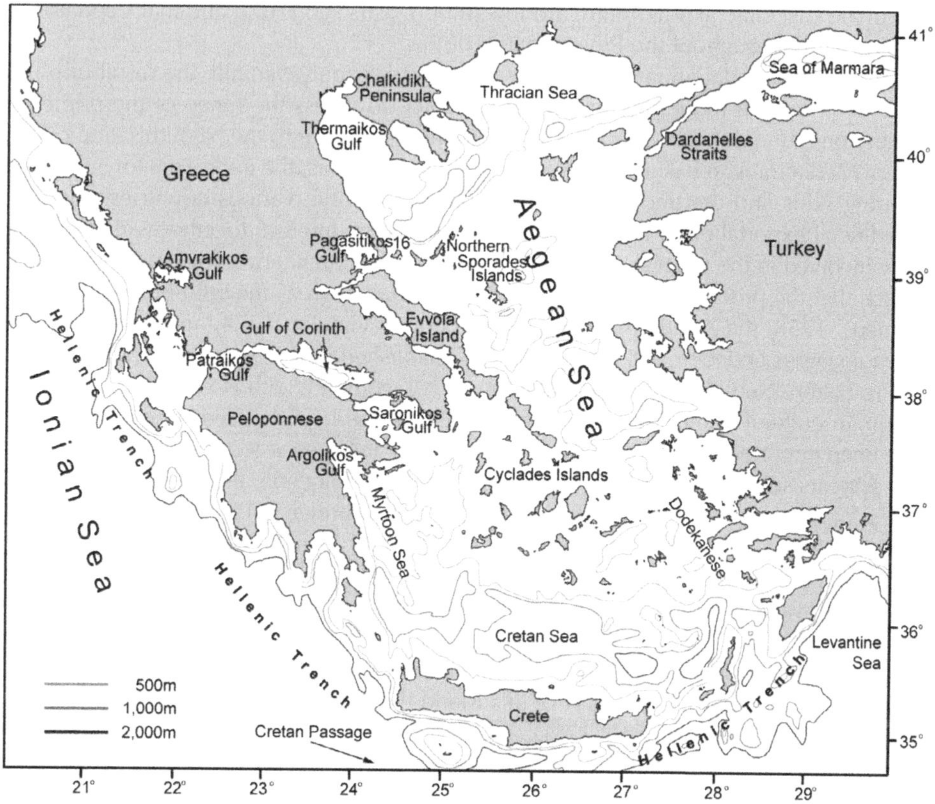

Fig. 5: Apart from a few islands that belong to Albania and Turkey, most of the territories surrounded by the Ionian and the Aegean seas are today comprised within the political borders of the Hellenic Republic.

to Crete, extending over 8 336 km^2 (it spans 260 km from east to west and is 60 km at its widest point), but the majority of those inhabited vary from 70 to 150 km^2.

From a geological point of view, the Ionian archipelago is largely made up of "continental islands", being the remains of the stable continental platform, now considered as the Pre-Apulian zone, which outcrops in the western parts of Cephalonia, Lefkada and Zakynthos [98]. In conceptual terms, "continental islands" are those close to a continental landmass, and geologically related to it, having been formed by separation from the nearest mainland (compare with [99]). The recent faunas of the Ionian islands have specific continental characters, having very likely originated from the southern Balkan peninsula opposite, from where they could easily have immigrated during the Late Glacial Maximum (24 000–18 000 years ago). Non-insular faunas are well known from Corfu and Cephalonia, the only Greek islands still characterised by the occurrence of the Balkan mole, *Talpa stankovici* V. Martino and E. Martino, 1931. On Cephalonia only a hippopotamus of normal size is known [100–101], whereas Corfu provided osteological remains of wild boars, deer and carnivores dating to the Upper Pleistocene [101–106], which did not substantially differ from their continental counterparts. Regarding the continental character of these islands, the evidence of their Palaeolithic occupation could also be very important if one takes into consideration the fact that Zakynthos and Cephalonia were connected, although not with mainland Greece, at the time of the Last Glacial Maximum, around 18 000 years ago [107], and that Cephalonia was an island throughout the Pleistocene [108].

The extant mammalian fauna of the Ionian islands is only partially the result of post-Pleistocene human-induced introductions. In fact, to assess the range of the primeval distribution of the different species, earlier chronologies prior to the Neolithisation of the Eastern Mediterranean basin should be considered. After this, the improvement in human seafaring skills and the trade networks set up between the various countries, enabled the artificial exportation of faunistic species of kinegetic interest, together with those already involved in the process of domestication [6, 109]. Archaeozoological investigation records that the process of domestication of mammals such as the wild boar, *Sus scrofa* Linnaeus, 1758, the Asiatic mouflon, *Ovis orientalis* Gmelin, 1774, and the wild goat, *Capra aegagrus* Erxleben, 1777, was already established in the Near East from the Middle Pre-Pottery Neolithic B period (MPPNB) ([110], compare with [111]). Thus, evidence of human cultural control over these mammals can be hypothesised, since they could have been imported onto the islands from mainland areas, even located very far afield. It now appears sufficiently credible that up to the early Holocene, the Mediterranean Sea increasingly represented less of a barrier and more of a bridge [112–116] promoting and multiplying in a relatively short time the circulation of ideas, merchandise, faunal elements, and human groups, which spread into new and different environments, and, over time and in various ways, became grafted onto the autochthonous substratum [23, 117] (see ▶Fig. 6). As far as is presently known, the first transfers of allochthonous mammals to Mediterranean islands have been documented from Cyprus and are chronologically referred to the human colonisation of this island that has been dated around the end of the 9th–8th millennium BC [118–121].

But the Eastern Mediterranean is also a huge seismic "hearth", and in many parts of it tectonic events have been accompanied by renewed bouts of volcanic activity [122]. Subduction of the eastern Mediterranean sea-floor beneath the Aegean began to produce volcanism about 5 million years ago along the South Aegean volcanic arc from

Fig. 6: The collections of the Goulandris Museum of Athens also comprise a Minoan fictile model of a vessel from early 2nd millennium BC. From the Mitsotakis Collection, catalogue number 91.

the north-eastern Peloponnese to the Dodecanese [98]. In the Cyclades, for example, volcanism started on Antiparos and continues now on Milos and Santorini (Thera). The eruption of the latter island, some 3000 years ago, around 1630 BC ([123], compare with [124]), was one of the most significant recent 'catastrophes' in the Earth's history, along with the eruption of the island of Krakatoa (off the south-western coast of Java, Indonesia) in 1883. Among other things, the Santorini eruption may have helped precipitate the collapse of the famous Minoan civilisation, which flourished in Crete and the surrounding Aegean area during the Bronze Age between 5000 and 3000 years BP [122, 125–126]. The Cyclades were named by the ancient Greeks, who saw them as a wheel, or cycle, placed round the sacred isle of Delos [98], whereas Sporades is the name given to all the other Aegean islands, which are scattered over the Aegean Sea all around the Cyclades. There are the Northern Sporades, including the islands and islets between Skiatos, Skopelos, Youra and Piperi, which are the uppermost parts of a horst extending eastwards from the Pelion peninsula [98]. Skyros, which is also comprised within the Northern Sporades, instead has a different tectonic setting. The Eastern Sporades include Lesbos, Chios, Ikaria and Samos, while Samothrace and Thasos are the northernmost Aegean islands. Twenty thousand years ago, when the sea level was much lower than it is now, Lemnos, located in front of the strait of the Dardanelles, was not an island but a peninsula of the Anatolian coast [98], with all the consequences on the natural dispersal of biological elements that such a phenomenon brought with it. The Southern Sporades, also incorrectly known as the Dodecanese, are instead a group of islands in the south-eastern Aegean Sea opposite the troubled coasts of south-western Anatolia. Due to their location, the latter islands, like most of those of the eastern Aegean, are politically part of Greece, but fall within the biogeographical range of Asia Minor [30, 127]. Lying between the Aegean Sea to the north and the Libyan Sea to the south, Crete is the fifth largest island of the Mediterranean, with an intermediate climate between the subtropical and the semi-desert except for the mountain ranges and its northernmost areas [97]. It has been characterised as a 'miniature continent' [128], since its great geomorphological variability is accompanied by a vast variety of habitats and climatic factors, ranging from the insular character of coastal areas to

Fig. 7: During the course of the 15th–16th centuries, the island of Crete was the destination of several Renaissance scientific missions, principally aimed at the study of the flora and collection of the precious dittany, *Origanum dictamnus* Linnaeus, which boasted such extraordinary pharmaceutical properties that it was considered a veritable panacea. It has even been conjectured that the Cretan deer and wild goats used to eat dittany to help heal themselves when wounded by hunters. According to Theophrastus, it was: "Said to be true, that, if goats eat it when they have been shot, it rids them of the arrow" (*Enquiry into Plants*, 9.16.1). Engraving from Dapper [132]: *Description Exacte Des Isles De L'Archipel. De l'Île de Crete ou Candie.*

a fully continental character of the higher parts of its mountains [129]. Consequently, the island is well known for its richness in endemic biological elements, mainly represented by plants (compare with [130]) (see ▸Fig. 7). The Cyclades are mountainous and largely barren islands, whereas Crete, Rhodes, Samos, Lesbos, Lemnos, Thasos, Corfu, Cephalonia and Zakynthos are the most extensively forested Greek islands (compare with [131]).

Within the unique range of biological elements always offered by the peculiar nature of the Aegean shores it could be interesting to recall the former existence of a number of endemic insular mammals. The late Quaternary island ecosystems were indeed quite different in some respects from the adjacent continental ecosystems [117]. As in some other islands of the Mediterranean basin, palaeontological and archaeozoological evidence demonstrates that several of the fossil faunas differed considerably from contemporary continental faunas, and were characterised by a very low taxonomic diversity [133–137]. The great number of islands in the Aegean Sea made this region highly favourable to the evolution of island endemics [9, 105–106, 138–139]). In this regard, examples from islands, such as Delos, Naxos, Seriphos, Kythnos, Milos, Amorgos, Crete, Kassos, Armathia, Karpathos, Rhodes, and Tilos are significant. From a biogeographical point of view, all these insular territories have to be regarded as "oceanic islands" (compare with [140–142]). In effect, within the Mediterranean basin it is almost impossible to ascertain the existence of islands of an authentically "oceanic type", which is the term generally adopted to indicate volcanic islands formed independently of continental land masses. In conceptual terms, oceanic islands are indeed the opposite of the above-mentioned "continental islands". Perhaps it would be more precise to refer to them as "oceanic-like islands" (compare with [143]), or authentic "true islands", which is an expression used to designate insular territories that have been separated from the closest mainland for a very long time, since at least the beginning of the Upper Pleistocene ([144]; compare with [145]). The islands of this type displayed very peculiar faunas: the existence of endemic mammals is perhaps the most obvious example of this. These mammalian assemblages were characterised by a very low taxonomic diversity. The most common trends of endemisation are the decrease in the size of macromammals, such as proboscideans and artiodactyls (deer and hippopotamuses), and the increase in the size of micromammals, such as soricomorphs and rodents. These modifications are generally considered to be essentially a consequence of genetic isolation from continental populations, a quantitative and qualitative reduction in food supply, an alteration of intraspecific competition, the absence of large carnivores, and as far as the micromammals are concerned, also of endothermic adaptations ([137] and references therein). Moreover, it is the absence of mammalian predators from these insular faunas that make it possible to define them as unbalanced. Dwarfism, for instance, was not only a suitable adaptive response to limited resources, but also an appropriate adaptation to living in the densely tangled vegetation of many islands [9], or island habitats (e.g., gallery or riparian forests, mountainous island habitats, etc.) at least at lower latitudes during the warm early–middle Holocene Climate Optimum. According to various authors, such as Felten and Storch [146] and Krapp [147], the issue of island gigantism too, as manifested by the micromammals, can only be understood as a consequence of genetic drift phenomena, including the so called "founder effect". The "oceanic-like islands" of the Aegean Sea provided endemic oligotypic associations and/or single species, the majority of which apparently vanished before any fully documented appearance of man.

4 The island endemics. The mammals of the Upper Pleistocene

Elephants are described from many Aegean islands, where they evolved in a rather peculiar way, often becoming dwarfed (compare with [148]) (▶Fig. 8). Pygmy elephants are in fact known among the Quaternary faunas of Rhodes and Tilos [104, 139, 149–157], Naxos, Delos and Seriphos [135, 158–163], Kythnos [106, 164–167], Milos [106, 135, 159, 165, 167–168], and Crete [135, 167, 169–172]. Other fossil proboscidean remains have also been discovered on the island of Astypalaia [8, 60, 108, 173] and Kalymnos [9, 173]. Most of these forms appear to derive from the straight-tusked elephant, *Elephas antiquus* Falconer and Cautley, 1847, a species dispersed up to the Late Pleistocene in the western Palaearctic (compare with [148, 174–177] (▶Fig. 9). Very recently the island of Kasos, in the Dodecanse, has also provided one molar

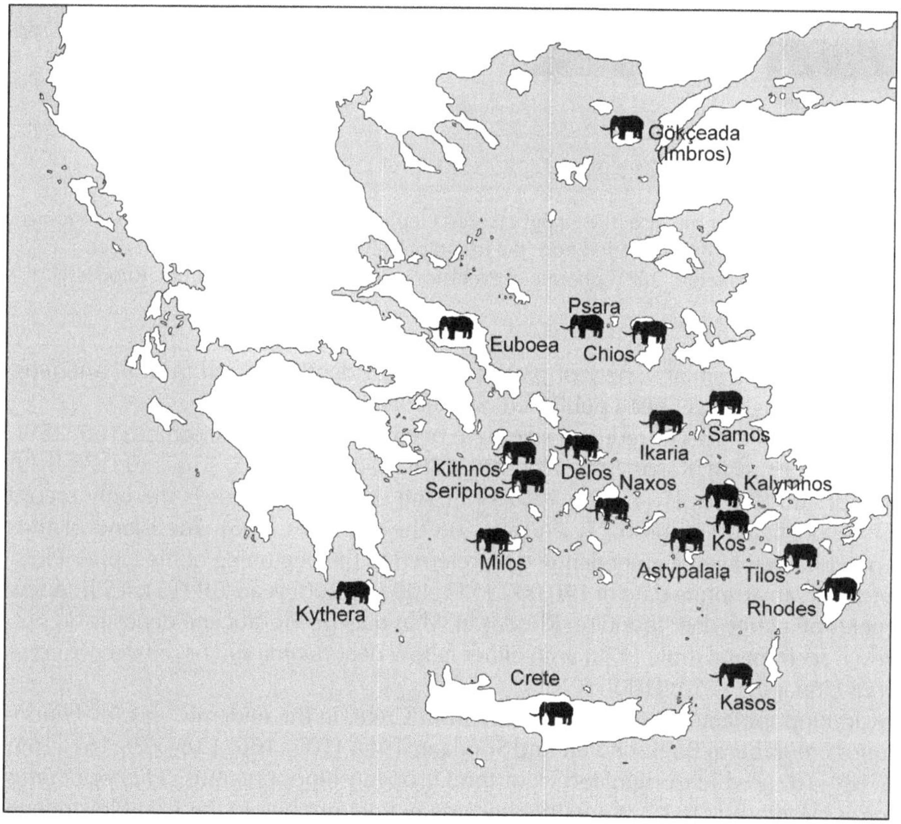

Fig. 8: Map of the Aegean basin showing the islands where Middle and Upper Pleistocene remains of proboscideans of the genus *Elephas* (*Palaeoloxodon*) Linnaeus, 1758 have been discovered.

Fig. 9: The third inferior molar of the straight tusked elephant, *Elephas (Palaeoloxodon) antiquus* Falconer and Cautley, 1847, reported from the island of Kalymnos, Dodecanese (Greece). Courtesy of Toula Marketou, 22nd Ephorate of Prehistoric and Classical Antiquities, Rhodes.

(left lower m_3) of a small variety of proboscidean, evidently derived from *E. antiquus,* although this has not yet been published (Sevket Sen, 2012: *in litteris*).

Late Pleistocene deer of reduced size have been reported from Amorgos [165, 168], Crete [106, 139, 178], Kasos [105, 139, 178–179], Karpathos [105, 138–139, 178–179] (►Fig. 10) and Rhodes [104, 106, 139, 149]. The deer of Amorgos is the only record of a Pleistocene cervid, probably endemic on the Cyclades [165]. The island of Tilos also provided remains of an endemic deer referred to the beginning of the Upper Pleistocene, with an absolute date of 140 000 ± (11 400 / 10 200) years BP [151, 154]. A few fragments of a large deer found on Rhodes in what may be Pleistocene deposits do not, however, seem compatible [178] with either fallow deer (*Dama* sp.) or red deer (*Cervus elaphus* Linnaeus, 1758) [180].

Dwarf hippopotamuses are known only from Crete, in the endemic species *Hippopotamus creutzburgi* Boekschoten and Sondaar, 1966 [105, 106, 138–139, 161, 165, 168, 180–182], which originated from the European hippopotamus, *Hippopotamus antiquus* Desmarest, 1822. It possibly became extinct not before the Late Pleistocene [183–184] (►Fig. 11). In fact, dwarf hippopotamuses from the Cretan Catharo basin are dated to about 12 135 ± 485 years BP [185]. A large Middle or Late Pleistocene

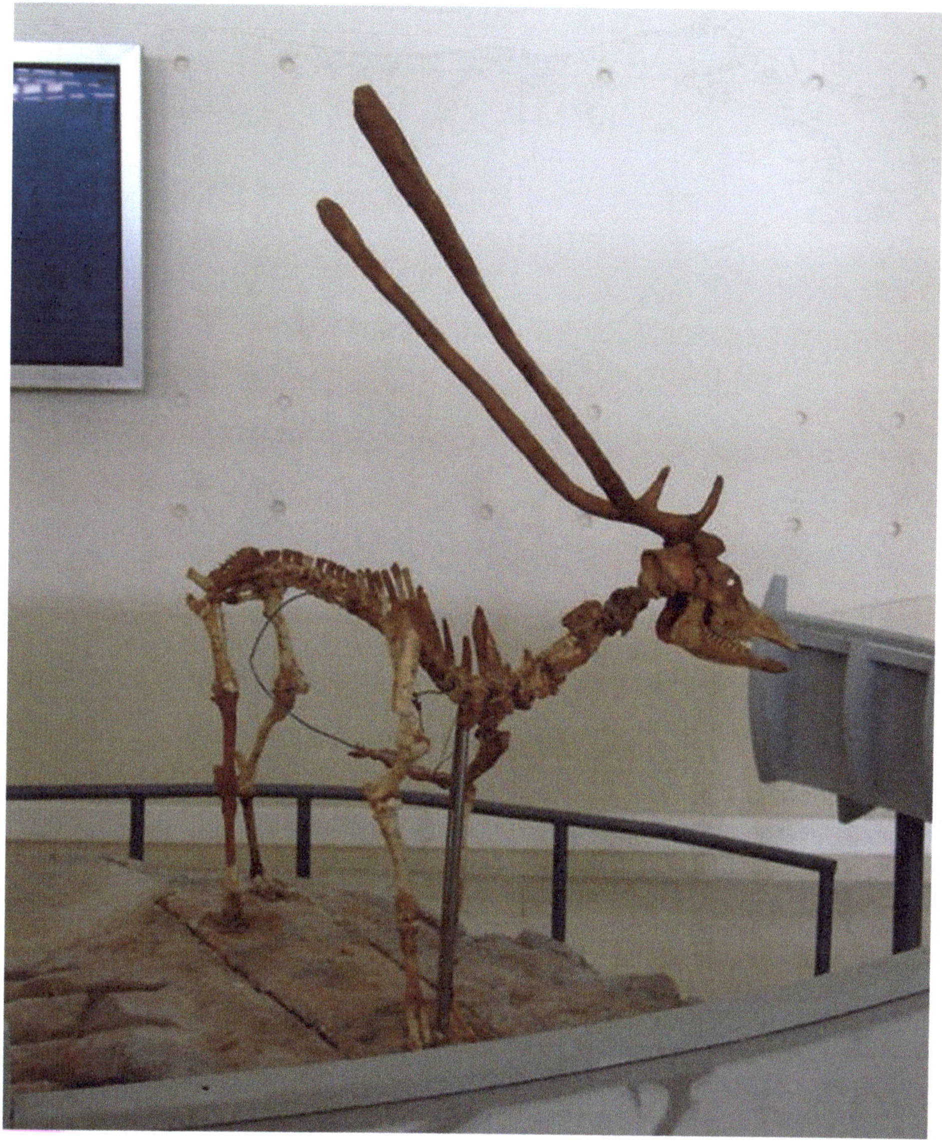

Fig. 10: The smallest of the Cretan pigmy deer was *Megaceroides (Candiacervus) ropalophorus* de Vos, 1979, which stood only 40 cm high. Photo by George E. Theodorou.

Apodemus sp. was found on Naxos [135, 186], whereas a few murid remains, possibly endemic, from the Charkadio cave of Tilos are still undetermined at species level. Dermitzakis and Sondaar [105] reported also the occurrence of an endemic murid of the genus *Mus* from the Upper Pleistocene of Karpathos.

Addressing the obvious difficulties faced by most organisms in trying to cross salt water, biogeographers used to postulate that most remote islands were once linked to continents by land-bridges now sunk from view [145]. It is in fact commonly

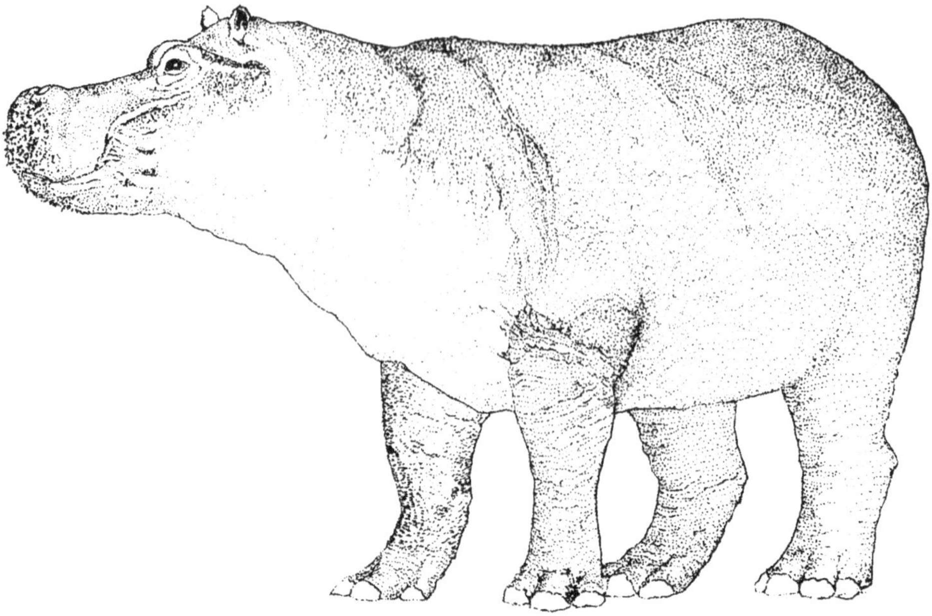

Fig. 11: Artist's reconstruction of the pygmy hippopotamus, *Hippopotamus creutzburgi* Boekschoten and Sondaar, 1966, of Late Pleistocene Crete. Drawing by Silvia Cantagalli Masseti.

assumed that, before becoming endemic, many of the mammals reached the islands from the mainland via temporary land-bridges that originated as a result of the lowering of the sea level during glacial episodes. On the basis of tectonic factors, Malatesta [187] observes, for example, that Crete may have been joined to continental Greece during the Pleistocene. However, as noted by Davis [163], it is then difficult to explain why its Pleistocene faunas are relatively rich compared to those of other "oceanic-like islands", such as Cyprus, for example. Today the general opinion is that most of the Mediterranean islands characterised by late Quaternary mammalian endemics could not have been reached over the last 130 000 years by non-volant vertebrates except by rafting or swimming [144].

For instance, Schüle [188] observed that most of the alleged Quaternary land-bridges to Mediterranean islands are geologically impossible. Thus, it is assumed that micromammals, such as shrews and rodents, may also have reached some islands by jumping onto floating logs and/or rafting.

Sondaar [136] has noted that endemic island faunas were dominated by good swimmer macromammals. Indeed, some authors have suggested that proboscideans may perhaps have swum to some of the islands [162]. New evidence today unequivocally demonstrates that elephants are excellent swimmers. Swimming speeds up to 2.70 km/h have been measured, with maximum distances estimated at 48 kilometres. For these reasons, Johnson [162] suggests reappraising the origin of land vertebrates on all the islands that held elephants, and reviewing the geological reconstructions that assumed that land-bridges once connected these islands to the mainland. There are also known records of the swimming abilities of deer [105, 189] (▶Fig. 12), and although not aquatic animals, cervids have long been known for their migratory water crossing [190].

Hippopotamuses are commonly believed to be excellent divers and swimmers, but they are poorly streamlined [191], not buoyant, and their feet are not sufficiently modified for swimming ([192] and references therein). Swimming hippopotamuses actually gallop short distances underwater [193], but they seldom lose contact with river and lake bottoms, and they tend to walk underwater rather than swim [194–195]. The feet are kept in contact with the bottom by controlling the gravity of the body and as a result of high bone density [195–196]. Animals that can migrate over long distances also include flying taxa, such as birds and bats [136].

This sporadic and highly selective dispersal from a continent to an island has been referred to as 'sweepstake migration', that is, possible only via the so-called *sweepstake routes* [105, 136, 197–199]. As we can imagine, the chances of such a dispersal are extremely low, and largely due to chance combinations of favourable circumstances. The sweepstake routes differ from biogeographical filters not merely in degree, but because the organisms that used the former were capable of exploiting them to disperse from island to island (compare with [200]). During the Pleistocene, the fluctuations of the sea levels caused by climate change favoured sweepstake migrations, and the subsequent evolution of endemic faunas, by altering the distances between the mainland and the islands. Low sea levels made the sea barrier easier to cross, but the subsequent high levels then rendered the channel crossing practically impossible, so that the founding population on the island would find itself isolated. Populations reaching the islands by

Fig. 12: Modern scientific investigation considers red deer, *Cervus elaphus*, as quite competent swimmers, yet they cannot survive a crossing of more than a few miles of open sea. In this regard, the description given by classical scholars, such as Aelian (170–235), is so detailed that it almost seems to be based on an eye-witness account: "Yet they [the deer] swim in single file, whereby the ones behind put their heads on the rears of the ones in front, and when the first one tires it seeks rest by lining up as the last one and brings up the rear". Photo by Franco Tassi, courtesy of Ente Autonomo Parco Nazionale d'Abruzzo, Italy.

such routes become genetically isolated from the respective continental population. The rapid evolutionary adaptation to the new insular environment usually affects body size and the pattern of locomotion [136, 199, 201]. Thus, the endemic mammalian assemblages of many of the Upper Pleistocene Mediterranean islands were oligotypic, characterised by a very low taxonomic diversity.

There are also several cases, however, in which the homogeneity with the continental Late Pleistocene fauna documents the continuing land connection with the nearest landmasses, as in the case of Kos (compare with [106, 202]) which was characterised by a Pleistocene continental non-endemic fauna, showing affinities with that of Asia Minor and the Near East [173]. In fact, cases of Middle/Late Pleistocene faunas are known also for the Aegean islands. Their distance from the nearest mainland did not allow the developments of endemic zoological varieties, while the connection with the closest mainland due to the lowering of the sea level in the Upper Pleistocene explains the balanced fauna which has been discovered there. Islands of this type are, for example, the Greek Kythera [108], Thasos [203], Kos, Kalymnos and the Turkish Gökçeada (Imbros) [105]. In the cases of Kos and Kalymnos, the homogeneity with the continental Late Pleistocene fauna documents the continuing land connection with the nearest landmasses [105, 179, 202, 204–205] (compare with [106, 206–208]), which show affinities with those of Asia Minor and the Near East. Moreover, the island of Kalymnos provided the remains of advanced *Hippopotamus antiquus* (compare with [209–210] (►Fig. 13), of the aforementioned straight tusked elephants [173], as well as several cervids (compare with [211]), which originated from Anatolia.

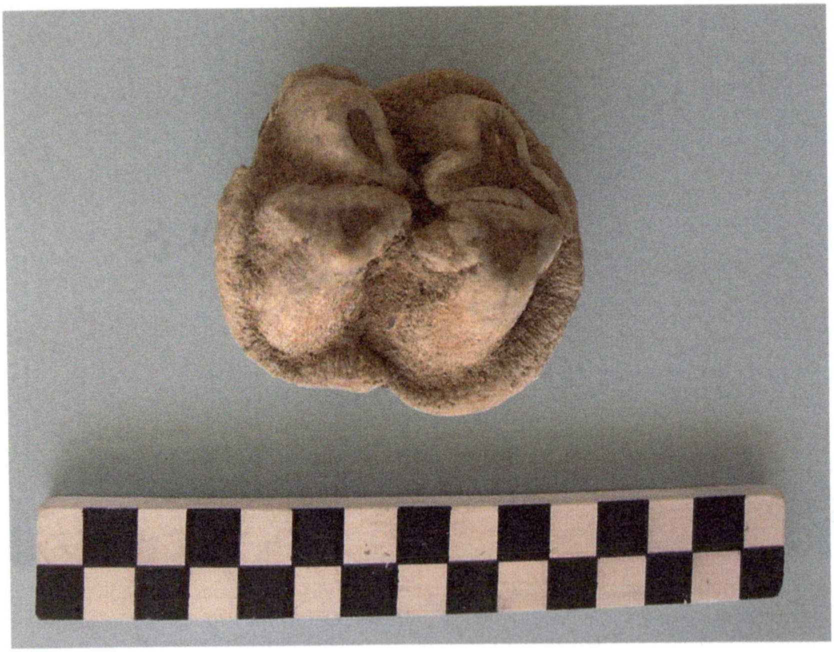

Fig. 13: The first upper right molar (M^1) belonging to an advanced *Hippopotamus antiquus* Desmarest, 1822, from the island of Kalymnos. Courtesy of Toula Marketou, 22nd Ephorate of Prehistoric and Classical Antiquities, Rhodes.

5 The disappearance of the endemics. Holocene relics of a vanished world

The situation of the insular endemics changed drastically around the end of the Pleistocene/Holocene, the most significant cases of extinction over the past few thousand years having been those of the insular species [212]. As far as is known today, the data available for the islands point to endemic faunal extinction being largely the result of human activities.

Human exploration of the shores and coastlines of the region began in ancient times, as man sought to exploit new natural resources and/or discover new geographical areas suitable for settlement. This was a lengthy process that began in pre-Neolithic periods and continued up to historical times. The available archaeological documentation, based on still quite fragmentary evidence, tends to indicate that the first relocations by sea in the Eastern Mediterranean basin were already carried out by hunter-gatherers in expression contexts of a Mesolithic type [117, 166, 213–218]. Evidence from the islands of Milos in the Western Cycladic archipelago [214, 218] and also possibly Kythnos [219] and Cyprus [217, 220–221], indicate improved seafaring capacities. In fact, from the late Mesolithic period onwards, the Mediterranean Sea can be considered as a preferential route for the exploration and subsequent colonisation of its coastlines and islands (compare with [214, 222–225]. There is also evidence of a pre-Neolithic human occupation of Crete (compare with [226]).

For some islands it would appear that the endemic oligotypic associations and/or single species vanished prior to periods for which there is evidence of permanent human occupation (compare with [117]). According to the 'theory of island biogeography', outlined by MacArthur and Wilson [227], island environments tend to be characterised by a reduced biodiversity, to which animal populations, like human communities, have to adapt. Thus, it has been suggested that the extinction of the insular endemics may have been related more to their inability to adapt further in the face of a basically unfavourable environment, than to the hunting prowess of pre-Neolithic man. In any case, this theory could prove more cogent for small and remote islets than for islands such as Crete, still characterised by a great variety of natural resources. In the light of the archaeological evidence, one might even conclude that the impact of hunter-gatherer communities on the island ecology was relatively limited. Nevertheless, as argued by Rackham and Moody [128], if in the Early Mesolithic trips to find obsidian, to Milos for example (►Fig. 14), were combined with hunting trips to other islands, many of the endemic mammals may have disappeared before there was any settlement such as to leave an archaeological record. In reality, in the case of Crete, for example, any overlap between the endemic Pleistocene fauna and the human occupation of the island has scant support from archaeological data, despite considerable efforts to uncover evidence of pre-Neolithic occupation (compare with [216, 228]). In any case, Kopaka and Matzanas [229] reported Palaeolithic tools *in situ*, dating back to perhaps circa 120 000–75 000 years BP, from the island of Gavdhos off the south coast of Crete, a first indication

Fig. 14: Located in the Western Cycladic Archipelago, the island of Milos (Greece) was exploited as a source of obsidian in Mesolithic times (circa 13 000 BP), but there is no evidence for permanent settlement on this island before the Neolithic era. Photo by Marco Masseti.

that humans may have colonised Crete before the Holocene, while Strasser et al. [230] reported Palaeolithic tools from the Plakias region, in southern Crete (▶Fig. 15).

It was particularly from the Early Neolithic period on that most of the Mediterranean island faunas began to reveal evidence of human exploitation, leading to a gradual extinction of the endemic species [231–232]. In the Aegean islands, very few of the endemic species persisted beyond the end of the Pleistocene to become extinct during the Holocene: these are the aforementioned Cretan shrew, the dwarf elephant of Tilos (see the following section "The island of Tilos in the Dodecanese: a case study"), and apparently one species of hamster. Effectively, a large endemic cricetid, described as *Mesocricetus rathgeberi* Pieper, 1984, was reported from the small islet of Armathia, off the north-western shores of Kasos (▶Fig. 16). According to Pieper [233], this represents a rather large species of the genus, and is thus comparable only with *Mesocricetus raddei* Nehring, 1894, from Dagestan; the similar size of the two species has been interpreted as possibly convergent. As far as is presently known, the Armathian golden hamster is the only representative of the family among the Upper Pleistocene faunas of the Aegean islands [101, 106, 233]. Rather surprisingly, the species was discovered within a typical human settlement context, comprising Holocene continental and anthropochorous species such as house mice, (*Mus musculus* Linnaeus, 1758), black rats, (*Rattus rattus* Linnaeus, 1758), and rabbits, (*Oryctolagus cuniculus* Linnaeus, 1758) (compare with [233]). Nevertheless, the archaeological context of its discovery raises doubts about the Armathian origin of this element [8].

Still dispersed on Crete, the endemic white-toothed shrew, *Crocidura zimmermanni* Wettstein, 1953, can be regarded as a relic in two senses: firstly, it is the direct descendant

Fig. 15: Palaeolithic quartz bifases from the site of Preveli 2, in the region of Plakias, in southern Crete. From Strasser et al. [230].

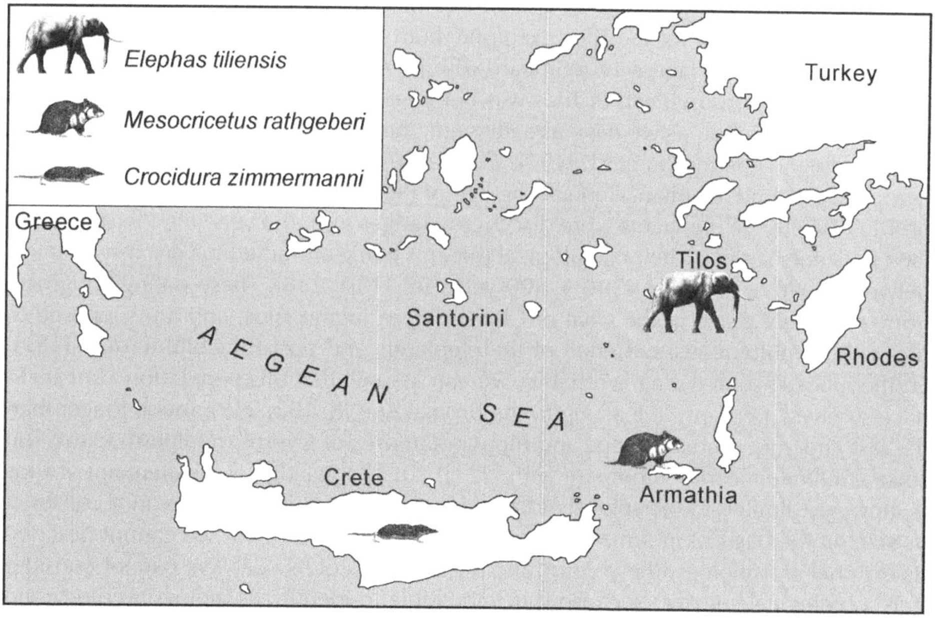

Fig. 16: Holocene endemic mammals of the Aegean islands.

of the Pleistocene group of *Crocidura* species that have disappeared from the rest of Europe and have been replaced by the current species; secondly, together with the Sicilian white-toothed shrew, *Crocidura sicula* Miller, 1900, it is the only known survivor of the wealth of endemic species that were found in many of the Mediterranean islands during the Pleistocene [15, 234]. Today this shrew is not known anywhere outside Crete [234] and it is therefore considered endemic to this island [17, 234].

5.1 The island of Tilos in the Dodecanese: a case study

Although this island lies only a few nautical miles off the western Anatolian coast, in the late Quaternary its native mammalian fauna was not characterised by continental taxa, but differed considerably from contemporary mainland wildlife. It was dominated by endemic dwarf elephants, regarded as belonging to the genus *Elephas* ([152–154], compare with [156]), only recently specifically described as *Elephas tiliensis* by Theodorou, Symeonidis and Stathopoulou [157] (▶Fig. 17). This form has often been compared to the *Elephas antiquus* or *Elephas falconeri* Busk, 1867, of Sicily [153], considered as the smallest elephant that ever existed, as well as for decades the last step of a progressive size reduction [176, 235]. Previously referred to two different forms, the endemic dwarf elephants of Tilos are now considered to belong to a single species with marked dimorphism. According to detailed comparisons by Theodorou [153, 236], this elephant also never arrived at the larger dimensions of some of its bigger endemic relatives of Sicily and Crete. The morphology of the new species points to a population with numerous changes concerning the postcranial material. It could be interesting to note that the height at the withers of the adult pygmy African elephants – possibly identifiable as *Loxodonta pumilio* Noack, 1906 – photographed by Böhme and Eisentraut [237] in the pluvial forest of northern Congo (▶Fig. 18) – is almost the same as that which it has been possible to reconstruct for the Tilos exemplar, both being adult specimens. This is symbolically illustrated in Figure 17 (compare with [8, 63]). However, the supposed ancestor of the dwarf proboscidean of Tilos was not a representative of the genus *Loxodonta* Anonymus, 1827, but, as we have already seen, the straight-tusked elephant.

According to Theodorou et al. [157], *Elaphas tiliensis* is the last Mediterranean and European endemic elephant. In fact, the age of the deposits of the cave of Charkadio, on Tilos, where the elephants were discovered ranges from the very late Pleistocene to the Holocene, some of their osteological remains being attributed to very recent times, between 7090 ± 680 and 4390 ± 600 years BP [150, 238]. These datings originate from a different place in the cave of Charkadio, in inland Tilos, and are supposed to prove the simultaneous existence of the elephants and post-Palaeolithic man [151]. Furthermore, if such dating is reliable, we can assume that this population appears to have survived, perhaps, at least up to the Bronze Age [8, 152, 239], much longer than the last endemic proboscideans and hippopotamuses of Cyprus, hunted to extinction several millennia earlier (compare with [221]). Our knowledge of the elephant of Tilos is, however, limited to the sphere of the palaeontological species. We can reconstruct its size on the basis of information derived from the skeleton, but we cannot describe its external morphology, the colours and the markings of its coat. We cannot consider it as a biological species, because it is impossible to identify its biology, ecology and behaviour.

Fig. 17: Artist's reconstruction of the extinct dwarf elephant, *Elephas tiliensis* Theodorou, Symeonidis and Stathopoulou, 2007, of Late Pleistocene-Holocene Tilos. The reconstruction is adapted from the osteological material in the Museum of Megalochorio (Tilos, Greece) and compared to the size of its supposed ancestor, *Elephas antiquus* Falconer and Cautley, 1847. Drawing by Alessando Mangione and Marco Masseti.

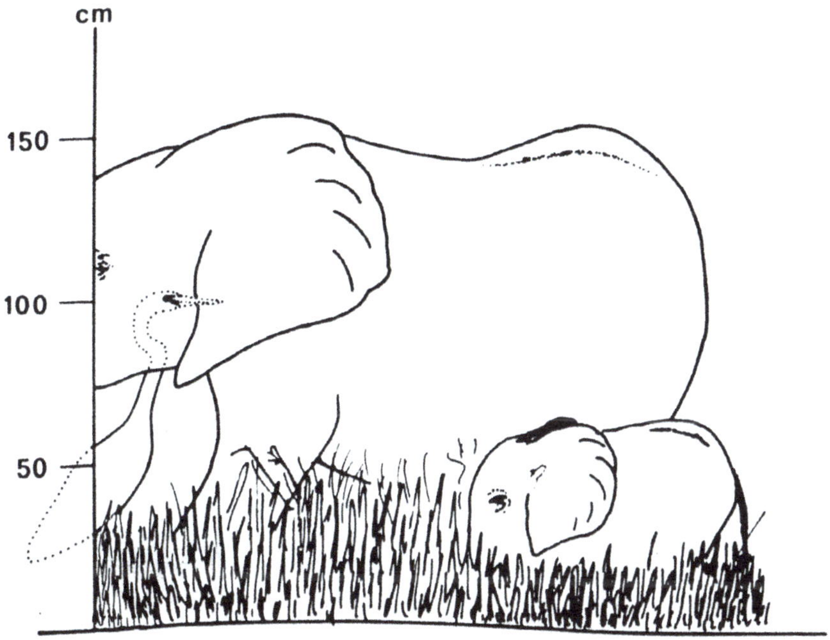

Fig. 18: Sketch reconstructing the size of an adult Congo forest pygmy elephant and a calf, compared with the dimensions of an individual of white egret, *Egretta alba* Linnaeus, 1758. From Böhme and Eisentraut [237].

Fig. 19: Detail of the wall-painting of the tomb of Rekh-mi-Rē, at Thebes (Egypt), showing a small-sized elephant borne by the Syrian tributaries. Photo by Nicole Douek.

Nevertheless, the survival of dwarf proboscideans in the circum-Mediterranean area up to the Late Bronze Age appears to be indirectly documented also by the artistic evidence of ancient Egypt. During the 1930s, a wall painting showing, among other animals, the figure of a small elephant was discovered in an Egyptian tomb of Thebes of the 18th Dynasty, dated around 1450 BC. It belonged to Rekh-mi-Rē, vizier of Thutmosis III and Amenhotep II (from about 1470 to 1445 BC) (compare with [240]), in Thebes (Egypt) (►Fig. 19). Several years ago, Rosen [241] and White [242] carried out a correspondence in the pages of *Nature* on the interpretation of the figure. Sparked off by the assumption of Lister [243] that Siberian dwarf mammoths lived up to the time of the Egyptian pharaohs, this correspondence led Rosen to suppose that the decoration effectively portrayed a dwarf mammoth, whereas White claimed that it was actually a small-sized African elephant. As already noted by the two correspondents, the image represents an adult specimen characterised by well developed tusks (►Fig. 20). According to other authors, such as de Garis [240], Scullard [244] and Osborne and Osbornová [245], it displays morphological patterns which might be referred to as Asian elephants, *Elephas maximus* Linnaeus, 1758, that possibly lived in the Near East at the time. According to pictorial, written and osteological evidence, it seems that wild herds of proboscideans lived in the ancient land of Niya, located in western Syria, between the late 2nd and early 1st millennium BC [246–248] (►Fig. 21). Regarding the peculiar

Fig. 20: The ideal reconstruction of the extinct Tilian elephant, compared to the detail of the wall-paintings of the 18th Egyptian Dynasty tomb of Rekh-mi-Rē at Thebes, which show a small-sized elephant borne by the Syrian tributaries. Drawings by Alessando Mangione and Marco Masseti.

Fig. 21: Elephant tusks of supposed Syrian origin displayed at the Archaeological Museum of Herakleion, Crete. Photo by Marco Masseti, courtesy of the Archaeological Museum of Herakleion.

size of the elephant in the Egyptian painting, as far back as 1935, Davies N. de Garis [240] remarked that the artists kept the animals small so that they would not dominate the tribute bearing procession, although the length of the tusks tends to suggest that they were thinking of an adult specimen. White [242] and Osborne and Osbornová [245] also noted that the differential scale of the human and animal figures is the result of stylistic convention rather than naturalistic representation (compare with [249–250]). It can be observed, however, that this conventional Egyptian method of portraying the size of animal and human beings is not consistently adopted in the wall-decoration of the Rekh-mi-Rē tomb. The tusks of the portrayed specimen are in fact much smaller and inconsistent with those shown being carried by the Syrian to the right ▶Fig. 19, and the Minoan and Nubian bearers pictured in other sections of the wall-painting (compare with [251]). This incongruence contrasts with the representation of other animals in the same picture. For example, the giraffe portrayed with the Nubian bearers takes up the entire available vertical space of the register. It therefore seems arguable that rather than being evidence of the stylistic convention to which the artists had to conform, we may actually be dealing with the portrait of a dwarf elephant. But, as already observed, there is no fully convincing evidence for the identification with the morphology of an Asian elephant. Thus, it may be possible to trace the morphological characteristics of the proboscidean to geographical species closer to ancient Egypt, possibly even among the Mediterranean islands where palaeontological evidence records the occurrence of dwarf and pygmy elephants from at least as far back as the Middle Pleistocene. But how could it come about that an insular dwarf elephant was brought to Egypt by Near Eastern tributaries? How did they get hold of it? And which Mediterranean island did it originally come from?

In the paintings of the Rekh-mi-Rē tomb, the Near Eastern tributaries bore the small-sized elephant, together with other goods which were to be offered to the pharaoh. In this context, an adult dwarf elephant would have cut a fine figure. Its value could have been related more to its curiosity appeal than to its effective economic worth, or even to both. On the other hand, there is considerable archaeological evidence for the circulation and trade of materials and ideas between the Mediterranean and south-eastern Asia. Cultural interaction between Crete, Cyprus, the Near East and Egypt increased markedly during the Late Bronze Age [252–253]. The main maritime route connecting the Aegean to the Near East during the Bronze Age passed between Rhodes and the peninsula of Bozburun [254], located a few nautical miles off the northern coast of the island of Tilos. The Minoan presence on Tilos and the other islands of the so-called "eastern Aegean string" is documented from at least circa 2000–1800 BC onwards [254–255]. This is not to say that the living proboscidean depicted in the Rekh-mi-Rē tomb is definitely the portrait of a Tilos elephant actually captured by the Aegean Bronze Age people on the island. It may have been a dwarf representative of the genus *Elephas* which survived on any Eastern Mediterranean island during the time of the Minoan–Mycenean control. It cannot be excluded that from this as yet unidentified island the dwarf elephant could have been exported to a mainland area where it could have represented a precious and rare curiosity to be exchanged as a costly gift between Aegean, Near Eastern and Egyptian rulers. While hopefully awaiting a revised dating for the elephants of Tilos, further investigations are also needed to better understand the significance of the Egyptian painting.

Except for one bat, the lesser mouse-eared bat, *Myotis blythii* Tomes, 1857, so far no other micromammalian osteological fragments have been found associated with the Tilos elephants, the stratigraphy of which also yielded the remains of marginate tortoise, *Testudo marginata* Schoeppf 1795, and brown bear, *Ursus arctos* Linnaeus, 1758, the latter presumed to be a hunter trophy [101, 106, 150]. Nonetheless, other as yet unknown representatives of the limited endemic mammalian fauna of Tilos may have survived much longer than on other Mediterranean islands, possibly thanks to the shelter afforded by the natural morphology of the island, particularly inhospitable and unsuitable for human settlement. On the basis of such a supposition, research was carried out several years ago by Masseti and Sarà [63] to investigate – for the first time – the current composition of the non-flying mammals of Tilos, also with a view to exploring the eventual relationship with species reported for the island for the previous Late Pleistocene-Holocene chronology. Several live local mammals were caught in traps and some remains have also been found in barn owl pellets. The outcome revealed that the extant insular non-volant mammals did not differ fundamentally from those of the great majority of the other Eastern Aegean islands, being almost exclusively characterised by continental taxa whose appearance on Tilos appears to be directly and essentially related to human activity (▶Tab. 1).

Only the local lesser white-toothed shrew revealed peculiar phenotypic patterns: while the external size and the colour of its coat fall within the variability range for the species, this shrew reveals an unexpected white tail tip [63]. Colour anomalies

Tab. 1: Holocene non-flying terrestrial wild mammals reported from the island of Tilos.

Taxon	Charkadio cave – late Pleistocene-Holocene fauna	Charkadio cave – sub-recent fauna	Present fauna
Erinaceus concolor			• [63]
Crocidura russula/suaveolens		• [152]	
Crocidura suaveolens			• [63]
*Oryctolagus cuniculus**			• [63]
Apodemus flavicollis/sylvaticus		• [152]	
Apodemus mystacinus		• [152]	• [63]
Rattus rattus			• [63]
Mus domesticus			• [63]
Vulpes vulpes		• [152]	
*Ursus compare with arctos***	• [101, 106, 150]		
Martes foina		• [152]	
Elephas tiliensis	• [101, 106, 150]		
Total = 12	**2**	**5**	**6**

*Imported in very recent times (about 1997–1998); **Very probably a hunter's trophy.

have been reported for several species of Soricomorpha [256–261]. Albinism and white spotting in shrews include *Crocidura suaveolens*, *Crocidura russula*, and *Crocidura leucodon* [262]. The light colouration of the coat has been recognised as a result of lack of pigment in the entire hair or hair fragments, and it appears that atypically coloured shrews occur more often in isolated populations whose gene transfer with neighbouring populations is limited [261]. Moreover, white tail tips do not appear to be uncommon in shrews of the genus *Sorex* [256, 263–266], but specifically as regards the representatives of the genus *Crocidura*, apart from the Tilos shrew, it is a characteristic reported to date only for the circum-Sicilian endemic *Crocidura sicula* [12, 63, 173, 267]. Also for all this, a more detailed study on the geographical variations of the lesser white-toothed shrews of the Eastern Aegean islands needs to be conducted.

If we look at the present non-flying terrestrial mammals of the Ionian and Aegean islands, we can hardly find any of the endemic elements that characterised the Pleistocene faunal structures. Further, it should be emphasised that, apart from those islands that were connected by land-bridges to the nearest mainland at different times, neither is the repertoire of the modern species traceable in Pleistocene deposits, nor, all things considered, does it seem likely that they have reached the islands by swimming, jumping onto floating logs or other so-called sweepstake routes.

6 Atlas of the non-volant mammals of the Ionian and the Aegean islands. The species account.

6.1 Northern white-breasted hedgehog, *Erinaceus roumanicus* Barrett-Hamilton, 1900

Greek name: ακανθόχοιρος (*akanthohiros*)
Greek vernacular name: ακανζόχιρος (*skanzohiros*)

Turkish names: *kirpi, doğu avrupa kirpisi*

The taxonomic classification of the hedgehogs of the genus *Erinaceus* Linnaeus, 1758, has recently been updated. This genus now includes at least four species (*Erinaceus amurensis* Schrenk, 1859, *Erinaceus concolor* Martin, 1838, *Erinaceus europaeus* Linnaeus, 1758, and *Erinaceus roumanicus* Barrett-Hamilton, 1900), with a geographical range extending over most of Eurasia, from western Europe to eastern China and Korea.

The northern white-breasted hedgehog, *Erinaceus roumanicus* Barrett-Hamilton, 1900, occurs throughout a large portion of this range, being widespread in Europe, from Poland to Austria and Slovenia, the Balkan states, Turkisk Thrace, and eastwards through Russia and Ukraine to the northern Caucasus, western Siberia and the basin of the river Ob [268] (▶Fig. 22). Former authors, such as Corbet [269], Giagia and Ondrias [68], Harrison and Bates [270], Reeve [271], Giagia-Athanasopoulou and Ondrias [272], Giagia-Athanasopoulou and Markakis [273], Lapini [274], and others, previously referred this range to the distribution of the eastern hedgehog or southern white-breasted hedgehog, *Erinaceus concolor* Martin, 1838, whose distribution has been instead recently restricted to Asia Minor and the Levant, including Syria and Palestine, Iraq, Iran and the southern Caucasus (compare with [268]). *Erinaceus roumanicus* occurs also in many of the Adriatic, Ionian and Aegean islands (compare with [268]), but according to von Wettstein [58], hedgehogs were absent from Pholegandros, Sikinos, Siphnos, Astypalaia, and Karpathos. Their absence from Astypalaia has been recently confirmed [61], but not from Karpathos, where it seems that hedgehogs were introduced in very recent times (over the last few decades) (Costas Papaconstantinou, personal communication). Hedgehogs have been reported from Corfu, Cephalonia, Zakynthos, Antikythera, Kythera, Skiathos, Skopelos, Alonissos, Skyros, Seriphos, Syros, Milos, Siphnos, Kythnos, Aegina, Euboea, Andros, Kea, Giaros (between Kea and Tinos), Tinos, Kymolos, Renia (west of Delos), Delos, Naxos, Folengandros, Ios, Amorgos, Astypalaia, Thera, Anaphi, Theodorou (off the north-western coast of Crete), Crete, Karpathos, Rhodes, Tilos, Kos, Kalymnos, Samos, Chios, Lesbos, Lemnos, Gökçeada (Imbros), Aghios Efstratios, Samothrace, and Thasos. The occurrence of the species was confirmed by A. Dimitropoulos (*in*

Fig. 22: Northern white-breasted hedgehog, *Erinaceus roumanicus* Barrett-Hamilton, 1900. Photo by Maria Dimaki.

verbis) for the islands of Syros, Andros, Tinos, and Naxos. During the present study, the northern white-breasted hedgehog was seen for the first time on the islands of Lefkada, Paros, Mykonos, Patmos, and Ikaria. *Erinaceus roumanicus* was also recorded from Sifnos by Martin Gaethlich and Yannis Ioannidis (*in verbis*), from Kymolos, Syros, Aegina, and Salamina by Y. Ioannidis (*in verbis*), and from Bozcaada (Tenedos) and Gökçeada (Imbros) by Beytullah Özkan (*in litteris*). The occurrence on Ithake of this representative of the Erinaceidae needs to be confirmed (C. Papaconstantinou, *in verbis*). Northern white-breasted hedgehogs have been introduced also on the island of Aghios Efstratios around 1986 (Niki Kardakari, *in verbis*). Ristow (2000, *in litteris*) confirmed their occurrence on Naxos. I have personally observed these hedgehogs on Zakynthos, Skyros, Euboea, Milos, Ios, Alonissos, Crete, Rhodes, Tilos, Kos, Kalymnos, Leros, Chios, Samos, Lesbos, and Thasos. They are also possibly present on Antiparos, but there is no recent or reliable confirmation. Their introduction on Santorini (Thera) must have occurred after the end of the 19th century, since Douglas [275] did not mention the species on the island. Two specimens from the islet of Theodorou (off the north-western coast of Crete), which are still preserved in the ZSM collections, were collected in April 1966.

According to a recent morphological study carried out by Kryštufek et al. [92], insular hedgehogs tend to be smaller on remote islands and larger on islands close to the mainland coast, which is their putative source of colonisation. However, the insular response is not uniform and large size close to the mainland is possibly a result of frequent introduction.

Systematic data

Although several subspecies have been described from the European range of the northern white-breasted hedgehog, not all of them are clearly distinguishable, because their ranges and systematic value are not yet sufficiently known (compare with [274]). Six subspecies were reported from continental and insular Greece:

1. *Erinaceus roumanicus roumanicus* Barrett-Hamilton, 1900. It was dispersed in most of continental Greece, from Peloponnese and Attica to Epirus [70]. On morphological grounds, the Cretan hedgehog was previously referred to by Bate [275–277] as *Erinaceus roumanicus.*
2. *Erinaceus roumanicus bolkay* V. Martino, 1930. More recently, morpological studies carried out on samples from the southern Balkan peninsula revealed that the hedgehogs dispersed in this area belong to the subspecies *bolkay* V. Martino, 1930 [68]. The same subspecies has been reported from Euboea [273].
3. *Erinaceus roumanicus drozdovskii* V. Martino and E. Martino, 1933. Its distribution is limited to Macedonia, north of Thessaloniki [70].
4. *Erinaceus roumanicus nesiotes* Bate, 1905. This subspecies is found on Crete, Tinos, Syros, Naxos, and the Ionian islands of Corfu and Cephalonia [58, 68, 70, 273]. Recognised as an insular subspecies on morphological grounds, the Cretan hedgehog is clearly distinguishable from the mainland subspecies [274]. It would, however, appear to be karyotypically identical to the mainland counterpart *Erinaceus roumanicus drozdovskii* V. Martino and E. Martino, 1933. Recent genetic analyses suggest that Cretan hedgehogs originated from mainland Greece or other places in the Balkans and not from Asia Minor or the Levant [278].
5. *Erinaceus roumanicus rhodius* Festa, 1914. This subspecies occurs in the eastern Aegean islands of Rhodes, Samos, and Chios. It has statistically discernible karyotypic differences from the continental counterpart [68].
6. *Erinaceus roumanicus transcaucasicus* Satunin, 1905. Dispersed in Asia Minor, it seems to occur also on Kos and Lesbos (compare with [55, 273]). As observed by Giagia-Athanasopoulou and Markakis [273], the position of hedgehogs from the island of Lesbos in canonical (discriminant) analysis (CA), and their similarity to the specimens from Asia Minor, support the opinion that it is also probably present on Lesbos.

Museum specimens

Corfu: BMNH 8.10.1.6, ZMUP 5600, 5612, 5613, 5842, 5844; Cephalonia: BMNH 8.10.1.7; SMF 55262, 556261; ZMUP 5816, 5817, 5820; Antikythera: GNHM 4042; Skyathos: SMF 55857; Skopelos: NMW 19213; Skyros: SMF 49998; Euboea: ZMUP 5420, 5421; Aegina: ZMUP 3887, 3888; 3889; Seriphos: ZMUP 3888, 3889; Milos: NMW 18950; Andros: BMNH 65.480, 66.5373–66.5375; Naxos: BMNH 66.5376–66.5377; Ios: NMW 19214; Crete: BMNH5.5.12.2.12–5.12.2.13; NHMC 80.5.26 (1–15); NMW 18942–18945, 18948; SMF 65768; ZMUP 5303, 5348, 5495, 5719; ZSM 1966/0465, 1977/1324–1977/1327, 1997/0004, NHMC 80.5.26. (1–15), ZMUP 5335, 5336, 5386, 5389, 5393, 5410, 5414, 5706, 5744, 5745, 5819, 5958; Theodorou (off the north-western coast of Crete): ZSM 1966/0184 and 1966/0185; Rhodes: NMW 18946–18947; ZMUP 5580, 5581, 5588, 5598, 5786, 5787, 5789, 5790, 5620; Kos: ZMUP 5993; Samos: ZMUP 5065, 5601, 5604, 5611, 5766, 5775, 5776; Chios: ZMUP 5594, 5599, 5930, 5943, 5944, 5976, 5990; Lesbos: ZMUP 5229, 5594, 5617, 5712, 5713, 5811, 5890, 5928, 5929, 5932, 5933, 5943, 5976; Aghios Efstratios: SMF 67305; Samothrace: NMW 19209–19212; Thasos: SMF 62885.

References

Corfu [49, 68, 70, 92, 273, 279–281]; Cephalonia [68, 70, 92, 273, 279–285]; Zakynthos [50, 285–286]; Kythera [52–53, 281, 287]; Skiathos [92]; Skopelos [58, 92, 288]; Alonissos [285, 289]; Skyros [58, 70, 92, 281, 290]; Kythnos [58, 70, 281]; Seriphos [58, 70, 281]; Milos [51, 285, 288]; Euboea [68, 273, 281, 285, 291–293]; Tinos [68, 273, 280–281]; Syros [58, 68, 70, 273, 280–281, 294]; Delos [280]; Naxos [92, 273]; Ios [58, 70, 92, 281]; Thera [295]; Theodorou [92]; Crete [58, 68, 70, 92,

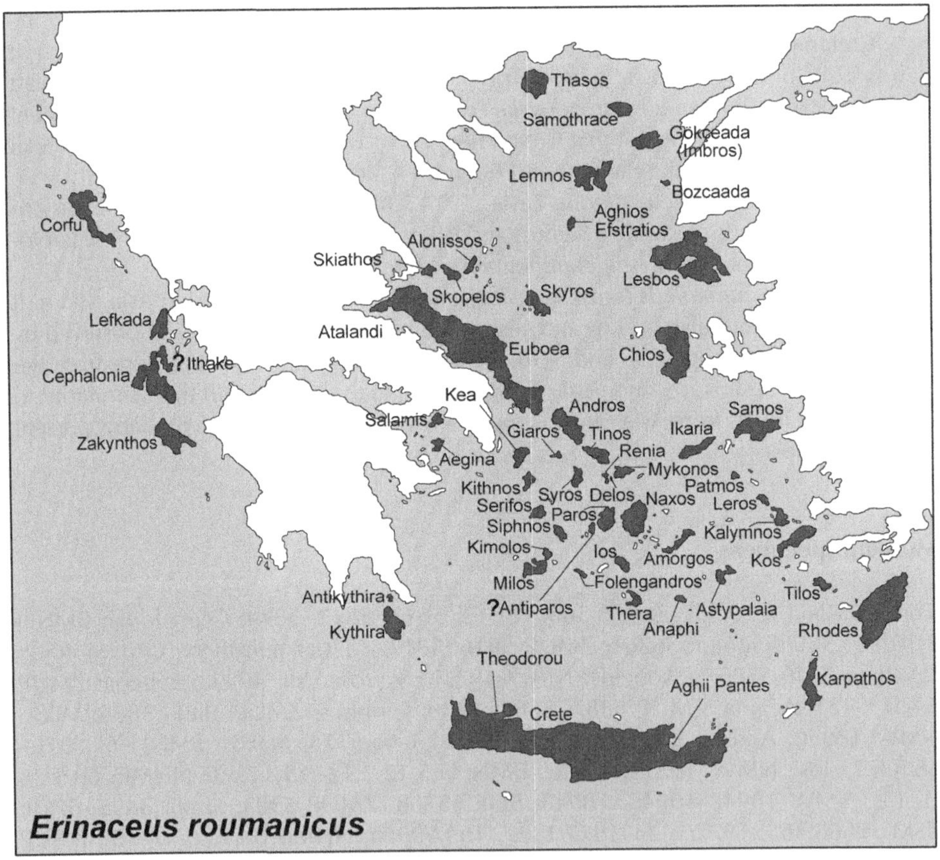

Fig. 23: Distribution of the Northern white-breasted hedgehog, *Erinaceus roumanicus* Barrett-Hamilton, 1900, in the Ionian and Aegean islands.

273, 276, 278–281, 285, 296–298]; Karpathos [285, 299]; Rhodes [55, 57–59, 64, 68, 70, 93, 141, 273, 280–281, 285, 300–301]; Tilos [60, 63, 302]; Kos [55, 58, 63–64, 70, 93, 140, 273, 281]; Kalymnos [58]; Leros [303]; Patmos [303]; Samos [65–66, 68, 93, 273, 280–281, 285, 304–306]; Chios [64, 67–69, 93, 273, 280–281]; Lesbos [68, 70–72, 93, 273, 280–281, 285, 304, 307]; Aghios Efstratios [92]; Lemnos [285]; Gökçeada (Imbros) [92–93, 308]; Samothrace [58, 70, 77, 92, 281, 285]; Thasos [92].

Ethnozoological and archaeozoological note

The extant occurrence of hedgehogs on many islands suggests a human-derived introduction. It appears, in fact, that these mammals have been artificially introduced onto several insular areas beyond their homeland to provide food supplies to the human communities since prehistory and throughout historical times [6, 309–310]. It is worth stressing, however, that not all of the zoological species provided by the archaeological exploration of ancient sites are necessarily edible, nor perhaps were they considered so at the time. Hedgehogs have been, in fact, also used as medicine, against snakes and/or for other purposes from prehistoric times on ([6, 9, 309–312]; compare with [304]). In antiquity, according to Pliny the Elder, their skin was also used in dressing cloth for garments: "*hac cute expoliuntur vestes*" (*Naturalis historia*, VIII: LVI 135) (see [313]; compare with [314]).

Hedgehogs are documented in the osteological record from early Neolithic settlements of the Eastern Mediterranean and Near East [315]. Noteworthy is their discovery in the prehistoric site of Skala Sotiros, on Thasos [316]. Early evidence of human consumption of these micromammals includes the osteological remains found in the Neolithic levels of the continental site of Lerna, in the Argolid, where cutmarks have been observed suggestive of carcass processing [317]. On several Aegean islands, such as Lesbos (Forest Department of Lesbos, 1997, *in verbis*) and Tilos, these animals are still appreciated in the local cuisine.

6.2 Eurasian pygmy shrew, *Sorex minutus* Linnaeus, 1766

Greek name: νανομυγαλίδα (*nanomygalida*)
Greek vernacular name: μυγαλή η μικρά (*migali i mikra*)
Turkish name: *cüce sivriburunu*

The distribution of the Eurasian pygmy shrew, *Sorex minutus* Linnaeus, 1766, ranges from the northern Iberian peninsula to Lake Baikal, in Siberia [318–319]. This representative of the Soricidae family is dispersed in forested zones of north-central Europe, occurring as far north as the Polar Circle, in Scandinavia, but it also penetrates further south than the larger *Sorex araneus* Linnaeus, 1758, in the Mediterranean region [93]. In Turkey the species is restricted to Thrace, where it is known from a single locality in the Istranca Mountains. It is regarded as a probably isolated population [93], whose karyotype is the same as that present in the rest of the species' European range: 2N = 42 [320–321].

The Eurasian pygmy shrew is absent from the Mediterranean coasts and islands, as it is from the Atlantic islands north of Scotland. One specimen, however, was captured alive while it was swimming in the sea in the vicinity of the mouth of Megalo Livari, in northern Euboea [293]. The species is also found temporarily on many islands in northern Europe [318]. Hence, it most probably colonises islands by active dispersal or passive transport.

Systematic data

According to Kryštufek and Vohralík [93], the pygmy shrew from the southern Balkans are larger than their central and northern European counterparts and show the phenotypic characteristics typical of *Sorex minutus gymnurus* Chaworth-Musters, 1932, described from the type in the locality of Mount Olympus, in Greece.

References

Euboea [293].

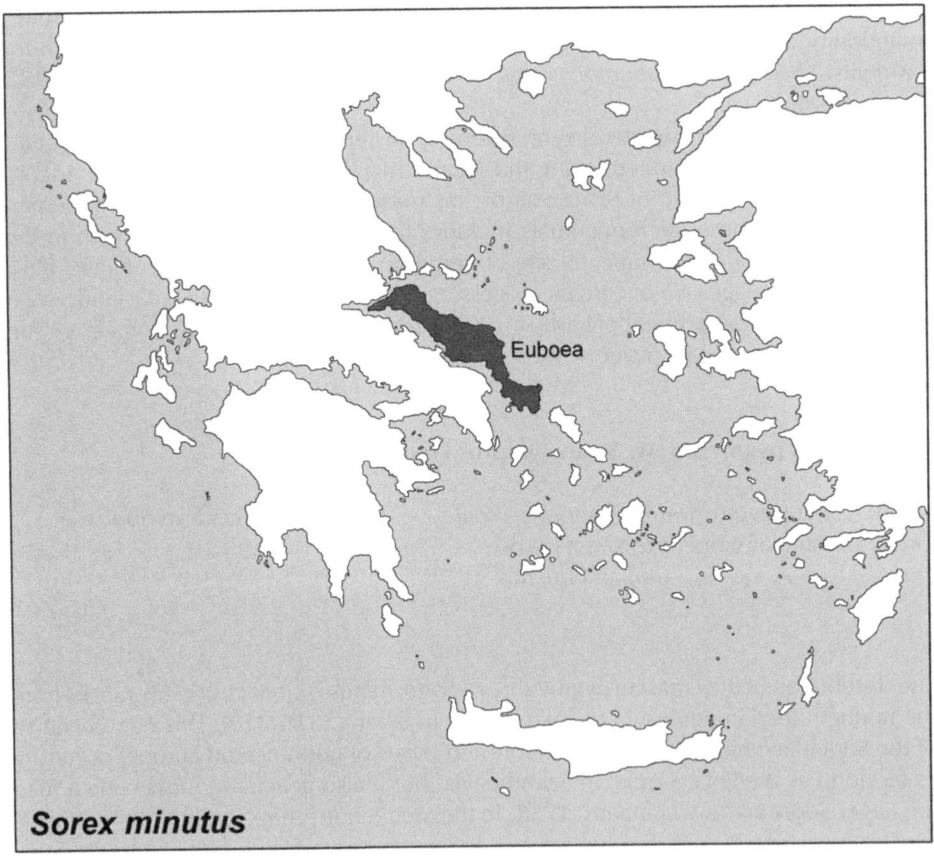

Fig. 24: Distribution of the Eurasian pygmy shrew, *Sorex minutus* Linnaeus, 1766, in the Aegean islands.

6.3 Mediterranean water shrew or Miller's water shrew, *Neomys anomalus* Cabrera, 1907

Greek name: βαλτομυγαλίδα (*valtomygalida*) Turkish name: *bataklik sivriburunu*
Greek vernacular name: μυγαλή η υδρόβιος (*migali i idrovios*)

The Mediterranean water shrew or Miller's water shrew, *Neomys anomalus* Cabrera, 1907, is dispersed in continental Europe between 37° and 55°N and Asia Minor, where it occurs in eutrophic riparian vegetation of still freshwater bodies, bogs and slow-flowing brooks and rivers from lowlands to altitudes of 1 850 metres [322–323]. It is common in northern Thrace, mainly along the Black Sea coast [93].

Although until today this species has never been reported from Mediterranean islands (compare with [323]), there may be some evidence of its occurrence in northern Euboea. In fact, remains of the shrew have been found in scats of otter, *Lutra lutra* Linnaeus, 1758, in the lagoon of Istiea [293]. It cannot be excluded, however, that we are dealing

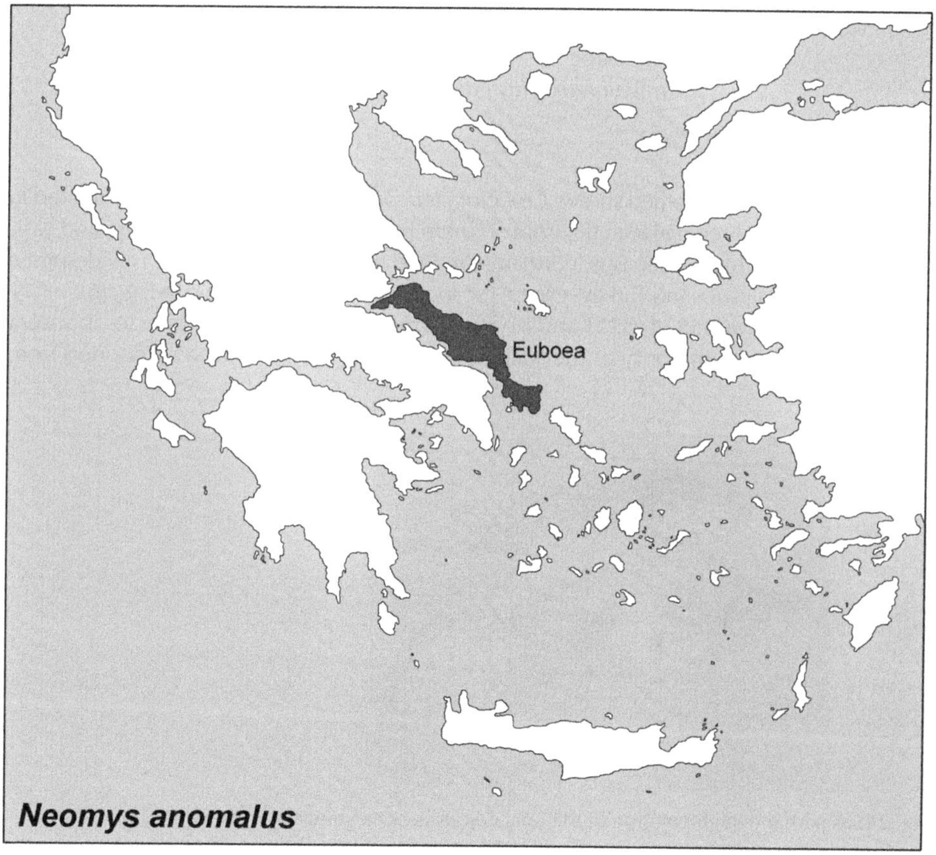

Fig. 25: Distribution of the Mediterranean water shrew or Miller's water shrew, *Neomys anomalus* Cabrera, 1907, in the Aegean islands.

in this case with an otter that imported onto Euboea the remains of a shrew previously hunted on the nearest mainland.

Systematic data

The Mediterranean subspecies of water shrew that lives in continental Greece is referred to as *Neomys anomalus milleri* Mottaz, 1907 [70]. Anatolian specimens are generally larger than their European counterparts, which are mainly sympatric with the water shrew, *Neomys fodiens* Pennant, 1777 [93].

References

Euboea [293].

6.4 Bi-coloured white-toothed shrew, *Crocidura leucodon* Hermann, 1780

Greek name: χωραφομυγαλίδα (*chorafomygalida*)
Greek vernacular name: μυγαλή λευκόδους η δίχρωμος (*migali leukodus i dichromos*)
Turkish name: *tarla sivriburunu*

The bi-coloured white-toothed shrew, *Crocidura leucodon* Hermann, 1780, is distributed in a vast area of the western Palaearctic, from northern France across central, southern and eastern Europe to the Volga, Caucasus, northern Iraq, and the Levant [93, 319]. It is widespread in the Balkan peninsula and Turkey, except for south-eastern Anatolia [93] (▶Fig. 26).

According to Krapp [324–325] and Kryštufek and Vohralík [93], this species is absent from Cyprus and the majority of the Mediterranean islands, whereas it is reported from

Fig. 26: A partial depigmentation of the coat colour is not so unusual in the representatives of the Crocidurinae subfamily, such as the bi-coloured white-toothed shrew, *Crocidura leucodon* Hermann, 1780. Photo by Massimo Del Guasta.

the Adriatic island of Cres and the Aegean island of Lesbos. In the course of the present research, however, bi-coloured white-toothed shrews were found in a significantly wider range. According to the data collected in the present survey, *Crocidura leucodon* is only distributed in the Eastern Aegean islands. In fact, besides Lesbos (A. Legakis, *in verbis*), it was found in five other islands: Karpathos [56], Leros [326], Chios [67]; Gökçeada (Imbros) and Bozcaada (Tenedos) (B. Özkan, *in litteris*). Cattaneo [303] reported from Leros only the occurence of *Crocidura* sp. The presence on Lesbos of a representative of the genus *Crocidura* different from the lesser white-toothed shrew, *Crocidura suaveolens* Pallas, 1811, was already noted by Ondrias [327].

Systematic data

Two subspecies are currently recognised: *Crocidura leucodon leucodon* Hermann, 1780, and *Crocidura leucodon narentae* Bolkay, 1925, whose respective ranges are however not completely understood [325].

References

Karpathos [56]; Leros ([326]; compare with [303]); Chios [67]; Lesbos [72–73, 93, 319, 327–330].

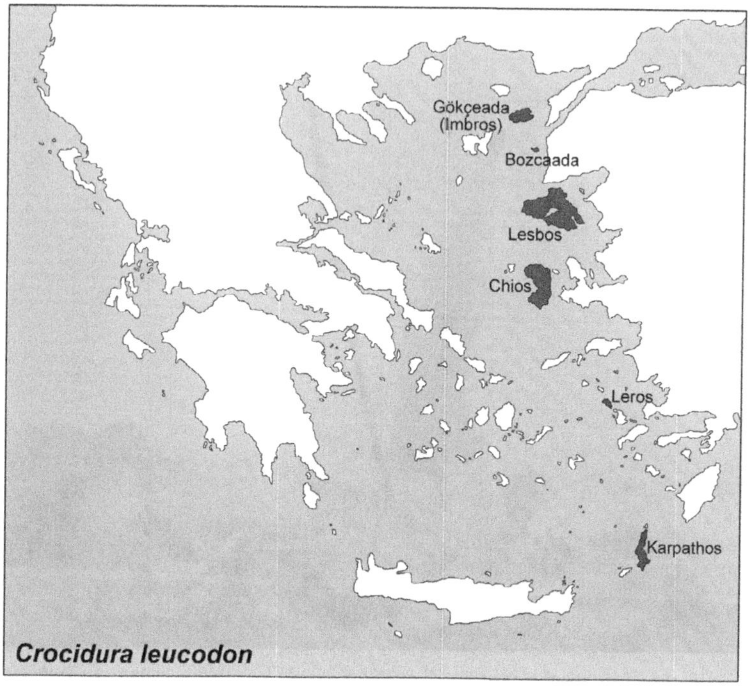

Fig. 27: Distribution of the bi-coloured white-toothed shrew, *Crocidura leucodon* Hermann, 1780, in the Aegean islands.

6.5 Lesser white-toothed shrew, *Crocidura suaveolens* Pallas, 1811

Greek name: κηπομυγαλίδα (*kipomygalida*)
Greek vernacular name: μυγαλή λευκόδους η μικρά (*migali leukodus i mikra*)
Turkish name: *bahçe sivriburunu*

The lesser white-toothed shrew, *Crocidura suaveolens* Pallas, 1811, is a synanthropic species dispersed in the southern Palaearctic, from northern Spain to Korea; it is also the most widespread and most common shrew in Turkey [93, 319].

Its distribution extends onto many Mediterranean islands. In the Ionian and Aegean seas it has been reported from Corfu, Zakynthos, Kythera, Euboea, Crete, the islet of Theodorou (off the northern coast of Crete), Rhodes, Kos, Samos, Psara, Chios, Lesbos, Samothrace, and Thasos. The biochemical similarity of Cretan populations with those from Anatolia reveals the origin of the extant lesser white-toothed shrew of this island [17, 331]. Bones recently found in barn owl pellets and individuals caught by trapping show that the species occurs also in Ithake, Cephalonia (H. Pieper, *in litteris*), and Tilos [63] (▶Fig. 28). Owing to archaeological finds from Konispol Cave, *Crocidura suaveolens* can be reported here for the first time on the Albanian coast off the Greek island of Corfu [332]. Angelici et al. [61] and Angelici and Riga [62] recorded a representative of the

Fig. 28: The lesser white-toothed shrew, *Crocidura suaveolens* Pallas, 1811, from the island of Tilos. Photo by Marco Masseti.

genus *Crocidura* Wagler, 1832, not yet recognised at the species level, from the island of Astypalaia, whereas Pieper [59] erroneously reported *Crocidura russula* Hermann, 1780 from Rhodes.

According to Poitevin et al. [333–334] and Catalan et al. [335], the almost allopatric distribution of *Crocidura* representatives on islands is explained by increased competition and mutual exclusion. In contrast, however, with this "mechanism of exclusion" theorised by Poitevin et al. [333], there are at least two Aegean islands wherein two species live sympatrically: *Crocidura suaveolens* and *Crocidura leucodon* on Lesbos, and *Crocidura suaveolens* and *Crocidura zimmermanni* in Crete [73]. In an analysis of a Holocene fauna of Chios, Besenecker et al. [336] found fossil remains of *Crocidura suaveolens* and *Crocidura leucodon* in the same cave but, as already observed, only the lesser white-toothed shrew lives on the island at present [67]. Moreover, in the central Mediterranean basin, *Crocidura leucodon* and *Crocidura suaveolens* seem to be sympatric on the Croatian island of Krk [337].

Systematic data

Crocidura suaveolens debeauxi Dal Piaz, 1924, is the taxon reported from Corfu [49]. Niethammer [52] suggested that the shrew from Kythera was possibly *Crocidura suaveolens mimula* Miller, 1901, which is the subspecies recorded by Ondrias [70] from Euboea. Crete is inhabited by *Crocidura suaveolens canae* Miller, 1909. However, according to Dubey et al. [338], *Crocidura suaveolens gueldenstaedtii* Pallas, 1811, is the variety occurring on Crete.

While Hoffmann [339] and Wolsan and Hutterer [330] considered the status of *Crocidua gueldenstaedtii* Pallas, 1811 still under discussion, according to Hutterer [319], *Crocidura suaveolens* is almost certainly conspecific with *Crocidura gueldenstaedtii* and must be regarded as a synonym of the latter.

Museum specimens

Corfu: BMNH 8.10.1.8; Crete: NHMC 80.5.20.4–80.5.20.23, 80.5.20.25–80.5.20.33, 80.5.20.39–80.5.20.59, 80.5.20.61–80.5.20.332, 80.5.20.359–80.5.20.371; NMW 36148, B5511–B5513; SMF 50537; Tilos: MZUPA ML137, TE928; Kos: SMF 56241–56278.

References

Corfu [49, 279, 330, 340–341]; Zakynthos [50]; Kythera [52–53, 330]; Euboea [70, 291, 293, 330, 340, 342–345]; Crete [17, 319, 330–331, 338, 341, 346–349]; Theodorou [348]; Rhodes [63, 93, 330]; Tilos [29, 63, 350–352]; Kos [59, 64, 330, 341]; Samos [65, 93, 330, 338, 341, 349, 353]; Psara [350, 352]; Chios [67, 330]; Lesbos [17, 73, 93, 329–330, 338, 349, 354]; Samothrace [77, 355]; Thasos [330, 356].

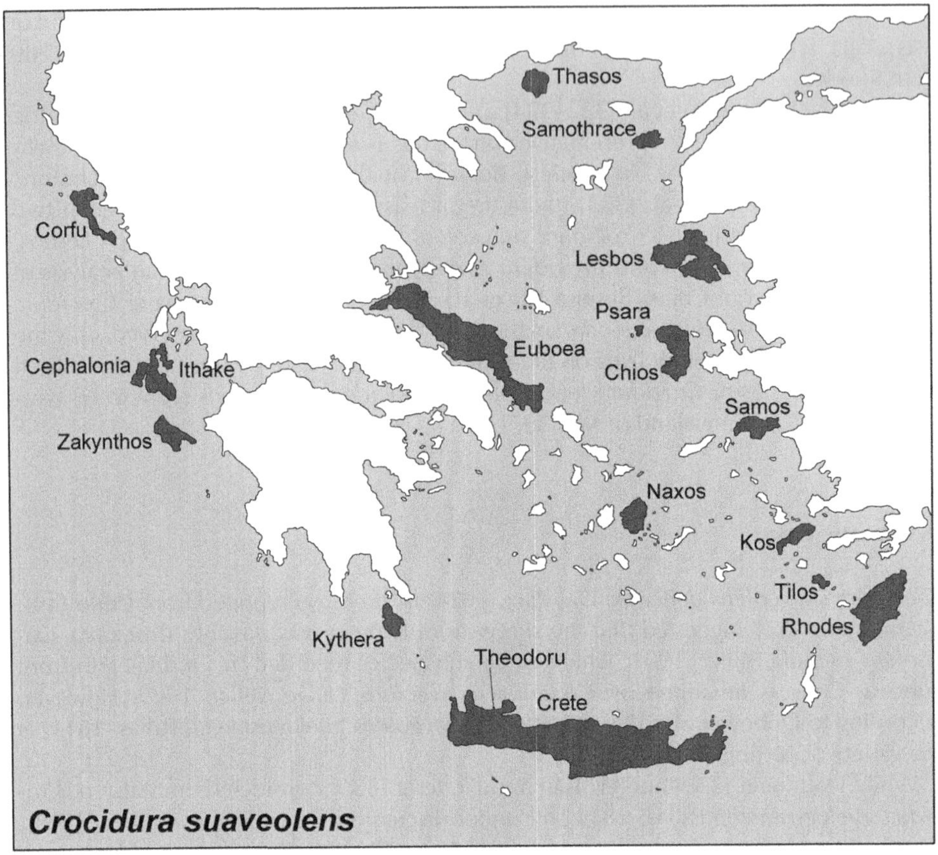

Fig. 29: Distribution of the lesser white-toothed shrew, *Crocidura suaveolens* Pallas, 1811, in the Ionian and Aegean islands.

Archaeozoological note

A medium-sized *Crocidura* was fairly common in Middle Minoan Kommos, in southern Crete [347]. These specimens were referred to as *Crocidura suaveolens canae* Miller, 1909, by Reumer and Payne [16]. According to Payne [347], the taxon reached Crete by the end of the Middle Minoan period.

6.6 Zimmermann's white-toothed shrew or Cretan white-toothed shrew, *Crocidura zimmermanni* Wettstein, 1953

Greek name: Κρητική σπιτομυγαλίδα (*Kritiki spitomygalida*)

Endemic to Crete, the Zimmermann's white-toothed shrew or Cretan white-toothed shrew, *Crocidura zimmermanni* Wettstein, 1953, has been recognised by Reumer [15] as a relic of the Early and Middle Pleistocene group of European *Crocidura* species, and of the Pleistocene Mediterranean fauna (▶Fig. 30). Fossils of this species have probably been present since the Early Pleistocene. *Crocidura* has been found in association with those endemic mammals that lived on Crete during the Pleistocene. It regressed after the humans introduced the species that characterise the current fauna [16]. *Crocidura zimmermanni* is considered endemic to Crete [17, 234, 319], because it is unknown today outside the island [17, 234]. More specifically, Reumer [15, 234] considers that the Cretan shrew is a relic for two reasons: firstly, it is the direct descendant of a group of *Crocidura* species that have disappeared from Europe already in the Pleistocene and that have been replaced by the present living species; secondly, it is the only known representative of the array of endemic shrews that were distributed in most of the Mediterranean islands during the Pleistocene.

The species appears to be limited to the Crete highlands [330, 357], but it was more widely distributed on the island during the Pleistocene [347]. In fact, the Cretan shrew is mainly distributed at high altitudes (over 1 200 metres), although Pieper [358] has collected pellets containing *Crocidura zimmermanni* bones at lower altitudes, and the shrew has also been collected in the gorge of Preveli, in southern Crete (P. Lymberakis,

Fig. 30: Endemic to Crete, the Cretan white-toothed shrew, *Crocidura zimmermanni* Wettstein, 1953, is today regarded as a relic of the rich Pleistocene endemic mammalian faunas that characterised the lost world of the Mediterranean islands. Photo by Apostolos Trichas.

Fig. 31: The subtropical gorge of Preveli, in southern Crete. Photo by Apostolos Trichas.

in verbis) (►Fig. 31). Its numerical dominance on the mountains and its absence from the fertile plains [331, 358] are probably symptomatic of direct competition with the invading lesser white-toothed shrew [17]. In fact, *Crocidura zimermanni* seems to be sympatric with *Crocidura suaveolens*, which was human-introduced, most probably during the Minoan period, about 1500 BC [16]. According to Vogel et al. [331], in the harsh ecological conditions of the mountains, to which *Crocidura zimmermanni* seems to be well adapted, the endemic shrew is more common than the introduced competitor. It is nonetheless interesting to note that the Cretan white-toothed shrew does not display the most common trends of endemisation reputed to affect endemic micromammals on islands: the well-known increase in size characteristic of fossil and/or subfossil shrews and rodents, but also documented in many extant populations [9, 30]. This fact raises a series of intriguing questions that are still far from being answered.

Systematic data

Crocidura zimmermanni was originally described as a subspecies of *Crocidura russula* Hermann, 1780, but Vesmanis and Kahmann [359] raised it to species rank on morphological grounds. The validity of this interpretation was then confirmed by the particular karyotype of 2N = 34, which is not found outside of this island [360–361].

Museum specimens

Crete: NHMC 80.5.20.24, 80.5.20.34–80.5.20.38, 80.5.20.60; NMW B5510.

References

Crete [15, 234, 330–331, 347–348, 357–360, 362].

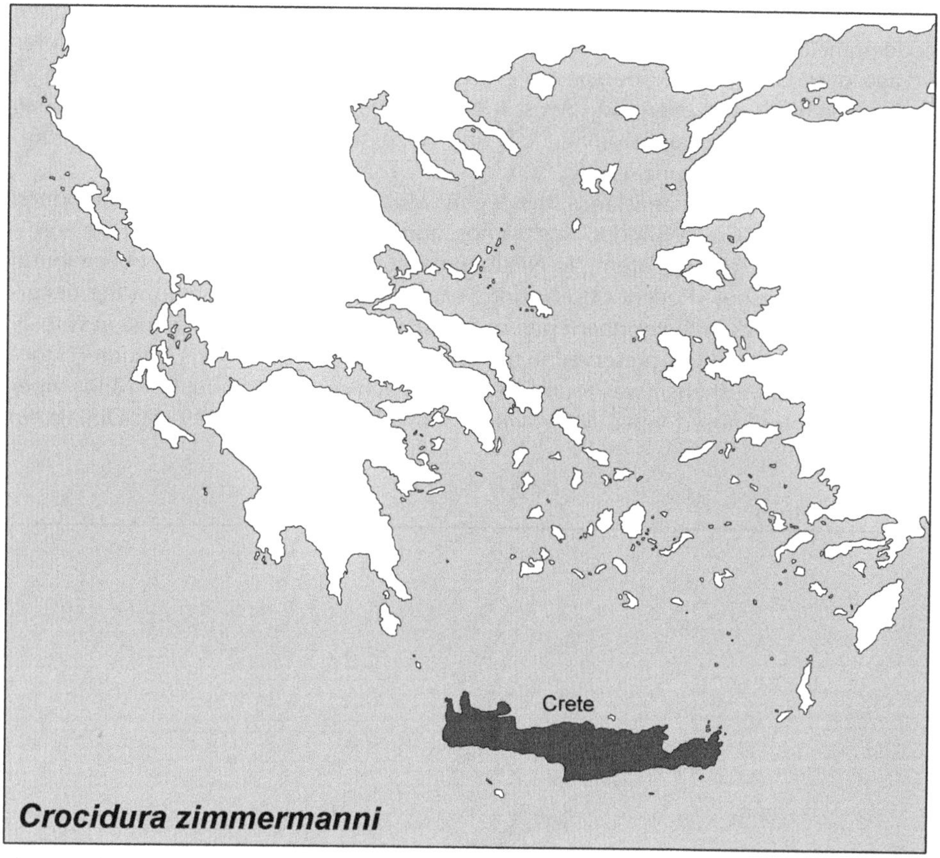

Fig. 32: The distribution of the Cretan white-toothed shrew, *Crocidura zimmermanni* Wettstein, 1953, is restricted to the fifth largest island of the Mediterranean Sea.

6.7 Pygmy white-toothed shrew, *Suncus etruscus* Savi, 1822

Greek name: Ετρουσκομυγαλίδα (*Etruscomygalida*) Turkish name: *Etrüsk sivriburunu*
Greek vernacular name: μυγαλή η πυγμαία
(*migali i pigmea*)

The pygmy white-toothed shrew, *Suncus etruscus* Savi, 1822, has a southern Palaearctic distribution, from the Iberian peninsula and the western Maghreb to Arabia, Asia Minor, Caucasus, Turkmenistan and Tajikistan [269, 319, 363]. Kock et al. [364] provide evidence of the occurrence of *Suncus etruscus* in eastern Anatolia (Norşum Tepe) in the period 3500–300 BC, and its continuous presence in this region was confirmed by Obuch [93, 365]. In Europe, it is restricted to the Mediterranean region, including several islands, where it prefers over-exploited or abandoned olive groves and vineyards with old dry stone walls and stonepiles, but occurs also in low maquis shrub and open Mediterranean forests [363]. Its distribution is confined to areas where July temperatures average over 20°C [363]. The species is also sympatric with humans, as it lives in gardens, orchards and even old towns. It also occurs on many of the Mediterranean islands, including Corsica, Sardinia, Sicily and Gozo, in the Maltese archipelago [366], Crete [297, 367] and Cyprus [368].

In the Ionian and Aegean islands, the pygmy white-toothed shrew has been reported from Corfu, Milos, Crete, Rhodes, Kos, Samos, and the islet of Theodorou, off the north-western coast of Crete. According to Niethammer [64], this species has not been found on Chios and Lesbos. Recent observations carried out in Euboea confirm the occurrence of the species in the northern part of the island ([291, 293]; Alivizatos, *in verbis*). The specimen from Milos, preserved in the collection of the Goulandris Natural History Museum (Kifissia, Athens), was found in the stomach of an individual of Milos viper or Cyclades blunt-nosed viper, *Macrovipera schweizeri* Werner, 1935 (M. Dimaki, *in verbis*).

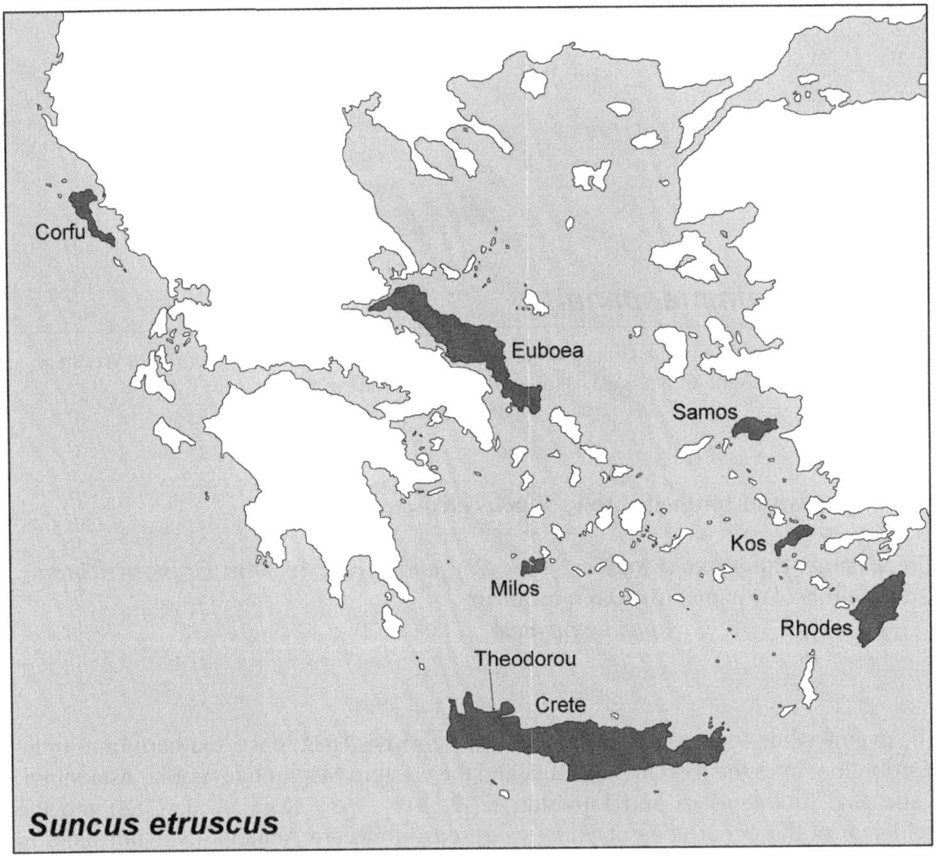

Fig. 33: Distribution of the Pygmy white-toothed shrew, *Sunucs etruscus* Savi, 1822, in the Ionian and Aegean islands.

Systematic data

On the island of Corfu, Niethammer [49] reported the taxon *Suncus etruscus etruscus* Savi, 1822.

Museum specimens

Milos: GNHM 90417; Crete: MZUF 10600; NHMC 80.5.21.3–80.5.21.8, 80.5.21.10–80.5.21.12, 80.5.21.14, 80.5.21.16–80.5.21.158, 80.5.21.164, 80.5.21.166–80.5.21.169; Kos: SMF 56210–56240.

References

Corfu [49]; Milos [51, 285]; Euboea [291, 293]; Crete [347–348, 367, 369–370]; Theodorou [285, 348]; Rhodes [58–59, 93, 348, 367]; Kos [59, 64, 93, 340, 371]; Samos [65–66, 93, 306, 340, 367].

Archaeozoological note

According to Payne [347], *Suncus etruscus* arrived on Crete at the end of the Bronze Age.

6.8 Balkan mole, *Talpa stankovici* V. Martino and E. Martino, 1931

Greek name: Βαλκανικός ασπάλακας (*Valkanikos aspalakas*)
Greek vernacular names: τυφλοπόντικας (*tiphlopontikas*)
ασπάλακας (*aspalakas*, Cephalonia)

The world distribution of the Balkan mole, *Talpa stankovici* V. Martino and E. Martino, 1931, is restricted to south-eastern Europe, where the species is confined to the Balkan peninsula [372]. It lives in fact in western Macedonia, as well as in the western and central part of Greece and probably in Albania [49, 372–376].

On the eastern Mediterranean islands, the species is known only from Corfu [49] and Cephalonia ([283–284]; C. Stamatopoulos, *in verbis*).

Systematic data

The Balkan mole was previously described as a subspecies of the Roman mole, *Talpa romana* Thomas, 1902, and treated as such until recently, when allozyme and morphometric analyses confirmed its specific independent status [372, 377]. According to

Mitsainas et al. [284], *Talpa stankovici* was recorded for the first time in Cephalonia: *"It was previously known that the genus* Talpa *is distributed on areas of Mount Ainos, however… it became possible to identify this* taxon *to species level"*.

Recent results of molecular phylogeny suggest a monophyletic origin of this taxon as sister species – together with *Talpa levantis* Thomas, 1906 – of other moles of western Europe, such as *Talpa europaea* Linnaeus, 1758, *Talpa occidentalis* Cabrera, 1907, *Talpa caeca* Savi, 1822, and the aforementioned *Talpa romana* [378].

Museum specimens

Corfu: ZFKM 1760–1765.

References

Corfu [49, 379]; Cephalonia [282–284, 380–381].

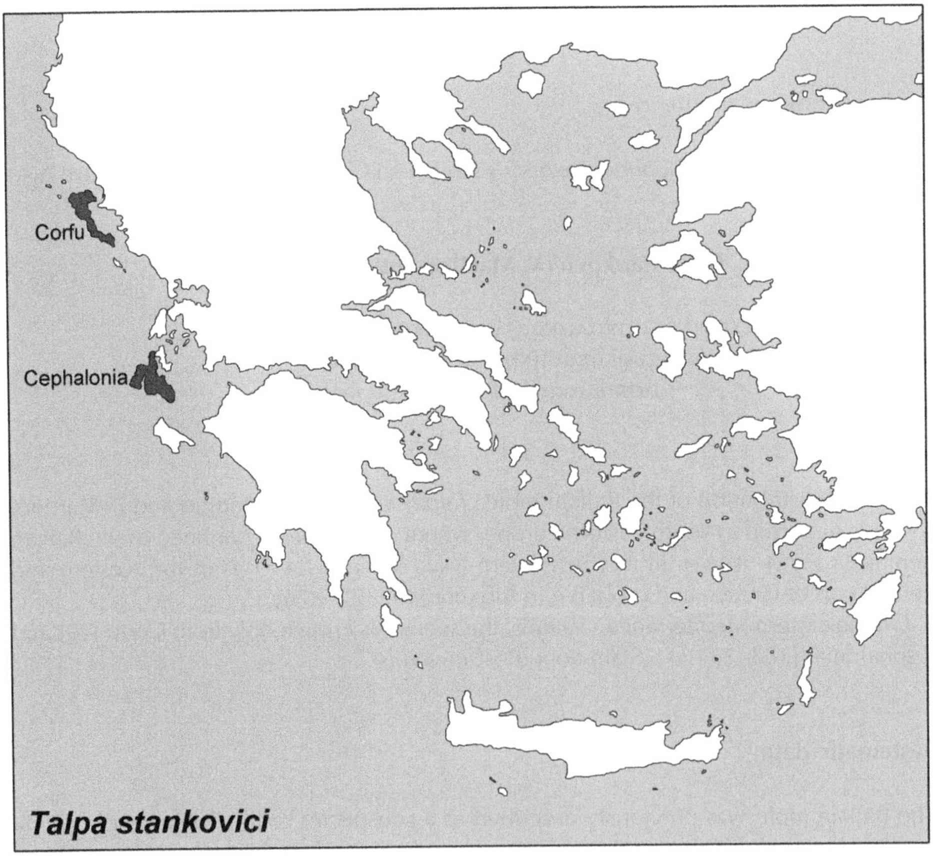

Fig. 34: Distribution of the Balkan mole, *Talpa stankovici* V. Martino and E. Martino, 1931, in the Ionian islands.

6.9 European or brown hare, *Lepus europaeus* Pallas, 1778

Greek name: λαγός (*lagos*) Turkish names: *avrupa tavşani, tavşan*

The European or brown hare, *Lepus europaeus* Pallas, 1778, is a continental faunal element characteristic of the Palaearctic Biogeographical Region (compare with [269]) (▶Fig. 35). Its range of distribution spans from western Europe to the west Siberian lowlands and the Near East, including Iran [382–383]. Brown hares occur practically in all kinds of grassland or open landscape, including coastal sand-dunes, cropland, pasture, clearings in scrub or forest, marshes and steppes, up to 1 500 metres above sea level.

This is one of the most versatile mammals dispersed in the Eastern Mediterranean region. In fact, it occurs virtually everywhere throughout the entire Balkan and Anatolian peninsulas, as well as on several islands [93, 269–270, 373, 384]. The presence of hares on islands in the Ionian and Aegean Seas is an indication of the vagility of the species [385]. It has been reported from Cephalonia, Lefkada, Zakynthos, Kythera, Milos, Siphnos, Aegina, Euboea, Kea, Andros, Tinos, Delos, Syros, Paros, Naxos, Crete, Dragonada, Paximada, Ghianissada, Gaidouronisi (Chrysi), Koufonisi, Gavdhos, Amorgos, Astypálaia, Karpathos, Rhodes, Kos, Ikaria, Samos, Chios, Lesbos, Leros, Samothrace, and Thasos. Data collected during the present study proves its occurrence, for the first time, on the islands of Kimolos (C. Antoniou, *in verbis*), Ithake (C. Papaconstantinou, *in verbis*), Fourni (M. Dimaki, *in verbis*), Amorgos (R. Tziakiris and C. Papaconstantinou, *in verbis*), Skyros (A. Dimitropoulos, *in verbis*), Gavdhopoula, off southern Crete (P. Lymberakis, *in litteris*), and Gökçeada (B. Özkan, *in litteris*). Hares possibly occur – or have occurred until very recent times – also on the Ionian islet of

Fig. 35: European or brown hare, *Lepus europaeus* Pallas, 1778. Photo by Marco Masseti.

Kalamos, east of Lefkada: there is in fact a very old report from local hunters. In the course of this research, it was possible to confirm its occurrence also on the following islands: Lefkada, Cephalonia, Zakynthos, Euboea, Syros (Y. Ioannidis, *in verbis*), Paros, Samos, Chios, Lesbos, Thasos (C. Antoniou, *in verbis*), Milos (M. Dimaki, *in verbis*; C. Antoniou, *in verbis*), Naxos (C. Antoniou, *in verbis*; in the year 1998, H. Alivizatos found remains of *Lepus europaeus* in pellets of Bonelli's eagle, *Hieraaetus fasciatus* Vieillot, 1822, *in litteris*), Gavdhos, Aghios Theodorou, Gaidouronisi (Chrysi), Koufonisi, Dragonada (P. Lymberakis and K. Paragamian, *in litteris*), Aghios Nikolaos, off the entrance of Souda Bay, Crete (P. Lymberakis, *in litteris*), Rhodes, Ikaria, and Samothrace. In September 1995, Dietrich Ristow [386] reported four carcasses of hare from the islet of Skandilion, west of Skantzoura in the Northern Sporades. The occurrence of the species on Skyros, reported by A. Dimitropoulis (*in verbis*), needs to be confirmed. M. Dimaki (*in verbis*) reported the occurrence of the species from Milos, Fourni, Ikaria and Samos, and according to A. Dimitropoulos (*in verbis*), young individuals were found on Syros at the end of the 1990s.

The hares of Delos became extinct during recent historical times, and, according to Bodson [387], they have been replaced by rabbits (compare with [26]). Angelici et al. [61] observe that the hare of Astypalaia is on the brink of extinction, if it has not become extinct already: only footprints have been observed on a sandy beach. According to D. Ristow (*in verbis*), however, these lagomorphs were no more present since the second half of the 1990s on the latter island, where only rabbits occur. Husband and Davis [38] reported the species also from the islet of Theodorou, off north-westen Crete, whereas Zimmermann [297] and Ondrias [70] confirmed the occurrence of only rabbits. There are no hares on the islands of Santorini (Thera), nor on Kalymnos. On Lemnos, these lagomorphs possibly vanished between the 1980s and the 1990s. Recently the European hare has become extinct also on Sifnos, in the western Aegean islands.

Systematic data

Geographical isolation may have often produced distinct varieties. In the literature, the various subspecies of insular hares are distinguished by the coat patterns as well as by the size of the body and skull. This logically led to a plethora of varieties which now need to be thoroughly revised [388]. On the basis of morphologic and geographic characters, for example, the hares of the Ionian and Aegean islands have been attributed to at least 5 or 7 different subspecies:

1. *Lepus europaeus creticus* Barrett-Hamilton, 1903, from Cephalonia [279], Crete [276–277] and several small islets off Crete, including Gavdhos, Gavdhopoula, Gaidouronisi (Chrysi), Koufonisi and the archipelago of the Dionysades [389]. Niethammer [389] observes that the hares of Paximada (Dionysades islands) and Koufonisi are smaller than those of Crete itself, but fall within the morphology of *Lepus europaeus europaeus* Pallas, 1778, and not of *Lepus capensis* Linnaeus, 1758.
2. *Lepus europaeus ghigii* De Beaux, 1927, from Astypalaia [55, 390].
3. *Lepus europaeus carpathous* De Beaux, 1929, from Karpathos [55].

4. *Lepus europaeus rhodius* Festa, 1914, from Rhodes [55, 58].
5. *Lepus europaeus cyrensis* Satunin, 1905, reported from the island of Kos, is the subspecies reputed to occur also in western Anatolia [55, 64].

According to Miller [279], the brown hare of Corfu might refer to *Lepus europaeus meridiei* Hilzheimer, 1906, whereas Ondrias [70] described it as *Lepus europaeus transsylvanicus* Matschie, 1901.

Although characterised by differtent varieties of coat colour, the sample of Euroean and Anatolian *Lepus europaeus* studied by Sert et al. [391] are closely related to one another genetically, and only some separate population pairs revealed increased genetic divergence. In fact, the recent study of Y DNA and mithocondrial genetic systems confirm the deep separation of Anatolian and Europen lineages of *Lepus europaeus* [392]. In the case of Rhodes, which was disconnected from mainland Anatolia during the evolutionary period of hares, the genetic lineages are phylogenetically relatively close to several central or south-western Anatolian haplotypes [393]. Most of the hares of the eastern Aegean islands are possibly to be described more as ecophenotypic varieties than true geographical subspecies. Nonetheless, as observed by Ghigi as far back as 1929, doubts may arise on their supposed endemicity and on the fact that the hares of the eastern Aegean islands might have been translocated by man.

Museum specimens

Cephalonia: BMNH 8.10.1.52–8.10.1.5256, 8.11.3.20–8.11.3.21; Zakynthos: NMW 42779; Andros: BMNH 66.5380; Crete: BMNH 2.11.9.1–2, 99.2.14.2; NHMC 80.5.39.6, 80.5.39.8, 80.5.39.23, 80.5.39.28; NMW 18622–18625; ZMUP, 6535, 6539–6540, 6542, 6548; Dragonada: NHMC 80.5.39.15–80.5.39.17; Paximada: ZFMK Niethammer collection 6022–6028, 6304–6305; Gaidouronisi (Chrysi): NMW 18627; Koufonisi: ZFMK Niethammer collection 10204; Rhodes: MCSTM 2327; MRSNT 2474, 2476–2486; NMW 43544; Samothrace: NMW 18626.

References

Corfu [49, 70, 279, 395]; Lefkada [396–397]; Cephalonia [70, 279, 282–283, 285, 304, 397]; Zakynthos [50, 285–286]; Kythera [53, 397–400]; Milos [51, 285, 401]; Siphnos [58, 401]; Kea [401–402]; Aegina [403]; Euboea [285, 292]; Andros [293, 401, 404]; Tinos [401, 404]; Delos [387]; Syros [401, 404]; Paros [401, 404]; Naxos [58, 290, 304, 400–401, 404–405]; Astypálaia [55, 58, 61, 70, 373, 390]; Theodorou [38]; Crete [58, 276–277, 279, 285, 297–298, 373, 384, 392, 397, 400, 406–408]; Gavdhos [70, 297, 348, 389]; Paximada [297, 389]; Ghianissada [348, 409]; Dragonada [348, 409]; Gaidouronisi [297]; Chryssi [389]; Koufonisi [297, 348, 389]; Karpathos [55, 58, 70, 285, 299, 410]; Rhodes [55, 57–58, 70, 285, 297, 301, 373, 384, 397, 400]; Kos [55, 58, 64, 70, 373]; †Leros [411]; Ikaria [58]; Samos [66, 305–306, 397, 400, 412]; Chios [64, 67, 69, 304, 397, 400]; Lesbos [72, 307, 397, 400]; Leros [411]; Samothrace [58, 304, 413]; Thasos [71, 285, 397].

Archaeozoological note. Early artificial appearance of *Lepus europaeus* on the Aegean islands

As we have already seen, brown hares are present on several Aegean islands. Many of the islands off the Anatolian coast had landbridge connections with mainland Anatolia when the sea level at the Late Glacial Maximum was lowered by about 120 m [414]. Thus, hares from islands such as Chios or Lesbos probably originated from mainland Anatolia, from were they could easily have immigrated some 15 000–10 000 years ago. Moreover, in the Late Pleistocene, hares lived in the whole of the Greek peninsula up to present-day Alonnissos [415] at a time when the Northern Sporades had not yet become islands. Indeed, genetic research carried out by Kasapidis et al. [400] showed that the eastern Aegean islands of the Asia Minor coast were most likely colonised naturally through a Late Pleistocene landbridge connection. However, hares from islands without Pleistocene landbridge connections to Anatolia or other continental regions, such as Crete or Rhodes, must have been introduced by humans in ancient times or more recently [414].

The brown hare is one of the species most adaptable to peculiar environmental conditions, including small and barren islands. These lagomorphs were brought by sailors and let loose on islands so that they could breed and provide a store of fresh meat that would be readily available for the ship passengers [6, 8]. Apparently this was the only way to exploit the natural resources of territories that could not be cultivated and that were generally economically unattractive for humans. The evidence providing the importation of hares on islands before the Early Bronze Age, however, is not fully convincing [8, 388]. In fact, in the light of recently found archaeozoological evidence, the earliest relevant documentation appears to come from the islands of Thasos [316], Santorini [416], Crete [417–418] and Amorgos [419] (▶Fig. 36). A few fragments of

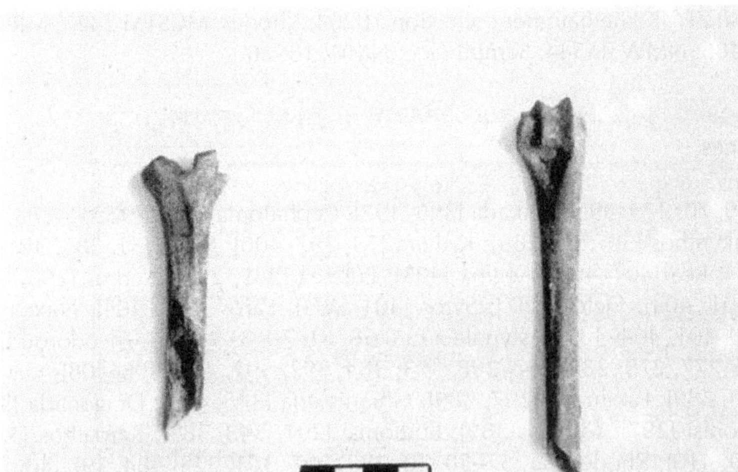

Fig. 36: Distal fragments of humerus and tibia of *Lepus europaeus* Pallas, 1778, from the Late Bronze Age site of Akrotiri, on Santorini. Photo by Y. Sarakinis, from Trantalidou [416].

hare were found in the Lower Mesolithic context of the cave of the Cyclops, on Youra (Northern Sporades) [415, 420], and in the Mesolithic levels of the site of Maroulas, on Kythnos (West Cyclades) [421], but very likely the lagomorphs reached both the islands naturally. In fact, it is still unclear whether Youra and Kythnos were already islands in the Late Pleistocene and/or the Early Holocene (compare with [166, 228, 420, 422]). Bones of hares have also been provided by the archaeological exploration of the layers of the Geometric period (8th century BC) of the site Minoa, on Amorgos [423]. The existence of hares on islands has been reported by classical authors, such as Homer *(The Odyssey,* IX, 116–124) and others. Xenophon (5th century BC), in his *Kinegeticon* (24–26), observed that hares were particularly abundant on islands because predators such as foxes and eagles are rarer than on the mainland. The ancient practice of releasing hares on islands, which are considered natural enclosures, was kept up into historical times in the *leporaria* of the Roman period (Varro, *De re rustica,* III, 12, 1; [387]) and the Middle Ages [8].

Fig. 37: Former and present distribution of *Lepus europaeus* Pallas, 1778, in the Ionian and Aegean archipelagos. Islands that are inhabited by the hare are shaded and those where they have become extinct are marked with a cross.

Hares can survive even on very small islands down to 3–3.5 km^2, as shown by the populations that still inhabit – or have inhabited until few years ago – several islets off Crete, such as Gavdhos, Gavdhopoula, Aghios Theodorou, Gaidouronisi (Chrysi), Koufonisi, Aghios Nikolaos, Paximada, Ghiannissada and Dragonada (Dhionysiades archipelago) (compare with [297, 389, 409]). These are barren-looking islands, short of water, very windy, and characterised by great sand-dunes. Apart from Dragonada, the majority of the hares which occur today on the circum-Cretan islets have been introduced at various times by the Forest Department of Crete (P. Lymberakis and K. Paragamian, *in litteris*). According to Karetsos et al. [409] and D. Ristow (*in litteris*), these lagomorphs became extinct on Paximada towards the end of the 1980s, whereas in September 2009 hunters again shot several individuals on Dragonada. Hares have been released even in very recent times on barren islets of the Ionian and Aegean seas mainly for hunting purposes and to some way exploit these territories, which are otherwise poor and unproductive. This is a way of simplifying management problems. Islands are considered natural enclosures where the wild game ought to derive its food supply directly from the carrying capacity of the environment [6]. We cannot overestimate the importance of the islands inhabited by free-ranging populations of herbivores; these represented living depositories of animal proteins available at any time along the marine routes of antiquity since prehistorical periods [6, 8].

6.10 Wild or European rabbit, *Oryctolagus cuniculus* Linnaeus, 1758

Greek name: αγριοκούνελο (*agriokunelo*)
Greek vernacular names: κόνικλος (*koniklos*)
κουνέλι (*kuneli*)

Turkish names: *ada tavşani, ada tavşcan*

The wild or European rabbit, *Oryctolagus cuniculus* Linnaeus, 1758, is the ancestor of all domestic varieties of rabbit (▶Fig. 38). The species is regarded as a post-glacial endemism of the Iberian Peninsula [388, 424–429]. Today, however, it figures as one of the world's most widespread mammalian species. Its spread in western Europe is the most important human-induced migration from its ancestral Iberian territories ([430]; compare with [431]). In fact, humans have exchanged rabbits from prehistory until the early Middle Ages; but, perhaps, the process is still far from over.

Both domestic and wild-type populations have been introduced into the wild all over the world for a variety of reasons: for sport, to farm for meat or fur production, as food for other animals or bait for lobster pots, as pets, to control vegetation, amuse tourists, and even to conserve representative stock from myxomatosis (compare with [26]). In Europe, the species is now distributed north to Shetland, southern Sweden and Gotland, east to the north coast of the Black Sea, south to the entire Iberian penisula, and part of continental and insular Italy; it is very localised in the Balkan penisula (compare with [426]). Several free-ranging populations are present

Fig. 38: European wild rabbit, *Oryctolagus cuniculus* Linnaeus, 1758. Photo by Marco Masseti.

in North Africa (compare with [432]), and in some islands off the Maghrebine coast [26, 433–435]. This is one of the mammalian species more widespread in the Mediterranean area, where it is regarded as a really successful island coloniser. In very recent times, the rabbit became extinct on a few islands of the western Mediterranean due to the spread of myxomatosis [26, 436]. To preserve the insular biological diversity of the Mediterranean area, the species has been subjected to several attempts of eradication in the course of the last decades (see [388, 437]). Rabbits can, in fact, damage vegetation, accelerate soil denudation and erosion, and may ultimately be responsible for the decline or extinction of various species of reptiles and birds, e.g., [9, 438–442].

Records from the literature and museum collections report it on Corfu, Arpia (Strophades), Antikythera, Kythera, Aegina, Euboea, Atalandi, Alonissos, Pappous (N. Sporades), Psathoura (north of Youra, N. Sporades), Grammiza, Kythnos, Kimolos, Milos, Giaros, Sikinos, Kardiotissa (between Pholegandros and Sikinos), Seriphos, Kea, Andros, Mykonos, Delos, Evreokastron (Paros), Levitha, Folegandros, Thera, Anaphi, Makria (Anaphi), Pachia (Anaphi), Crete, several satellite islets off Crete, such as Theodorou, Dhia, Aghii Pantes, Gaidouronisi (Chrysi), Koufonisi, Elasa, and Makroulo, Armathia, Kassos, Karpathos, Tilos, Kos, Gyali, Leros, Arki, Agathonisi, Chios, and Lemnos. Data collected during the present survey confirm its occurrence on the following Ionian and Aegean islands: Arpia (Strophades) and Milos (M. Dimaki, *in verbis*), Syros (A. Dimitropoulos, *in verbis*), Kimolos and Aegina (Y. Ioannidis, *in verbis*), Astypalaia (D. Ristow, *in verbis*; A. Sioulas, *in verbis*), Armathia, Kassos, Nissiros, Telendos (off Kalymnos), and Lipsoi (M. Mandragos, *in verbis*), Lesbos and Fourni [388]. The species was personally observed on Milos, Mykonos, Delos, Antiparos, Paros, Leros, Lemnos, Tilos Kos, and Rhodes. On the latter island, the rabbit is present today where I had reported it – at least from the year 2005 – in the suburban park of Rhodini, possibly after a very recent human introduction. The coat colour of all the specimens observed in Rhodes falls within the wild phenotypes, which are of the "agouti" type without any variation towards partial depigmentation. Rabbits were also imported in very recent times on Tilos (over the last few years) for hunting [63]. This leporid was first reported on the islands of Vromonas (between Ithake and the mainland) (C. Papaconstantinou, *in verbis*), Stamphani (Strophades) (M. Dimaki, *in verbis*), Velopoula (east of south-eastern Peloponnese), Astakida (north of Skopelos), Cristianà (south of Thera), Zaforà (south of Astypalaia), Astakida (north-west of Karpathos), Sirna (south-east of Astypalaia), Leros (C. Constantinou, *in verbis*), Lipsoi, Patmos (C. Costantinou, *in verbis*; personal observations), Seriphos, Kythnos, Kymolos, Aegina (Y. Ioannidis, *in verbis*), Prasouda (between Euboea and Skyros) (A. Dimitropoulos, *in verbis*), Petaloi (off Euboea) (Y. Ioannidis, *in verbis*), Giaros (between Kea and Tinos), Tinos (M. Dimaki, *in verbis*), Gavrionissia (off Andros) (M. Dimaki, *in verbis*), Andros, Renia (west of Delos), Folegandros, Thera, Anaphi, Aghii Pantes (off of Crete), Nissiros (M. Mandragos and Y. Ioannidis, *in verbis*), Kalymnos (personal observation), Telendos, west of Kalymnos (personal observation), Leros (personal observation), Alatonisi (west of Fourni) (M. Dimaki, *in verbis*), Psara (off Chios), Andipsara (off Chios), Ionusses (off Chios), Aghios Stefanos (off Chios), Aspronissos (off Lesbos), Panaghias (off Lesbos), Barbalias (off Lesbos), Aghios Efstratios (N. Kardakari, *in verbis*), Bozkaada (Tenedos) (B. Özkan, *in litteris*). They have been observed also on the islets of Tria Nisia, south of Sirna.

Although the species is officially considered to be absent from Crete, Niethammer [395] and Pieper [369] inform that rabbits formerly inhabited few areas of the island, while Dietrich Ristow (*in litteris*) reported its presence in western Crete. Masseti and De Marinis [388] confirm the occurrence of rabbits on several islets offshore, such as Theodorou, Gadaironisi, and Dhia. On the latter island rabbits have been reported since 1906 (the specimen BMNH 5.12.2.35 was captured 5 April 1906). In the course of a HOS (Hellenic Ornithological Society) survey on Lipsoi, in the years 1997 and 1998, H. Alivizatos (*in litteris*) found remains of *Oryctolagus cuniculus* in pellets of Bonelli's eagle, *Hieraaetus fasciatus* Vieillot, 1822. The coexistence of the European hare and the European rabbit has also been reported from several islands, such as Milos (M. Dimaki, in verbis), Aegina (Saronic Gulf), Gadaironisi, Theodorou, and Koufonisi (circum-Cretan islands), Paros, Andros, Milos, Leros and Tinos (Cyclades), Kos (Dodecanese), Lesbos and Fourni (north-eastern Aegean Islands) (see also [388]). In this regard, however, Stephanides [443] quotes a local Delian informant: "… *you will only find hares on the islands north of Delos and rabbits on those to the south* " (see also [26]).

Only the two islets of Theodorou and Gadaironisi, off the northern and the southern shores of Crete, respectively, are less than 5 km^2. Rabbits can survive on even smaller islets, as shown by the populations that still inhabit Delos (3 km^2), Dhia (1.35 km^2), Aghii Pantes (0.4 km^2), the already mentioned Theodorou (0.7 km^2), and many others ever smaller ones [8, 388], where they have been introduced at an unknown date, presumably for meat and fur (compare with [128]). According to Kock [67], Pauli's [444] record of rabbits on Chios probably refers to hares (compare with [26]). The species is also present even on islets only 0.088 km^2 in area (i.e., Makroulo, north of Koufonisi, off south-eastern Crete), and even smaller. In Turkey, feral individuals occur only on some offshore islands [93], such as Yassica, Göcek, and Delitli, in the gulf of Fethyie, and Koca, in the Gulf of Selymie, Buzburun peninsula (personal observation).

Systematic data

Wild rabbits of the Aegean archipelago are referred to the taxon *Oryctolagus cuniculus cnossius* Bate, 1905. Specimens which fall within the phenotypes of this subspecies have been reported from the islands of Kardiotissa (Sikinos) and Evreokastron (Paros) [58, 70], Theodorou (Crete) [70, 297], Dhia (Crete) [26, 276, 297] and Kos [64, 394]. Moreover, according to Miller [279] and Zimmermann [297], the colour of the coat of the rabbits of Dhia is particularly typical of this subspecies. The populations of several Greek islands, such as Zafora, Sirna and Tria Nisia, as well as those of many other islets off the Greek and Turkish coasts, are characterised by coat colours of the domestic phenotypes, showing tendencies towards melanism, partial depigmentation, and/or albinism.

Museum specimens

Arpia (Strophades): GNHM 4951; Antikythera: GNHM 90512; Pappous (N. Sporades): SMF 55786–55788; Psathoura (north of Youra, N. Sporades): SMF 55789, 55856; Sikinos: NMW 18620; Kardiotissa (between Folegandros and Sikinos): NMW 18621; Dhia: BMNH 5.12.2.35; NHMC 80.5.39.4–80.5.39.10, 80.5.39.18–80.5.39.20, 80.5.39.22–80.5.39.28; NMW 18616–18619, B1259–B1261; Aghii Pantes: NHMC 80.5.39.21; Gaidouronisi (Chrysi) (south of Ierapetra, Crete): NHMC 80.5.39.14; Koufonisi (south of south-eastern Crete): NHMC 80.5.39.1–80.5.39.3, 80.5.39.11; Makroulo (north of Koufonisi): NHMC 80.5.39.12–80.5.39.13; Kos: MZUF 12685; Gyali (between Nisiros and Kos): SMF 34972.

References

Corfu [49]; Arpia (Strophades) [445]; Atalandi [43, 446]; Psathoura [447]; Grammitza [289]; Kythera [398–399]; Antikythera [350, 448]; Aegina [403]; Euboea [292]; Alonissos [285], Kea [449]; Kythnos [401]; Kimolos [401]; Milos [51, 285]; Seriphos [401]; Sikinos [58]; Giaros [401]; Kardiotissa (near Sikinos) [58, 70, 340]; Andros [401, 404]; Mykonos [401]; Delos [26, 443]; Evreokastron (Paros) [58, 70]; Levitha [58]; Folegandros [401, 404]; Thera [275, 295, 404]; Anaphi [58, 450]; Makria (Anaphi) [26, 58, 70, 450]; Pachia (Anaphi) [26, 58, 70, 450–451]; Crete [369, 395]; Theodorou (Crete) [70,

340, 362]; Dhia (Crete) [26, 276–277, 297, 340, 348, 452]; Aghii Pantes (Crete) [348, 452]; Gaidouronisi (Chrysi) [128, 348, 362]; Kouphonisi [128, 348]; Elasa, east of Crete [26]; Armathia [233]; Kasos [233]; Karpathos [285]; Tilos [63]; Kos [64, 70, 340, 371, 394]; Leros [141, 303, 411]; Patmos [303]; Agathonisi [453]; Arkios [453]; Kavolos [453]; Makronissi [453]; Smineronissi [453]; Strongyli [453]; Agrelloussa [453]; Chios [26]; Lemnos [285, 454–455].

Archaeozoological note

On the Greek island of Lemnos, located in front of the Turkish Dardanelles strait, stands the settlement of Poliochni, which is considered the first town in Europe [456]. Remains of rabbits were found in the entire stratigraphy, from the Final Neolithic period to Early Helladic III period (3100–2000 BC) [457]. The occurrence of these lagomorphs in such

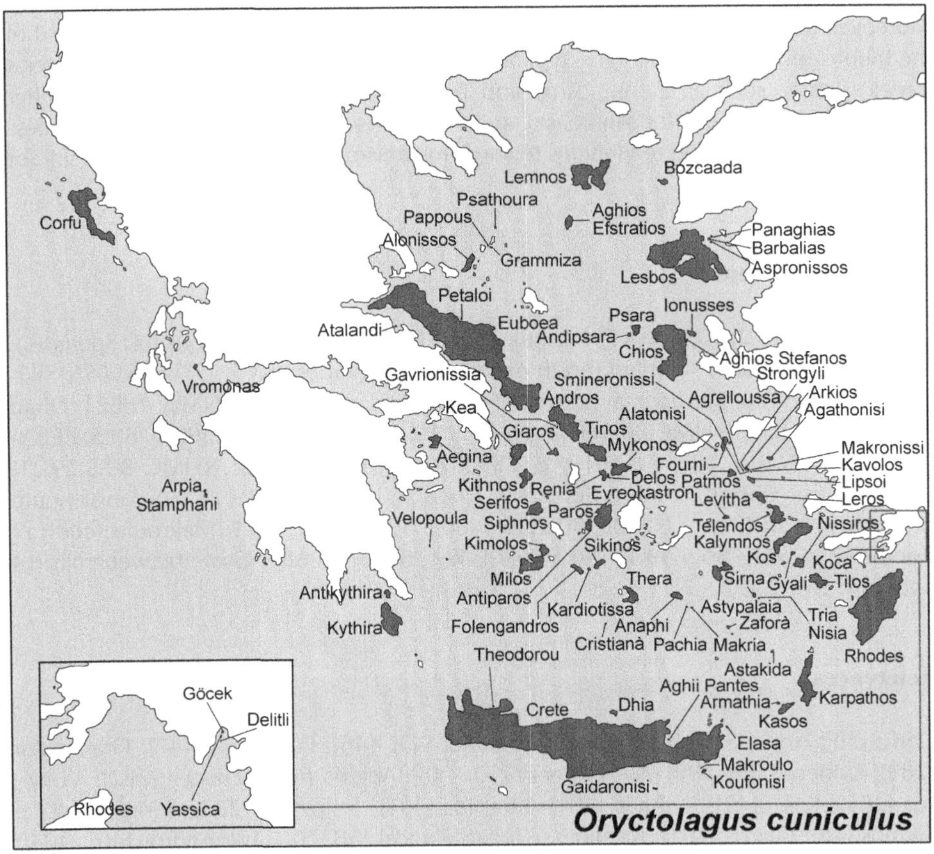

Fig. 39: Distribution of the European wild rabbit, *Oryctolagus cuniculus* Linnaeus, 1758, in the Ionian and Aegean islands.

ancient layers is however quite surprising, since *Oryctolagus cuniculus* appears to be unknown in the post-glacial Mediterranean region east of the Iberian peninsula [127, 388].

According to Nobis [458], *Oryctolagus cuniculus* makes its entrance in the history of human culturally controlled animals around 1100 BC, when the Phoenicians colonised the Iberian peninsula. This assumption is based on the knowledge that Phoenicians first discovered the wild rabbit in Spain and populated the coasts and islands along their routes with this self-replenishing supply of fresh meat so as to ensure regular provisions at strategic points of their routes (compare with [459]). The extant toponym *España* seems in fact to derive from *Hispania*, the Latin corruption of the ancient semitic expression *i-shephan (shaphan)-im* which means "island of hyraxes" [460]. The hyrax or dassie, *Procavia capensis* Pallas, 1766, is a medium-sized mammal naturally dispersed from South Africa to the Levant, through the Afro-Syrian Rift valley. Its external aspect may recall that of a rabbit, although the systematic position of the hyrax is very far from lagomorphs which are taxonomically closer to elephants and dugongs. When the Phoenicians landed for the first time onto the Iberian peninsula they noted the rabbits, these hyrax-like creatures, which at the time, however, they did not yet know. Thus, they called the new land "island of hyraxes". Moreover, quoting several classical scholars, such as Varro (*De Re Rustica*, III, 12, 6), Martial (*Epigrams*, XIII, 60), Pliny the Elder (*Naturalis historia*, VIII), Catullus (XXXVII, 18), and Strabo (III, 2, 6), [461]) summarises the literary evidence of the close association of rabbits with Spain in the conception of the ancient Romans. In reality, the history of the human-induced diffusion of *Oryctolagus cuniculus* beyond its natural distribution began many centuries before. In fact, the earliest recorded introductions of rabbits onto the Mediterranean islands date back to the second half of the 2nd millennium BC (1400–1300 BC), when the lagomorph was apparently imported to Menorca, in the Balearic Archipelago, by ancient settlers from the Iberian Peninsula [3]. In the Central Mediterranean basin, however, the subfossil evidence of the introduction of the species may date to no earlier than the 3rd–2nd century AD, since remains of rabbits have been found during the archaeological exploration of the islands of Nisida and Capri, in the Gulf of Naples (Italy) and Zembra in Tunisia [26, 427–428, 431, 434, 462–463]. Recent archaeozoological evidence, however, suggests even earlier importation, comprised between the 1st century BC and the 1st century AD, in the case, for example, of the sanctuary of Juno at Tas Silg on Malta [388].

Rabbits were probably not imported onto the Eastern Mediterranean islands before the end of the Classical period and/or the beginning of the Middle Ages. This can be also shown by the lack of toponyms related to their name in the Eastern Mediterranean basin. Among the other elements supporting the theory of a recent historical introduction of the rabbit onto the latter territories, we can consider the fact that Turkish people still refer to the species as the *ada tavscan* ("hare of the islands") [388]. Therefore, the occurrence of *Oryctolagus cuniculus* among the stratigraphic levels of the prehistoric town of Poliochni, on Lemnos, could be related to the animal's digging capacities [127]. Rabbits are in fact capable to excavate whole networks of underground holes. In archaeological sites, it is therefore possible to find osteological remains of the species in deeper layers than one might expect.

6.11 Persian squirrel, golden squirrel or Caucasian squirrel, *Sciurus anomalus* Gueldenstaedt, 1778

Greek name: Ασιατικός σκίουρος (*Asiatikos skiouros*)
Greek vernacular name: γαλιά (*galià*, Lesbos)
Turkish name: *Anadolu sincabi*
Turkish vernacular names: *dereek, kallay*

In the Palaearctic region, the genus *Sciurus* Linnaeus, 1758, is represented by two species: the European red squirrel, *Sciurus vulgaris* Linnaeus, 1758, in Europe and northern Asia, and the Persian or golden squirrel, *Sciurus anomalus* Gueldenstaedt, 1778, in south-west Asia. The latter is a medium-sized arboreal squirrel, which occurs in Asia Minor, Northern Arabia, including Lebanon, Syria, Jordan and northern Israel, Transcaucasia and western Iran [270, 464–468]. Its westernmost continental distribution extends to far eastern Europe and western Anatolia [466]. On the Mediterranean archipelagos, it has been reported only from the Turkish island of Gökçeada (Imbros) [74–75, 466] and the Greek island of Lesbos [71, 465–466, 469] in the Eastern Aegean sea. As far as is presently known, and maybe with the only exception of *Sciurus vulgaris* on Veli Brijuni (Croatia) [470–471] and perhaps Euboea, these two populations seem to represent the only contingents of the genus *Sciurus* Linnaeus, 1758, today dispersed on the eastern Mediterranean islands. During the present research, *Sciurus anomalus* was observed on both the islands of Gökçeada (B. Özkan, *in litteris*) and Lesbos (personal observation). Its natural colonisation of the latter two islands cannot be excluded.

Museum specimens

Lesbos: GNHM 8951; ZMUP 3153, 3615, 3618, 3619.

References

Lesbos [71, 93, 465–466, 468–469]; Gökçeada (Imbros) [74–75, 93, 466].

Ethnozoological note

In the light of modern ethnozoological enquiry, it would appear that Persian squirrels figure among those mammal species that have human interest for a variety of cultural reasons. In the Levant, for example, people still eat these squirrels and live specimens are regularly sold in the markets [467]. According to the rangers of the Forest Department of Lesbos, local people still feed on Persian squirrels. This is the last remnant of a tradition that survived in the culture of the elderly persons which sometimes seems to be practised also by the young. This is not surprising, however, since other animals are unexpected food in the Aegean islands, such as the shags, *Phalacrocorax aristotelis* Linnaeus, 1761, which are still eaten by local people on the not far away Leros (N. Kardakari, *in verbis*). In the towns along the Mediterranean coast of Turkey, until few years ago, Persian squirrels were still dressed by hawkers for tourist amusement and appreciation (▸Fig. 40).

Fig. 40: In the towns along the Mediterranean coast of Turkey (such as Antalya), until few years ago, Persian squirrels were still dressed by hawkers for tourist amusement and appreciation. Photo by Marco Masseti.

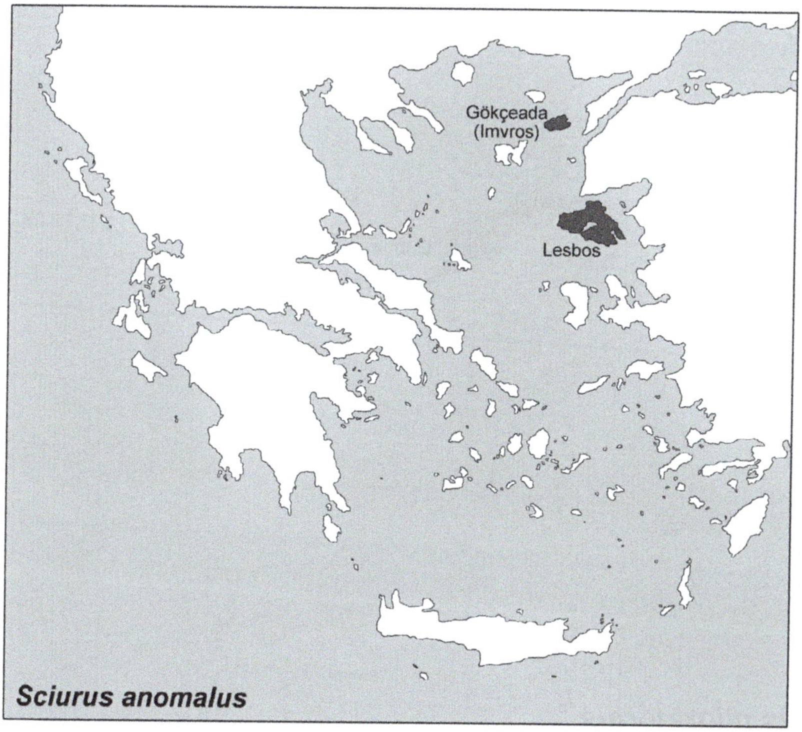

Fig. 41: Distribution of the Persian squirrel, *Sciurus anomalus* Gueldenstaedt, 1778, in the Aegean islands.

6.12 Grey hamster, *Cricetulus migratorius* Pallas, 1773

Greek name: νανοκρικετός (*nanokriketos*) Turkish name: *gri hamster*

The grey hamster, *Cricetulus migratorius* (Pallas, 1773), occurs in eastern Europe, east to Pakistan and Turkestan, and south to the Levant and the northern Arabian peninsula [270, 467, 472]. Its European distribution ranges from the eastern Balkan peninsula to Russia, through European Turkey and Ukraine [473]. The species exhibits considerable habitat plasticity. It prefers steppes and stony places covered with sparse vegetation. Common in fields and dry pastures, it inhabits also deserts and semideserts [473]. The grey hamsters distribute from the sea level to high mountain slopes. In part of its eastern range of diffusion (e.g., Central Asia and Transcaucasia), the species is synanthropic, and often becomes commensal with man [270, 474]. Remains of *Cricetulus migratorius* have been provided by the archaeological excavation of Konispol Cave, on the Albanian coast off the Greek island of Corfu [332].

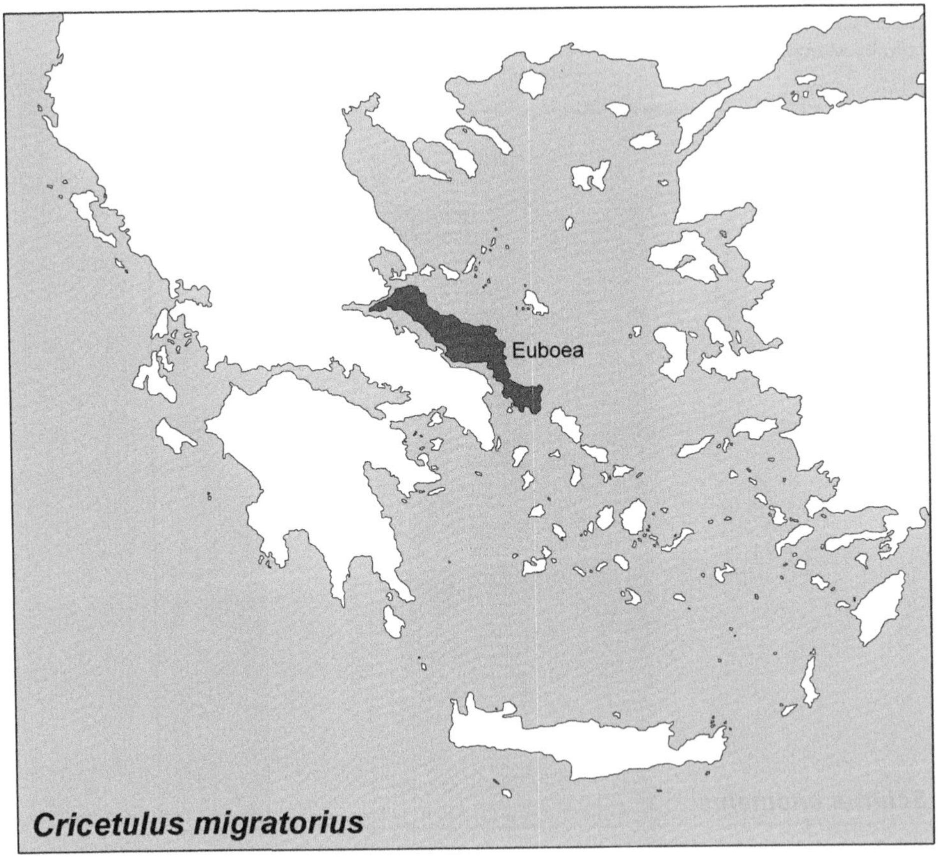

Fig. 42: Distribution of the grey hamster, *Cricetulus migratorius* Pallas, 1773, in the Aegean islands.

Cricetulus migratorius is absent from the islands off Turkish coasts [94]. The only Mediterranean island from which the species has been reported is Euboea: Sfougaris [293] recorded this rodent from the northern part of the island, while Akriotis [291] found its remains in pellets of the eagle owl, *Bubo bubo* Linnaeus, 1758.

Systematic data

Twenty-five subspecies of grey hamster have been described, mostly distinguished on the basis of differences in size and colour [473]. *Cricetulus migratorius atticus* Nehring, 1902, is the subspecies which is reputed to occur in continental Greece [70–71, 475]. Insignificant differences in the body and the skull size seem to distinguish the specimens from Greece and Asia Minor [71].

References

Euboea [291, 293].

6.13 Common vole, *Microtus arvalis* Pallas, 1778

Greek name: αρουραίος (*aroureos*) Turkish name: *kisa kuyuruklu adi tarla*

The common vole, *Microtus arvalis* Pallas, 1778, occurs in a continuous range from the Atlantic coast of Europe to central Russia, where it inhabits open cultivated agricultural land, grazed pastures, and shortgrass meadows [476]. It is found from sea level up to 2 600 metres. Insular populations are also known in western Europe, on the Orkney islands (Channel islands, but not the British isles), where it was introduced prehistorically, and Yeu, France (compare with [472, 477]).

Zima [476] records the species as absent from Mediterranean islands, but its occurrence on Euboea was reported by Cheylan [304], and on Lesbos by Pieper [72] and Niethammer [64]. Moreover, a label without specimen, written by Ioannis C. Ondrias himself, in the mammal collection of the University of Patras (collection number 3158) reports the occurrence of the common vole from the area of Mytilene, in south-eastern Lesbos [478]. Furthermore, specimens of *Microtus* sp. (not identified at the species level) were collected by trapping on Lesbos during field research carried out by the University of the Aegean in the spring of 1999 (A. Legakis, *in verbis*). However, according to Stella Fraguedakis-Tsolis (*in litteris*, 13th July 2006), the species does not appear to exist or to have ever existed on Lesbos: thus, sheet ZMUP 3158 would appear to be wrong. Furthermore, in the course of a new examination of the ZMUP collection of mammals carried out by the author in 2006, the specimen to which the label referred unfortunately proved to have been lost.

Fossil remains of *Microtus arvalis* have been found in a late Middle Pleistocene fauna of the island of Chios [106, 204].

Systematic data

In spite of numerous attempts to distinguish *Microtus arvalis* from other species on the basis of morphological criteria, the results continue to be problematic. Taking the karyotype as the sole character enabling a reliable determination, for example, Vohralik and Sofianidou [77, 356], assigned the Macedonian and Thracian populations of common voles to the taxon *Microtus epiroticus* Ondrias, 1966, whereas Zima [476] considers the latter as a synonym of *Microtus rossiameridionalis* Ognev, 1924.

Microtus arvalis levis Miller, 1908, is the form recognised as occurring in northern Greece (Thessaly, Macedonia and Thrace) [71]. Specimens of *Microtus arvalis brauneri* Martino, 1926, have also been reported in eastern Macedonia and in Thrace [475]. Ondrias [71] described the new subspecies *Microtus arvalis epiroticus* Ondrias, 1966 from Epirus, north-western Greece.

Museum specimens

Lesbos: ZMUP 3158.

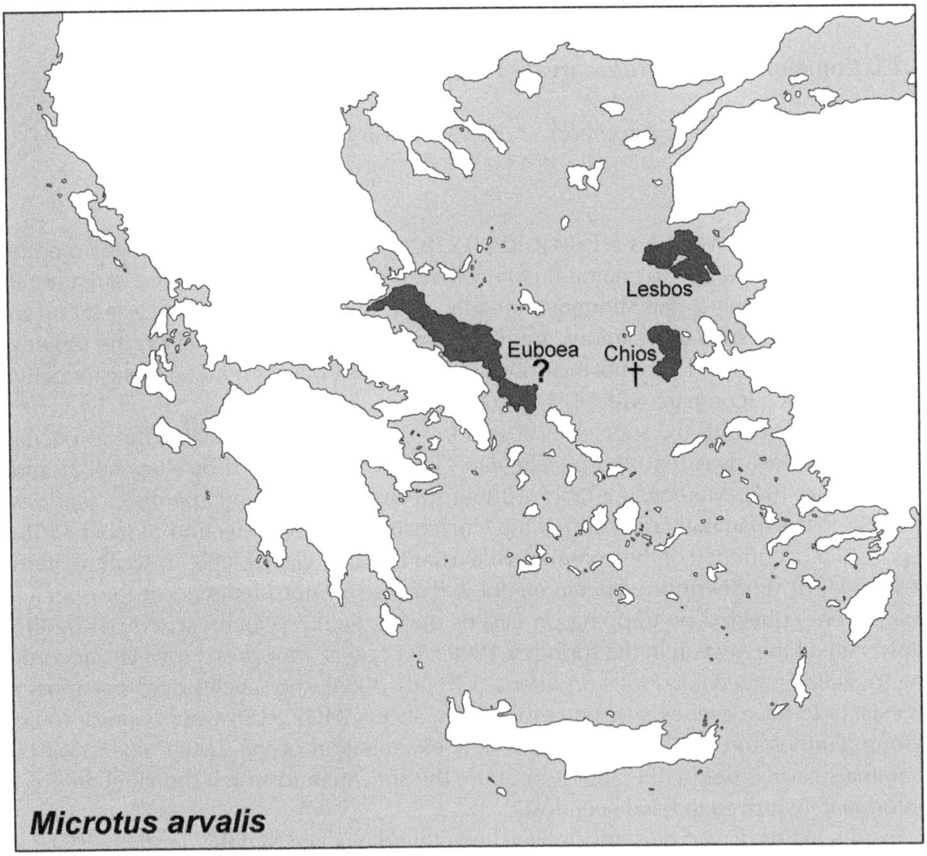

Fig. 43: Distribution of the common vole, *Microtus arvalis* Pallas, 1778, in the Aegean islands.

References

Euboea [304]; †Chios [67]; Lesbos [64, 72].

6.14 Guenther's vole, *Microtus guentheri* Danford and Alston, 1880

Greek name: αρουραίος της Ανατολικής (*aroureos tis Anatolikis*)

Turkish name: *tarla faresi*

The Guenther's vole, *Microtus guentheri* Danford and Alston, 1880, is the most southerly species of the genus in the Old World [467] (▶Fig. 44). In south-eastern Europe, this rodent inhabits dry meadows and pastures with sparse vegetation on well drained soil, from lowlands up to an altitude of between 150 and 500 m (compare with [356, 479]). An isolated population in Cyrenaica in Libya, also known as *Microtus mustersi* Hinton, 1926, closely resembles *Microtus guentheri* but its exact taxonomic identity is not established with certainity [94, 480]. In the Balkan peninsula, it extends northward from eastern Greece to southern Bulgaria, Macedonia, and southern Serbia [71, 77, 356, 472,

Fig. 44: Guenther's voles, *Microtus guentheri* Danford and Alston, 1880, are diurnal rodents. Photo by Andrea Bonetti.

479]. The species is reported at present from the islands of Lesbos [64, 72, 94, 479, 481], off the eastern Anatolian coast, which is so far the only record from any Mediterranean island [94]. It also occurs on the island of St. Thomas, in the Black Sea [472, 479, 482].

The species, however, has been documented among the subfossil Holocene fauna of Chios [67, 336]. Local specimens were very large in size, possibly a sign of endemisation.

Systematic data

Microtus guentheri hartingi Barrett-Hamilton, 1903, is the typical subspecies of Greece (northern Attica, western Boeotia, Phtihotis, Thessaly, Macedonia and Thrace) [71, 475].

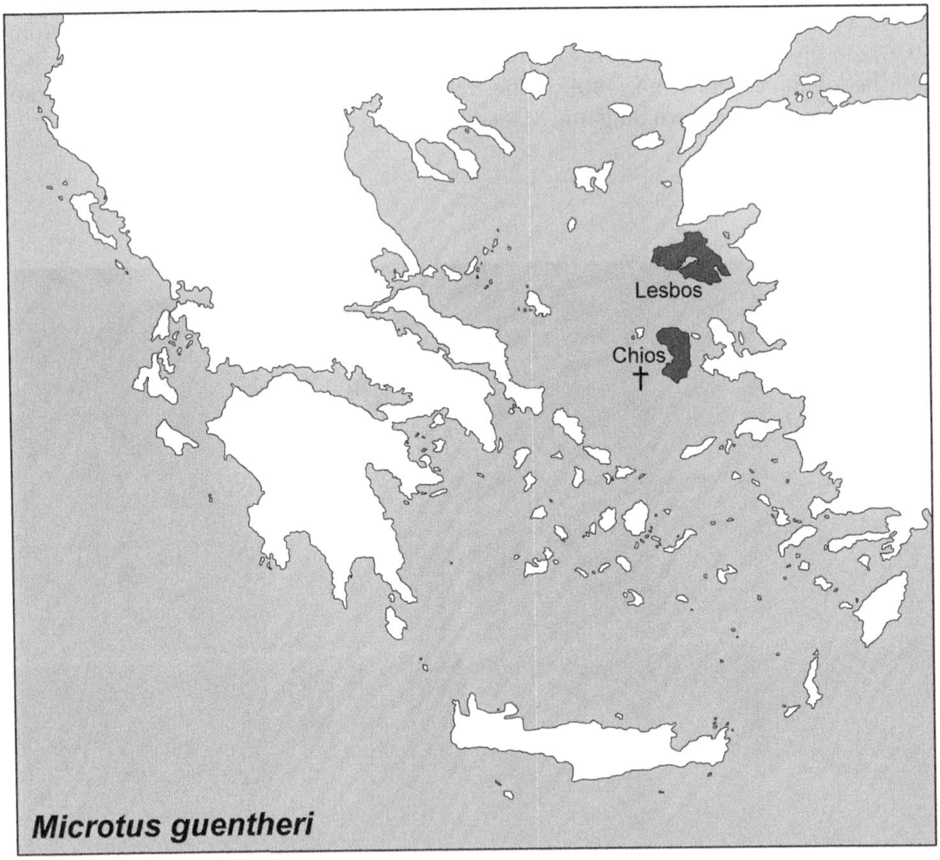

Fig. 45: Former and present distribution of Guenther's vole, *Microtus guentheri* Danford and Alston, 1880. Islands and areas that are inhabited by the vole are shaded and those where they have become extinct are marked with a cross.

References

Lesbos [64, 72, 93–94, 479, 481, 483]; †Chios [67, 336].

6.15 Thomas' vole, *Microtus thomasi* Barrett-Hamilton, 1903

Greek name: σκαπτοπόντικος του Thomas (*skaptopontikos tou Thomas*)

The Thomas's vole, *Microtus thomasi* Barrett-Hamilton, 1903, is a taxon endemic to the south-western Balkans, where it is the most widely distributed vole, occurring from the Neretva river to the Peloponnese [483–485], through Macedonia [356], and the island of Euboea [486–488].

This is the most fossorial species among the Balkan voles and its distribution is restricted to deeper soil. It prefers open habitats and its vertical range extends from 40 to 1 700 metres. In Greece, the species occurs in high mountain pastures [485]. Remains of *Microtus thomasi* have been provided by the archaeological exploration of Konispol Cave, on the Albanian coast opposite the Greek island of Corfu [332]. With regard to the rest of the Balkan peninsula, it has been reported from Albania, the former Yugoslavian Republic of Macedonia, Montenegro and south-eastern Bosnia-Herzegovina [80, 84, 488].

Specimens of *Microtus thomasi* were collected on Euboea both at sea level [489] and on Mount Dirphys, between 900 and 1 000 metres [490].

Systematic data

The proposal of Kratochvíl [491] to recognise two species, *Microtus thomasi* Barrett-Hamilton, 1903 and *Microtus atticus* Miller, 1910 from Attica, was not accepted, and the latter taxon is generally considered as a subspecific variety [485].

According to Giagia [489, 492], the *Microtus atticus* karyotype variety is distributed in eastern Greece, the northern Peloponnese and Euboea, while the *Microtus thomasi* karyotype form occurs in western Greece and was found as a mountain relic in Mount Taygetos, in the southern Peloponnese.

Museum specimens

Euboea: ZFMK 128, 145, 152, 153; ZMUP 394–416, 417–444, 446–492, 1190–1210, 1211–1230, 1232–1260, 2326–2808.

References

Euboea [71, 269, 293, 483, 486–490, 492–496].

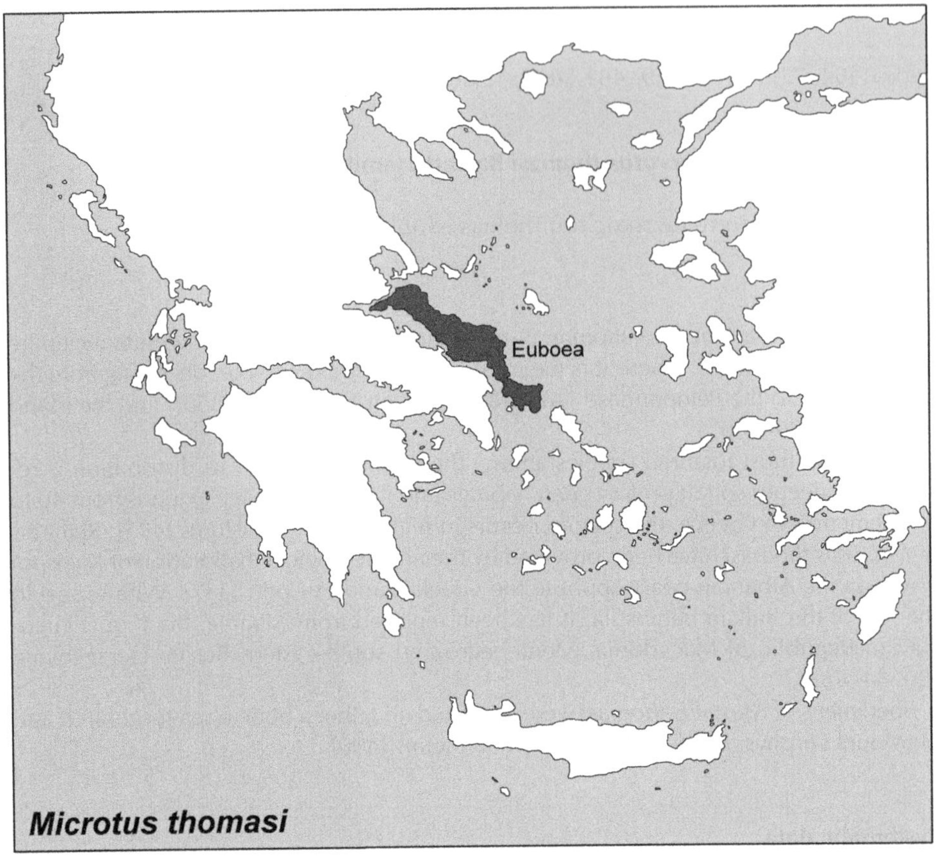

Fig. 46. Distribution of Thomas' vole, *Microtus thomasi* Barrett-Hamilton, 1903, in the Aegean islands.

6.16 Snow vole, *Chionomys nivalis* Martins, 1842

Greek name: χιονοπόντικος (*chionopontikos*) Turkish name: *kar faresi*

The snow vole, *Chionomys nivalis* Martins, 1842, is a nocturnal and strictly alpine species, which has a discontinuous distribution, ranging from the mountains of northern Spain, the Alps, Apennines, Tatra, Carpathian and Balkan ranges, to Asia Minor, southwestern Turkestan and Iran [270, 497–498]. It is, in fact, a typical mountain-top dweller living above the tree-line between 1 000 and 4 700 metres. Nonetheless, in some areas it has also been found in woodless and rocky biotopes at lower altitudes [498]. Ondrias [71] trapped a specimen at just 600 metres above sea level on a rocky hillside along the river Nestos (Drama, Macedonia).

The species is found only on the Mediterranean island of Euboea, where it was reported from Mount Dirphys, at an altitude of 1 700 metres above sea level [70–71, 342, 490, 499].

Systematic data

Chionomys and *Microtus* have recently come to be considered as separate genera on the basis of fossil records and genetic differentiation [498]. *Chionomys nivalis nivalis* Martins, 1842 is the western European subspecies, whereas *Chionomys nivalis pontius* Miller, 1908 has been reported from Turkey and *Chionomys nivalis hermonensis* Miller, 1908 is recorded from the Levant and Arabia [270].

In Greece, however, the subspecies of the snow vole are still imperfectly known. According to Ondrias [71], the two varieties – *Chionomys nivalis malyi* Bolkay, 1925 and *Chionomys nivalis aleco* Paspalev, Martino and Peschev, 1952 [500] which were described, respectively from Herzegovina and Mount Vitosha, in the Balkan peninsula – are valid since their type localities lie closer to Greece than other known varieties of snow vole. Specimens from Mount Dirphys (Euboea) were referred to as the subspecies *Chionomys nivalis aleco* by Peus [342].

Museum specimens

Euboea: ZFMK 77.114, 77.118.

References

Euboea [70–71, 304, 342, 490, 498, 499].

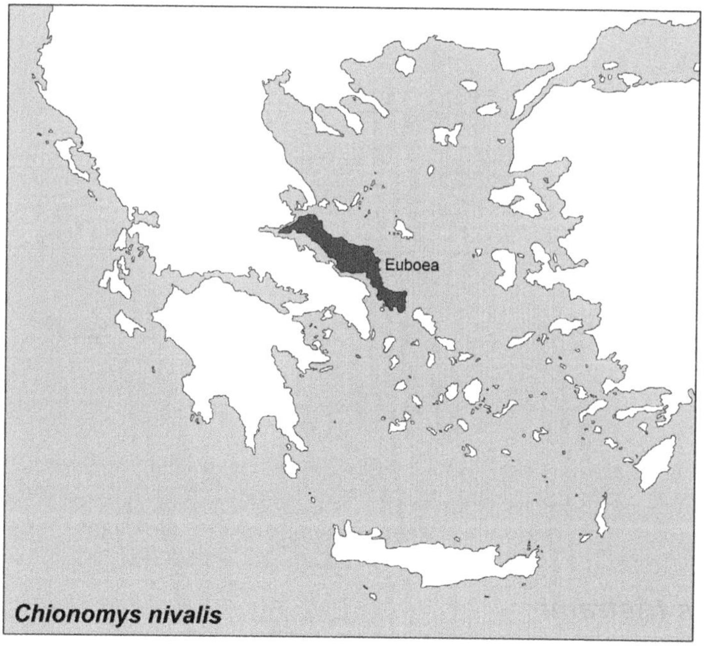

Fig. 47: Distribution of the snow vole, *Chionomys nivalis* Martins, 1842, in the Aegean islands.

6.17 Tristram's jird, *Meriones tristrami* Thomas, 1892

Greek name: γέρβιλος (*gervilos*)
Greek vernacular name: θυσανοποντικος (*thusanopontikos*)

Turkish name: *çöl siçani*

The Tristram's jird *Meriones tristrami* Thomas, 1892, is a nocturnal, crepuscular jird characterised by a wide habitat tolerance [270]. Its distribution ranges from the northern Arabian peninsula to Asia Minor, Transcaucasia and north-western Iran. In its Near Eastern dispersion, it seems to prefer desert and dry steppes with more than 100 mm of annual precipitation and it is distributed up to 1 550–2 000 m above sea level [501]. In Asia Minor, the species has widespread distribution from the surroundings of Izmir, in the Aegean coast, across central Anatolia as far east as the eastern side of the river Tigris [95].

In Europe, the Tristram's jird is of very marginal occurrence, being restricted to the Aegean island of Kos [501].

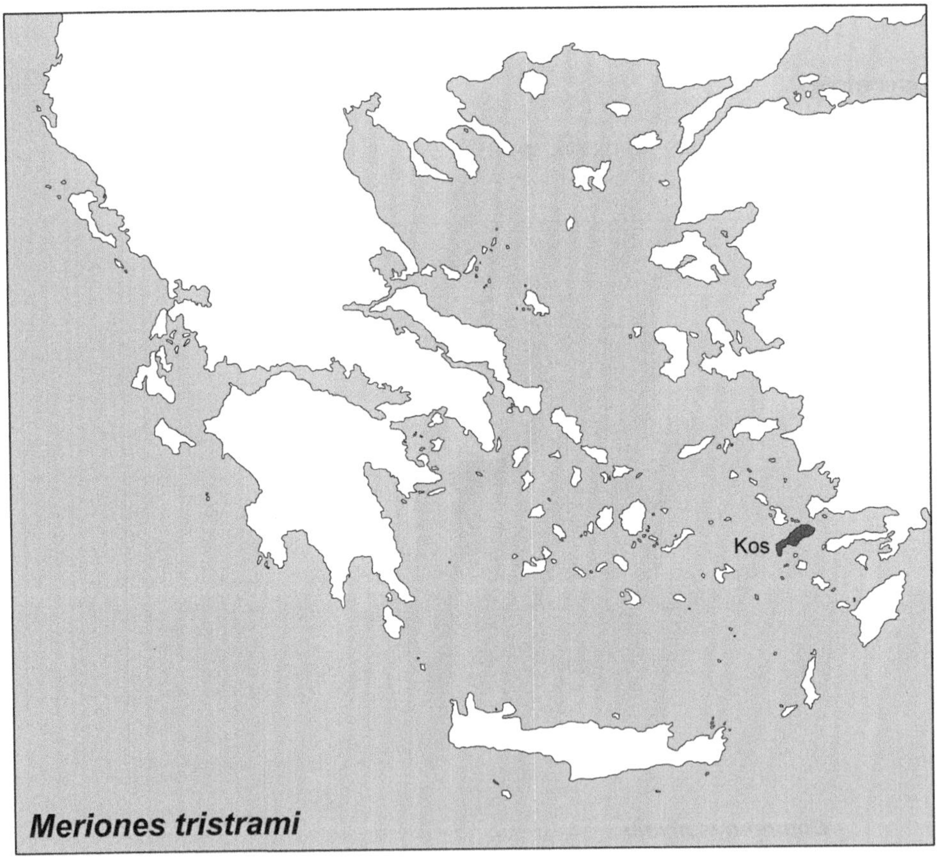

Fig. 48: Distribution of Tristram's jird, *Meriones tristrami* Thomas, 1892, in the Aegean islands.

Museum specimens

Kos: SMF 42276.78, 51694–51701.

References

Kos [59, 64, 501].

6.18 Lesser mole rat, *Nannospalax leucodon* Nordmann, 1840

Greek names: μικροτυφλοπöοντικος (*mikrotiflopontikos*)
Ασπάλαξ (*aspalax*, Aristotle: *De Animalibus Historiae*)
Greek vernacular name: ασφάλακας (*asfalakas*, Eastern Aegean islands)

Turkish names: *kör fare, köstebek*

The lesser mole rat, *Nannospalax leucodon* Nordmann, 1840, extends from Libya to the Arabian peninsula, Asia Minor, Transcaucasia, and south-eastern Europe, where it spends most of its life underground, although occasionally emerging at night, when it is not infrequently preyed upon by owls [270]. It inhabits steppes, meadows and pastures, from sea level to 2 400 m above sea level [502]. In the Mediterranean basin, its range spans only in the Aegean archipelago, where it occurs on some of the islands that have been connected to the mainland by continental land-bridges until early Holocene times (compare with [223]).

During the present research, the species has been observed on the islands of Euboea (Ioannidis, *in verbis*), Lesbos, Lemnos (personal observation), Bozcaada and Gökçeada (Özkan, *in litteris*), and Samothrace. The species is not known in the late Middle Pleistocene fauna of Chios, which shows continental Anatolian patterns yielding instead the Nehring's blind mole rat, *Spalax nehringi* Satunin, 1898 [106, 204].

Systematic data

According to Kryštufek and Griffiths [503], the systematics of fossorial rodents is particularly confusing, and in particular that of Spalacinae: a number of different chromosomal varieties of *N. leucodon* have been described from the Balkans [504] and at least some of these may represent distinct allopatric or parapatric species. Taxonomic splitting of *Nannospalax leucodon* would increase the degree of endemism in an area already recognised as a hot spot for endemics. Moreover, of necessity, the approach often adopted ignores the fact that real ranges are mosaics.

The subspecies recognised for the Eastern Aegean islands is the same described for Lemnos, *Nannospalax leucodon insularis* Thomas, 1917.

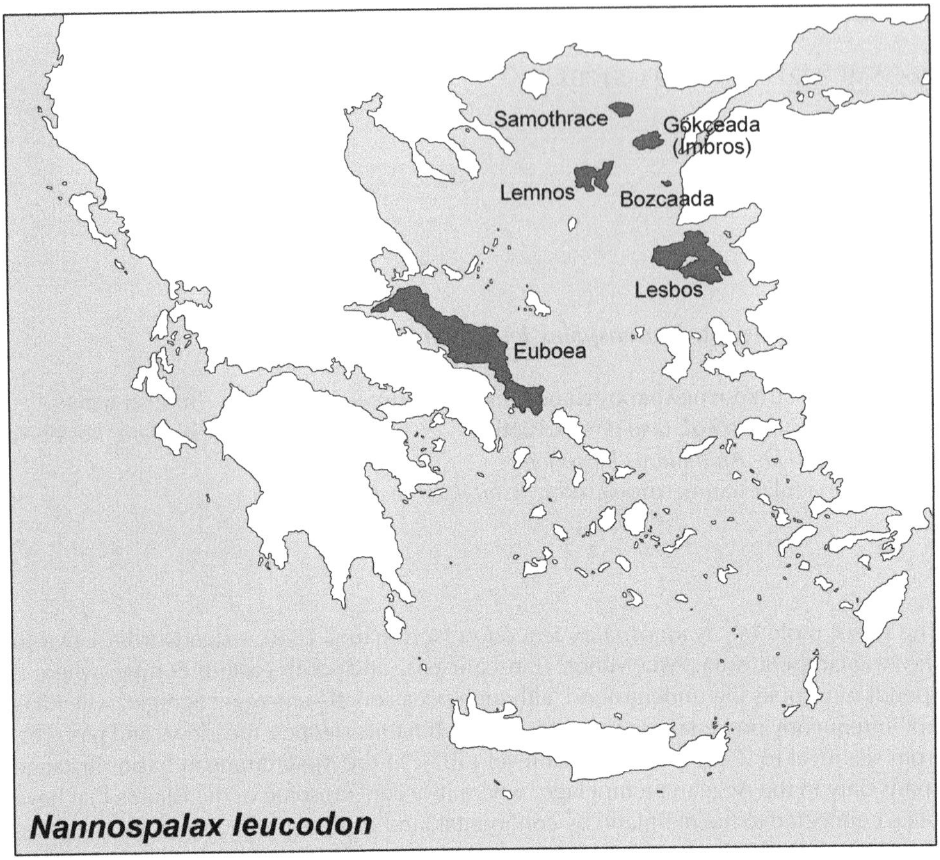

Fig. 49: Distribution of the lesser mole rat, *Nannospalax leucodon* Nordmann, 1840, in the Aegean islands.

Museum specimens

Lesbos: ZMUP 5988, 6221; Lemnos: BMNH 16.6.27.1.

References

Euboea [505–506]; Lesbos [72, 502, 505–507]; Bozcaada [75]; Lemnos [70–71, 304, 454, 505–506, 508–511]; Gökçeada [75]; Samothrace [77, 505–506].

Biogeographical note

As already seen, the extant distributional range of *Nannospalax leucodon* on the Aegean islands comprises several Greek and Turkish islands of the north-eastern Aegean Sea, but it cannot be explained as the result of a human-induced introduction, since it is a fossorial mammal adapted to digging and life underground. The presence of the rodent in these territories can only be explained through a direct, natural colonisation at the time when, about 20 000 years ago (compare with [98]), the islands were connected to the nearest mainland (▶Fig. 50). In fact, as observed by Kurtén [148], the similarity

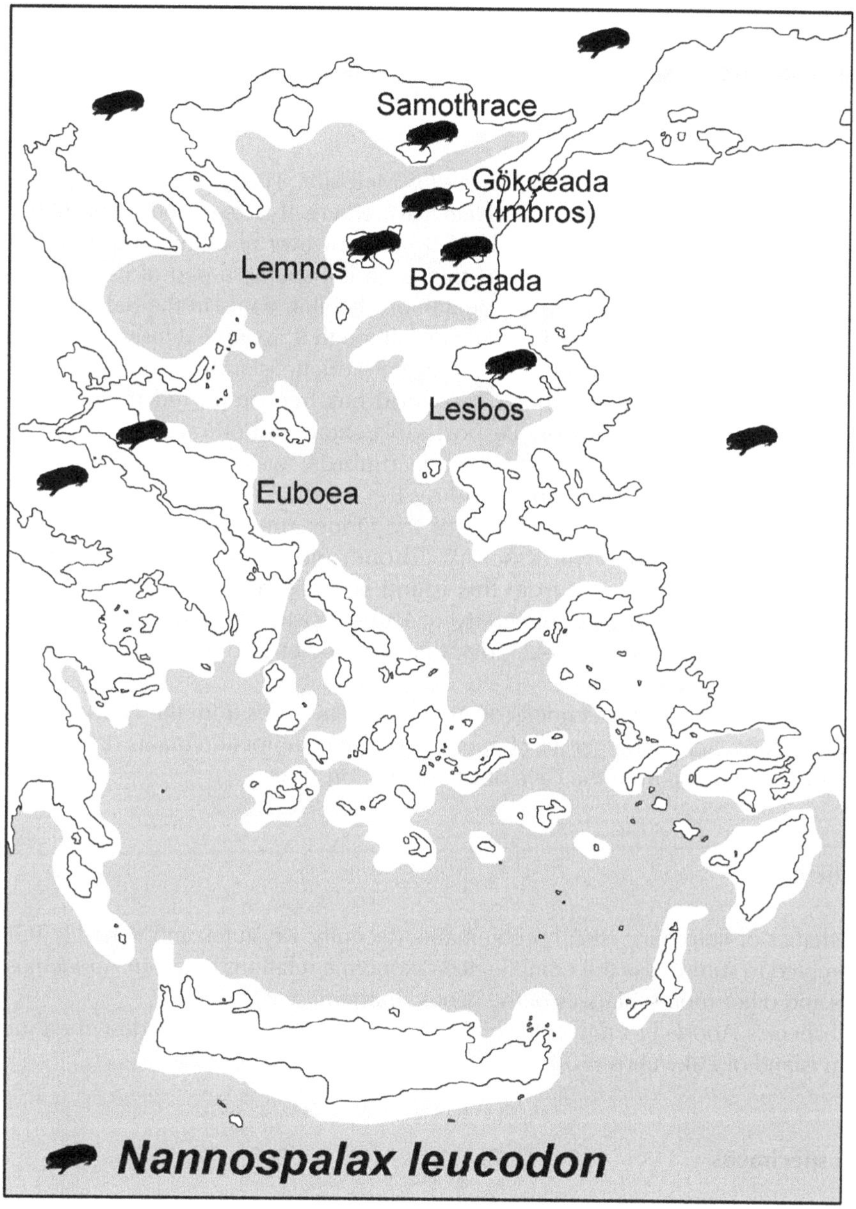

Fig. 50: Present occurrence of the lesser mole rat, *Nannospalax leucodon* Nordmann, 1840, in the Aegean region. The pale grey area represents land lost during Holocene sea level rises.

between the fossil and recent European ranges of these animals is striking and can presumably be correlated to their subterranean habitat, making the nature of the soil the most important environmental factor in their survival. Mole rats feed mainly on roots and bulbs, and require fertile, dry steppe or garden soil.

6.19 Yellow-necked mouse, *Apodemus flavicollis* Melchior, 1834

Greek name: κρικοπόντικος (*krikopontikos*) Turkish names: *orman faresi*

The yellow-necked mouse, *Apodemus flavicollis* Melchior, 1834, is dispersed in the western Palaearctic, from Europe to the Near East, where it is regarded as primarily a species of mature deciduous woodland [512]. It is however also known from other deciduous woodland types and coniferous forests. In the southern part of its range of distribution, the species prefers a more mountainous habitat, while in the Balkans and western Anatolia it is also common in lowlands, although it avoids extensive stretches of cultivated land [95, 512]. It also occurs on several Adriatic islands.

On the Ionian and Aegean islands, this murid has been reported from Corfu, Zakynthos, Kythera, Euboea, Kasos, Lesbos, Gökçeada, Samothrace, and Thasos. Its occurrence on Zakynthos and Gökçeada (Imbros) was recently confirmed respectively by Y. Ioannidis (*in verbis*), and by B. Özkan (*in litteris*); on Lesbos by A. Legakis (*in verbis*). Among the islands of the Dodecanese, the yellow-necked mouse appears to occur only on Kasos (V. Chondropoulos and S. Fraguedakis-Tsolis, *in verbis*). One specimen from this island is preserved in the collection of the Zoological Museum of the University of Patras (ZMUP 6805). According to F. Fladerer *(in litteris)*, it cannot be excluded that the species also occurs on Skiathos, in the Northern Sporades.

Kock [67] described fossil specimens of *Apodemus flavicollis* from the Holocene of the island of Chios. Spitzenberger and Bauer [513] found subfossil remains of the species in a Holocene fauna from the cave of Charkadio, on Tilos.

Systematic data

The systematics of field mice is still problematic. Recently, Kryštufek and Vohralík [95] have attempted to summarise the complicated taxonomic relations between *Apodemus flavicollis* and other representatives of the genus *Apodemus*.

The subspecies *Apodemus flavicollis braunei* Martino, 1927 has been described for the Ionian island of Zakynthos [50].

Museum specimens

Zakynthos: ZMUP 6467–6475, 6477–6480; Euboea: ZMUP 400–484, 2319–2783; Kasos: ZMUP 6805; Lesbos: ZMUP 3126–3182; Thasos: SMF 62882–62884.

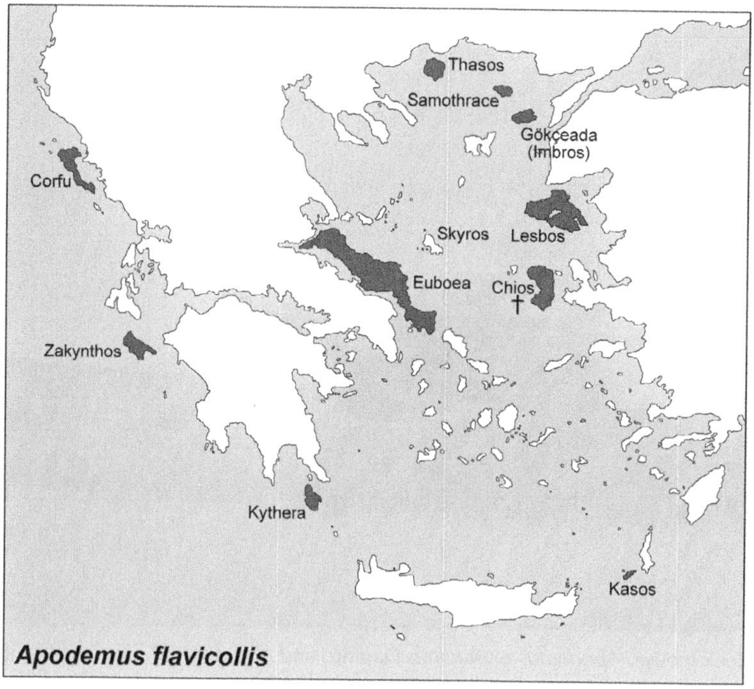

Fig. 51: Former and present distribution of the yellow-necked mouse, *Apodemus flavicollis* Melchior, 1834, in the Ionian and Aegean archipelagos. Islands and areas that are inhabited by the mouse are shaded and those where they have become extinct are marked with a cross.

References

Corfu [49, 514]; Zakynthos [50]; Kythera [52–53]; Euboea [71]; †Chios [67]; Lesbos [64, 72, 76, 329]; Gökçeada [75–76, 95]; Samothrace [76–77, 95]; Thasos [76, 95, 356].

6.20 Rock mouse or eastern broad-toothed field mouse, *Apodemus mystacinus* Danford and Alston, 1877

Greek name: βραχοπόντικος (*brachopontikos*) Turkish names: *kaya faresi*

The rock mouse or eastern broad-toothed field mouse, *Apodemus mystacinus* Danford and Alston, 1877, is a western Palaearctic species whose European distribution is restricted to the southern and western Balkan peninsula, including various Adriatic (Korčula and Mliet), Ionian and Aegean islands [472, 515]. In the latter islands, as far as is presently known, rock mice have been found on Corfu, Cephalonia, Zakynthos, Kythera, Alonissos, Peristera (east of Alonissos), Euboea, Salamis, Naxos, Crete, Kasos,

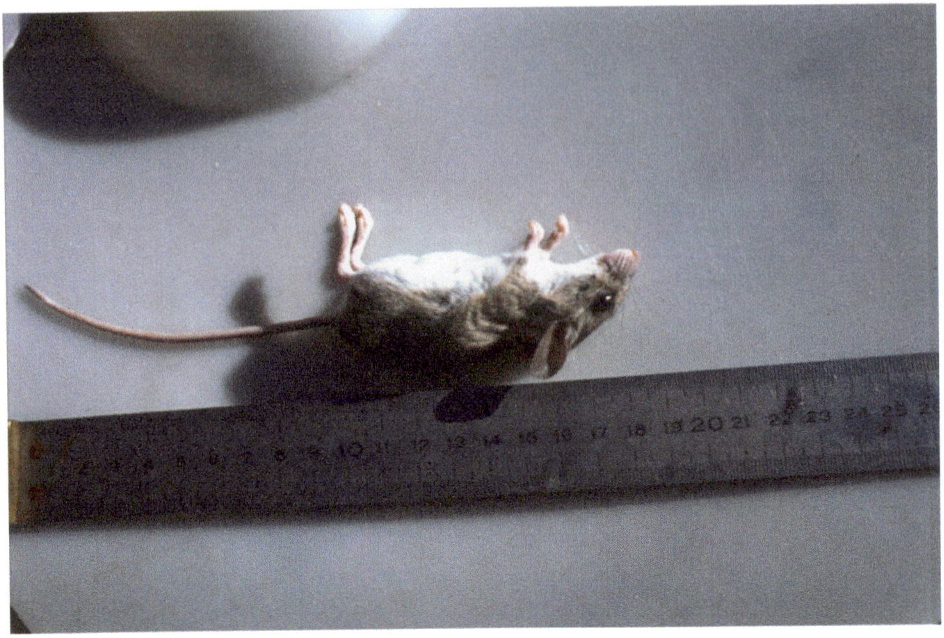

Fig. 52: Rock mouse, *Apodemus mystacinus* Danford and Alston, 1877, from the island of Tilos. Photo by Marco Masseti.

Karpathos, Rhodes, Tilos, Kos, Ikaria, Samos, Chios, Lesbos, Gökçeada, Samothrace and Thasos (▶Fig. 52). Other data collected during the present survey confirm the presence of the species on Aegina (Y. Ioannidis, *in verbis*), Lesbos (A. Legakis, *in verbis*), and Cephalonia where remains of the species were found in barn owl pellets (H. Pieper, *in litteris*). Bones of *Apodemus mystacinus* have been provided by the archaeological exploration of Konispol Cave, on the Albanian coast off the Greek island of Corfu [332].

Systematic data

Apodemus mystacinus, was described by Danford and Alston, 1877, around the end of the 19th century, on the basis of the examinations of specimens from western Anatolia (compare with [516]). Four are the subspecies described for the Ionian and Aegean islands:

1. *Apodemus mystacinus mystacinus* Danford and Alston, 1877, on Kos (SMF 51702–51717).
2. *Apodemus mystacinus epimelas* Nehring, 1902 on Corfu [49, 71], and other Ionian islands [516], Kythera [52], Euboea [71], and Peristera, in the Northern Sporades (SMF 55817–55818).
3. *Apodemus mystacinus smyrnensis* Thomas, 1903, on Samos [65, 71].
4. *Apodemus mystacinus rhodius* Festa, 1914, on Rhodes [57, 71], and Crete [71, 516].

According to Storch [517], however, the islands of the Eastern Aegean (Lesbos, Chios, Samos, Ikaria, Kos, Rhodes and Karpathos) and Crete are inhabited by the Anatolian subspecies *Apodemus mysaticinus mystacinus* Danford and Alston, 1877, whereas *Apodemus mystacinus epimelas* Nehring, 1902 is the variety occurring in the Western Aegean islands, such as Salamis.

Museum specimens

Zakynthos: ZMUP 2525; Peristera: SMF 55817–55818; Euboea: ZMUP 2776–2802; Crete: NHMC no 80.5.41.12, 80.5.41.18, 80.5.41.43, 80.5.41.(59–60), 80.5.41.62; Tilos: MZUPA ML138, TE926; Kos: SMF 51702–717; Chios: ZMUP 3119; Lesbos: NMW no 11147–11150; ZMUP 3138–3143.

References

Corfu [49, 71, 517]; Cephalonia [71, 95, 282, 380]; Kythera [52–53, 518]; Alonissos [289, 355]; Salamis [186, 517]; Euboea [71, 76, 95, 291, 293]; Naxos [519]; Crete

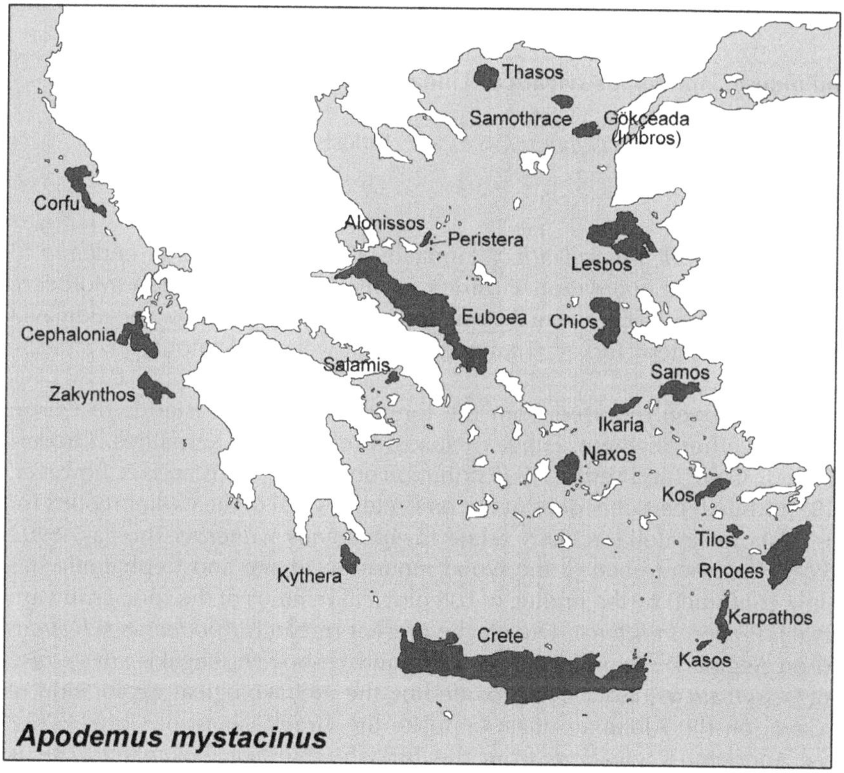

Fig. 53: Distribution of the rock mouse, *Apodemus mystacinus* Danford and Alston, 1877, in the Ionian and Aegean islands.

[71, 76, 95, 279, 297, 328, 347–348, 516–518]; Kasos [299]; Karpathos [56, 76, 95, 517–518]; Rhodes [57, 59, 76, 95, 297, 328, 516–518]; Tilos [63, 351]; Kos [64, 76, 95, 517, 518]; Ikaria [71, 76, 95, 328, 517, 518]; Samos [65, 71, 76, 95, 306, 517, 518]; Chios [67, 95, 517, 518]; Lesbos [72, 76, 95, 329, 517–518, 520]; Gökçeada [75–76]; Samothrace [518]; Thasos [518].

Palaeontological and archaeozoological note

The eastern broad-toothed field mouse has been known in the Eastern Mediterranean since the Lower Pleistocene, at least [206]. Kock et al. [364], for example, reported it in a subfossil and recent micromammalian fauna from western Anatolia. Fossil and subfossil remains of the species were respectively uncovered from Middle Pleistocene [204] and Holocene [336] sediments of the island of Chios.

A single M2 of the species was found in Middle Minoan III of the archaeological site of Kommos, on Crete. Payne [347] notes that the measurements of this tooth are very close to those given by Mayhew [521] for Late Pleistocene *Apodemus mystacinus* from mainland Greece. According to Payne [347], the species reached Crete by the end of the Middle Minoan period.

6.21 Wood mouse, *Apodemus sylvaticus* Linnaeus, 1758

Greek name: δασοπόντικος (*dasopontikos*) Turkish names: *adi tarla faresi*

The wood mouse, *Apodemus sylvaticus* Linnaeus 1758, is practically endemic to Europe, with an isolated population in Africa, in the northern Maghreb (Morocco, Algeria and Tunisia) [95]. Its European distribution spans to the north from Scandinavia and south to north-western Turkey (Thrace and north-western Anatolia) [472, 522–523].

The species has been reported from the Ionian and Aegean islands of Corfu, Cephalonia, Zakynthos, Kythera, Euboea, Naxos, Crete, Kasos, Karpathos, Rhodes, Kos, Agathonissi, Chios, and Lesbos. Its distribution on the Aegean islands is, however, not clear [95]: while it certainly does occur on Crete, several of the earlier reports for the islands offshore Anatolia actually relate to *Apodemus witherbyi* Thomas, 1902 [76, 95, 524]. The occurrence of the wood mouse on Ithake and Cephalonia has been recently confirmed by the finding of osteological remains of the species in barn owl pellets (H. Pieper, *in litteris*). During the present research *Apodemus sylvaticus* was found on Aegina (Y. Ioannidis, *in verbis*), and Lesbos (A. Legakis, *in verbis*). Remains of *A. sylvaticus* have been provided by the archaeological exploration of Konispol Cave, on the Albanian coast opposite the Greek island of Corfu [332]. A report of *Apodemus sylvaticus* from the Turkish island of Gökçeada [75], is erroneous [76, 95].

Systematic data

Two varieties of *Apodemus sylvaticus* have been recognised within the Ionian and Aegean islands:

1. *Apodemus sylvaticus creticus* Miller, 1910, on Corfu [49], Crete [277, 525] and Chios (BMNH 6538).
2. *Apodemus sylvaticus dichrurus* Rafinesque, 1814, on Samos [65].

Museum specimens

Cephalonia: SMF 55258–60; Euboea: ZMUP 397–462, 2325–2358; Naxos: NMW AB86/365–AB86/376; Crete: BMNH 1938.11.25.6, 5.12.2.25, 5.12.2.29; NHMC 80.5.41.1, 80.5.41.4, 80.5.41. 5–80.5.41.15, 80.5.41.20–80.5.41.42, 80.5.41.44–80.5.41.50,

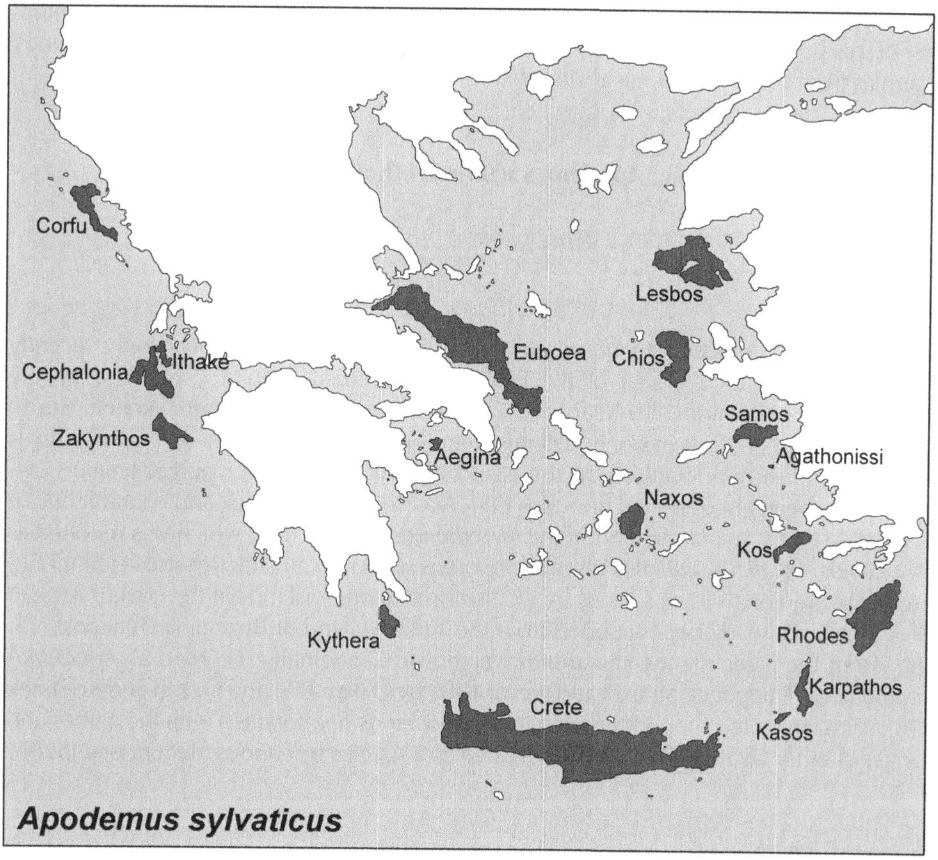

Fig. 54: Distribution of the wood mouse, *Apodemus sylvaticus* Linnaeus, 1758, in the Ionian and Aegean islands.

80.5.41.56–80.5.41.58, 80.5.41.61, 80.5.41.63, 80.5.41.65; NMW 36152–36157; ZSM 1977/0380a, 1977/0381–1977/0391; Kasos: MZUF 12687; Chios: BMNH 6583.

References

Corfu [49, 71]; Cephalonia [279, 283, 526]; Zakynthos [50]; Kythera [52–53, 76, 527]; Euboea [71, 76, 291, 293]; Crete [71, 76, 95, 186, 276–277, 279, 298, 347–348, 525]; Karpathos [56]; Rhodes [59, 76]; Kos [64, 76]; Agathonissi [326]; Samos [65–66, 76, 306]; Chios [67, 69, 186, 336]; Lesbos [329].

Palaeontological and archaeozoological note

Two size-based morphotypes are reported from the Early Pleistocene of the Dodecanese island of Kalymnos [205], and the Pleistocene of other Eastern Mediterranean islands [106].

According to Payne [347], the species reached Crete not before the Iron Age. A number of specimens of medium-sized *Apodemus* cf. *sylvaticus* are present in contemporary samples from the archaeological site of Kommos, southern Crete [347].

6.22 Steppe field mouse, *Apodemus witherbyi* Thomas, 1902

Greek name: στοεποποντικος (*stepopontikos*)

The steppe field mouse, *Apodemus witherbyi* Thomas, 1902, has been only recently recognised as a full member of the European mammalian fauna [524]. It is mainly distributed in Asia Minor, the Near East, Caucasia, Crimea and southern Ukraine, east to the river Dnepr [472]. It has only recently been found in Greece, where few individuals were trapped in two localities on the island of Rhodes (Kameiros and Kattavia) [524, 528]. It has been also reported from Kos [95]. According to Kryštufek and Vohralík [95], it is possible that a re-examination of the material from Samos [65], which was reported as *Apodemus sylvaticus,* will also reveal *Apodemus witherbyi*. In fact, Rovatsos et al. [528], are of the opinion that the species might also exist in other islands of the eastern Aegean Sea, as it has already been reported from the Turkish island of Bozcaada (Tenedos) [75, 93]. From the latter islands, this murid was however originally recorded as *Apodemus hermonensis* Filippucci, Simson and Nevo, 1989 by Özkan [75] and Özkan and Kryštufek [76]. Allozymic data support the identity of *Apodemus hermonensis* with the names and ranges of other taxa which have been treated as a species by various authors (see [528]).

Systematic data

The taxonomic position of this species is rather controversial. Zagorodnyuk [530], Zagorodnyuk et al. [531] and Mezhzherin [532] elucidated the morphological and distributional boundaries of *Apodemus witherbyi* (as *arianus*) and their definition also incorporates the name and range of *Apodemus hermonensis* Filippucci, Simson and Nevo, 1989. As observed by Musser and Carleton [472], the latter species was described from the alpine "tragacanthic" belt at about 2 000 metres in Mount Hermon (Israel) and was considered to be morphologically closely related to *Apodemus flavicollis*, which it displaced on Mount Hermon (and possibly also Lebanon and Antilebanon mountain ranges) at elevation above 1 900 metres [533]. Some of the earlier reports of *Apodemus sylvaticus* for the islands offshore Anatolia actually relate to *Apodemus witherbyi* [76, 524].

In fact, the taxonomy of the East European field mice of the genus *Apodemus* started to change considerably in the late 1980s, following the application of electrophoretic analyses of allozymes [533–534]. It appeared that, contrary to previous belief [269], *Apodemus sylvaticus* Linnaeus, 1758 is mainly absent from the Asiatic coasts of the eastern Mediterranean, being represented instead by a newly described *Apodemus hermonensis* Filippucci, Simson and Nevo, 1989 [524]. It was shown subsequently

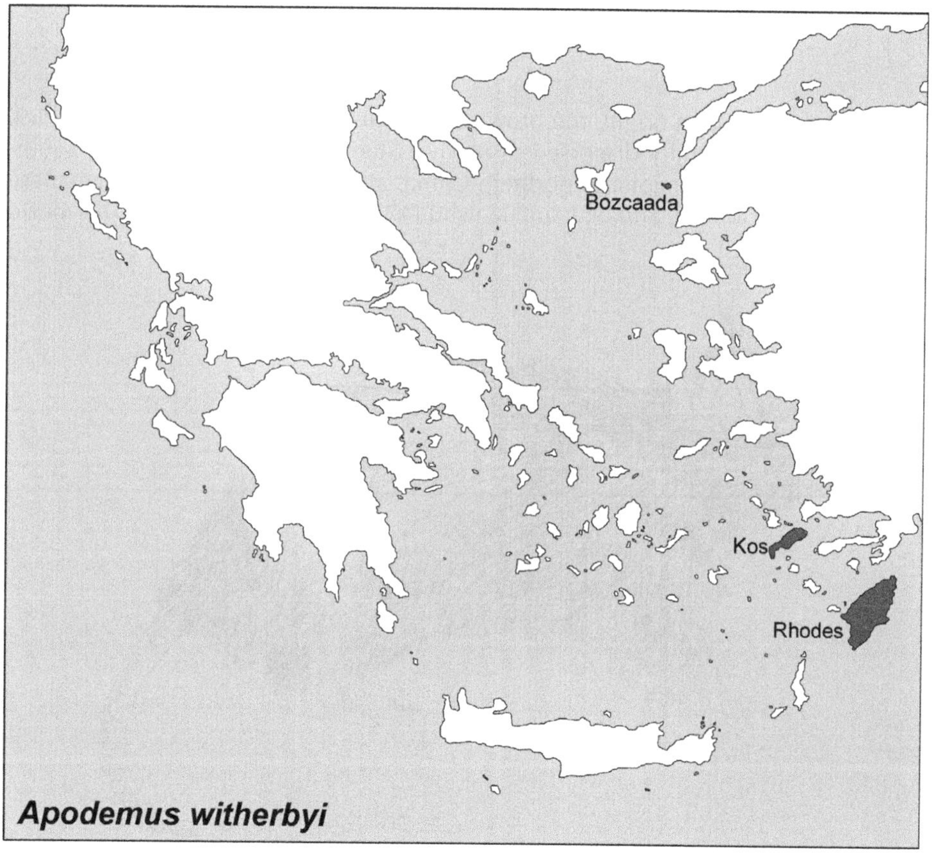

Fig. 55: Distribution of the steppe field mouse, *Apodemus witherbyi* Thomas, 1902, in the Aegean islands.

[525] that *Apodemus hermonensis* is predated by several names, of which *Apodemus iconicus* Heptner, 1948 (*nomen novum* to replace *tauricus* Barrett-Hamilton, 1900, not Pallas, 1811) poses no doubt about its taxonomic identity. The oldest name for *Apodemus hermonensis* is possibly *Mus sylvaticus witherbyi* Thomas, 1902, from Iran, which, however, differs from the Anatolian *Apodemus iconicus* in smaller bullae [525]. The above nomenclatoral solution has been mainly ignored (compare with [529]; for the opposing view see [524, 535]).

References

Bozcaada (Tenedhos) [75–76, 93, 528]; Kos [95]; Rhodes [95, 528].

6.23 Brown rat, *Rattus norvegicus* Berkenhout, 1769

Greek name: δεκατιστής (*dekatistis*)
Greek vernacular name: αρουραίος (*aroureios*)
Turkish name: *göçmen siçani*

A species of Far Eastern origin, the brown rat, *Rattus norvegicus* Berkenhout, 1769, is reputed to be naturally dispersed in north-eastern Palaearctic Asia, from south-eastern Siberia and Mongolia to northern China, and perhaps some of the Japanese islands, such as Honshu, Shikoku, and Kyushu [472, 536–540] (►Fig. 56). This murid

Fig. 56: Brown rat or Norway rat, *Rattus norvegicus* Berkenhout, 1769. Drawing by Silvia Cantagalli Masseti.

has been introduced worldwide. However, although some records in Europe date back to the Middle Ages, it did not begin its permanent invasion of western Europe before the end of the 17th and the18^{th} century [212, 541–543]. A brown rat skull was found in a galleon shipwreck of the18^{th} century AD discovered off the northern shores of Corsica [544]. This example shows, in practical terms, how marine navigation was a vector for the passive dispersal of small commensal mammals through time [545].

In the Mediterranean islands, *Rattus norvegicus* is not so widespread as its relative *Rattus rattus* Linnaeus, 1758, because it is presumably unable to successfully colonise the dry environments, due to its dependance on water (compare with [546–548]). In the Ionian and Aegean islands, as far as is presently known, brown rats have been found on Corfu, Cephalonia, Zakynthos, Kythera, Euboea, Atalandi (north Euboean Gulf), Crete, Karpathos, Rhodes, Kos, Ikaria, Chios, and Gökçeada. Data collected during the present study allowed to report for the first time the occurrence of the species from the islands of Skiathos (F. Fladerer, *in litteris*), and Lemnos (personal observation), but also to confirm its presence on Cephalonia (H. Pieper, *in litteris*). There is, however, the suspicion that brown rats occur on other islands as well, especially in harbour areas. Norway rats are reported to have swum up to 2 km [549]. They can float for 72 hours at 34°C [550], or up to 15 minutes in the cold Aleutian waters [551]. Very curiously, Husband and Davis [38] and Koster and Grether [452] reported this species respectively from the islet of Theodorou, and of Dhia off the northern coast of Crete, both characterised by a dry and barren environment apparently quite unsuited to the species which requires moister habitat conditions.

Systematic data

Beside the nominate subspecies *Rattus norvegicus norvegicus* Berkenhout, 1769, which inhabits Europe, *Rattus norvegicus caraco* Pallas, 1779, lives in easternmost Eurasia [542].

Rattus norvegicus norvegicus Berkenhout, 1769, has been reported from Corfu [49, 71], Cephalonia [71] and Crete [71].

Museum specimens

Corfu: BMNH 8.10.1.30; Cephalonia: BMNH 8.10.1.38; Zakynthos: ZMUP 6476; Crete: NMW B983; ZMUP 6650; two specimens are present within the collections of the Zoological Museum University of Athens (ZMUA 291 and 308, collected by Krüper), whose origin is generically referred from Sporades.

References

Corfu [49, 71, 518]; Cephalonia [71, 279, 518]; Zakynthos [50, 71, 518]; Kythera [53]; Atalandi [446]; Euboea [291, 293]; Crete [71, 95, 297, 347–348, 369, 518]; Theodorou

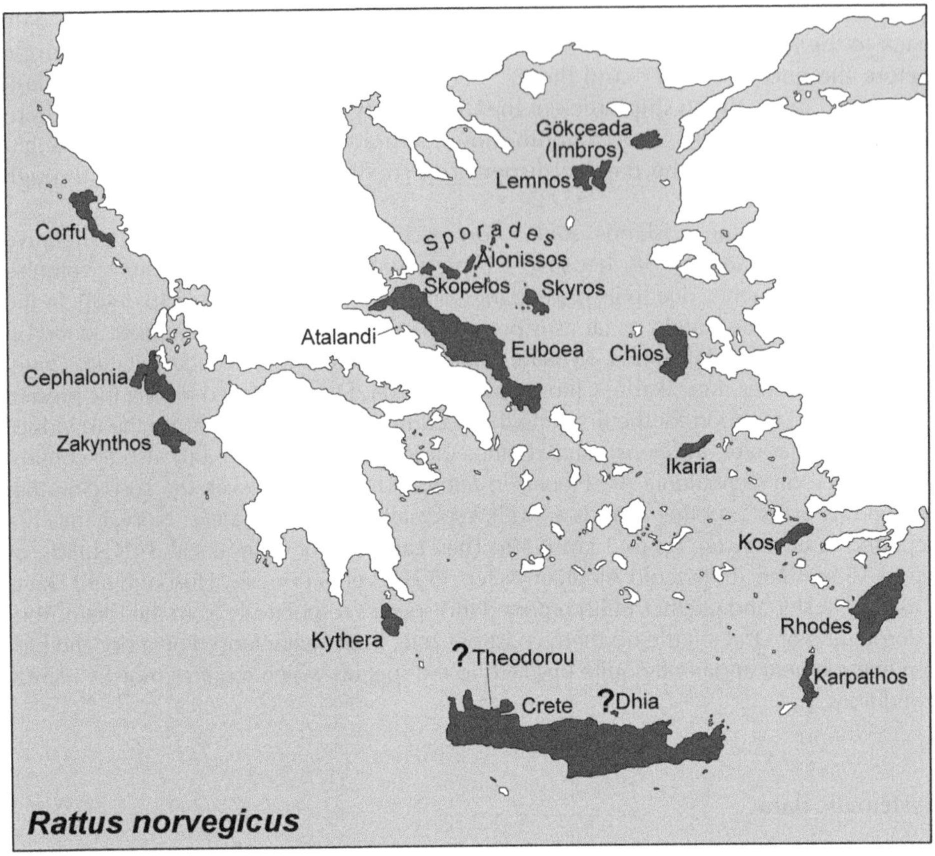

Fig. 57: Distribution of the brown rat or Norway rat, *Rattus norvegicus* Berkenhout, 1769, in the Ionian and Aegean islands.

[37], Karpathos [95, 518]; Rhodes [95, 518]; Kos [59, 95, 371, 518]; Ikaria [71, 95, 509, 518]; Chios [69]; Gökçeada [75, 95].

6.24 Black rat or ship rat, *Rattus rattus* Linnaeus, 1758

Greek name: μαυροπόντικος (*mavropontikos*) Turkish name: *ev siçani*
Greek vernacular name: αρουραίος (*aroureios*)

Black rats or ship rats, *Rattus rattus* Linnaeus, 1758, most likely originated in the equatorial Indo-Malayan region [472, 552] (▶Fig. 58). They have been introduced worldwide and today, they are found from northern Europe to Oceania [472, 541, 553–554]. This species is very likely the most widely distributed member of the Muridae family in the

Fig. 58: The black rat, *Rattus rattus* Linnaeus, 1758, is one of the most adaptable species of rodent that has ever existed. Drawing by Silvia Cantagalli Masseti.

Mediterranean region, where it distributes from the sea level up to mountainous areas. It largely behaves as a commensal of man, associated with human settlements, but it is able to originate "wild" vital populations.

Rattus rattus seems to be so far the most dispersed non-flying terrestrial mammal on the Ionian and Aegean islands. It can survive even on very small, isolated and uninhabited rocks and/or islets of up to about 0.2 ha of extent (compare with [555]), and where the rodent is often the only mammal present. Records from the literature report it on the Albanian island of Saseno, or Sazan [556], Corfu, Paxos and Antipaxos (south of Corfu), Lefkada, Cephalonia, Zakynthos, Antikythera, Kythera, Antimilos, Milos, Skiathos, Alonissos, Kassidis, Lachanou and Polemika (east of Alonissos), Kyra-Panaghia, Grammiza, Youra, Psathoura, Pappous, Piperi, Skantzoura, Peristera, Skyros, Euboea, Andros, Tinos, Mykonos, Delos, Paros, Naxos, Folegandros, Sikinos, Thera, Theodorou, Dhia, Crete, Paximada, Ghianissada, Dragonada, Gavdhos, Koufonisi, Astypalaia, Armathia, Kasos, Karpathos, Rhodes, Tilos, Gyali, Nissiros, Kos, Patmos, Arkios, Kavolos, Psathonissi, several satellite islets of Lipsoi, Agathonissi, Ikaria, Kiseria (Fourni), Chios, Lesbos, Bozcaada, Gökçeada, Samothrace and Thasos. According to Cattaneo [303], *Rattus* sp. occurs on the Dodecanese islands of Leros and Patmos (▶Fig. 59). During the present research it has been found for the first time on the islands of Stamphani (Strophades), Sapientza, Astakida, Salamis, Aegina, Hydra, Siphnos, Kymolos, Polyaigos, Syros, Kea, Amorgos, Levitha, Gavdhopoula and Aghii Pantes (off Crete), Chamilonisia, Zafora, Chalki and Simi (N. Theodoridis, *in verbis*), Macronisi, Frangos, Petrocarabo, Anidros, Leros, Agrelousa, Farmakonissi, Lipsoi, Arki, Anthropofas (C. Papaconstantinou, *in verbis*) and Lemnos (personal observation).

Fig. 59: Sub-adult black rat on the island of Leros, Dodecanese. Photo by Marco Masseti.

The species was found to occur on every island visited in the course of the present survey: Lefkada, Sapientza, Hydra, Poros, Aegina, Kea, Alonissos, Kira Panagia, Youra, Skyros, Euboea, Antimilos, Milos, Antiparos, Paros, Naxos, Delos, Mykonos, Ios, Amorgos, Thera, Crete, Theodorou, Dhia, Aghii Pantes, Rhodes, Simi, Tilos, Nisiros, Kos, Kalymnos, Leros, Lipsoi, Patmos, Ikaria, Samos, Chios, Lesbos, Lemnos, and Thasos. Black rats were also found on Zakynthos (F. Barbagli, *in litteris*), Aegina, Hydra, Kymolos and Syros (Y. Ioannidis, *in verbis*), Milos and Poyaigos (M. Dimaki, *in verbis*), Naxos (Dietrich Ristow, *in litteris*), Skiathos (Florian Fladerer, *in litteris*), Lesbos (A. Legakis, *in verbis*), Bozcaada (Tenedos) and Gökçeada (Imbros) (B. Özkan, *in litteris*). Their presence on Ithake and Cephalonia was confirmed by osteological remains of the species contained in barn owl pellets (H. Pieper, *in litteris*). In the course of a HOS (Hellenic Ornithological Society) survey on the island of Lipsoi, in the year 1998, H. Alivizatos (*in litteris*) found remains of *Rattus rattus* in pellets of Bonelli's eagle, *Hieraaetus fasciatus* Vieillot, 1822. The occurrence of the murid on the latter island had already been noted in the course of the present research. Black rats seem to have been reported from Dhia not earlier than the 1920s (Ristow, 2000, *in litteris*). According to Dietrich Ristow (*in verbis*), they are still rather common on Paximada, Dragonada and Giannissada, as well as on the eastern coast of Naxos.

Original data collected during the present study allowed to ascertain for the first time the presence of the black rat on Vromonas and Atokos, Pontikonissi, Astakida (north of Skopelos), Gavdhos, Gavdhopoula, Koufonissi (off Ierapetra, southern Crete), Lipsoi, Arki, Agathonissi, Patmos, Petrokaradon (north of Patmos), Chamilonisia (Hamili) and Zaforà (between Kassos and Astypalaia) (C. Papaconstantinou, *in verbis*), Siphnos (Y. Ioannidis, *in verbis*), Kalymnos (personal observation), Stamphani (Strophades), Andidragonada and Prassonisi off Kythera (east of Kythera), Polyaigos (Milos) (M. Dimaki, *in verbis*), Prasonisi (Antikythera), Aghia Pelagia (Euboea), Polemika, Lachanou, Kasidis, Skantili, and Korakas (Skantzoura, Northern Sporades), Kastronissia (north of Skiathos, Northern Sporades), Gaidaros (Peristera, Northern Sporades), Strongilo (Naxos), Theotokos (Andros), Chtapodia and Tragonisi (Mykonos), Prasonisi, Drakonisi, Dysvato, and Kalogeri (Tinos), Aghios Andreas and Choironisi (south of Tilos) (Jakob Fric, *in litteris*). Direct and indirect evidence of the presence of black rats was also recorded from several islets of the gulf of Fethiye, in the south-western Turkish coast, such as Turgut, Kameriye, Göcek, Delitli, Kizil, Yassicalar, Terzane, and Domuz.

A program of eradication was launched in the Northern Sporades in 2005, within the frame of a project for the protection of sea birds. That year the rodents were eliminated from the islets of Polemika and Lachanou (Skantzoura). In 2006, it was the time for Kastronisi 1 and Kastronisi 2 (Skiathos), whereas the eradication of black rats from Aghios Andreas (south of Tilos) was implemented in December 2006 (Jakob Fric, *in litteris*). Apart from *Rattus rattus*, on the latter islet, no other mammal has been found.

Systematic data

There are three known colour varieties, one for each of the subspecies *Rattus rattus rattus* Linnaeus, 1758, *Rattus rattus alexandrinus* Geoffroy, 1803, and *Rattus rattus frugivorus* Rafinesque, 1814.

The "*rattus*" morph is mostly black, with a black back grades to a slate grey belly; young rats may show a bluish tinge.

The "*alexandrinus*" phenotype has a grey-brown back and a slate-grey and/or creamy-white belly. The dorsal pelage consists of brown-tipped grey hairs overlain by long, black guard hairs.

The "*frugivorus*" phenotype also has a grey-brown back, but ventrally, from the chin to the anus, it is pure or lemony-white. These three morphs interbreed freely [557–559]. They all occur on the Ionian and Aegean islands.

Rattus rattus alexandrinus has been reported on Corfu [49, 71, 279, 518]; Cephalonia [71, 279, 518]; Euboea [71, 518]; Crete [277]; Paximada, Ghiannissada and Dragonada (Dietrich Ristow, *in litteris*); Rhodes [57]; Tilos and Leros (personal observation).

Rattus rattus frugivorus is known on Corfu [49, 71, 27, 518]; Cephalonia [71, 279, 518]; Zakynthos [50, 71, 518]; Euboea [71], Andros [71, 518]; Delos [71, 518]; Sikinos [58, 71], Pholegandros [58, 71], Astypalaia [58, 61, 71, 518]; Crete [71, 300, 518]; Karpathos [56, 58, 71, 518]; Rhodes [58, 71, 518]; Gyali [71]; Samos

[65, 71, 518]; Kiseria [58, 71]; Ikaria [58, 71, 518]; Chios [67] and Samothrace [58, 71, 355].

Museum specimens

Lefkada: NMW 23461, 45865; Cephalonia: BMNH 8.10.1.31–32, 8, 10.1.34–37, 8.10.1.40–41; NMW 45866; SMF 55250–55257; Zakynthos: ZMUP 6484–6484; Skiathos: BMNH 66.1468–1471; SMF 45204–45211; Psathoura: FMS 55819–55821; Pappous: SMF 55822–55839, 55855; Piperi: SMF 55840–55841, 55854; Skantzoura: SMF 55842; Peristera: SMF 55646–55649, 55843–55845; Skyros: SMF 49541–49548, 49999; Sikinos: NMW 24059–24060; Andros: BMNH: 66.5382–66.5386; Mykonos: BMNH 66.5381; Pholegandros: NMW 24058; Crete: BMNH no 8.10.24.4–5; MZUF 16436–16438; NMW B978–B980, 20.677–20.683, 36158–36163 ; SMF 49540, 66603–66604; ZMUP no 6687, 6710, 6731–33, 6759–6778, 6792–94; Paximada: ZFMK, Niethammer collection 6029; Astypalaia: NMW 24062–24064; Karpathos: NMW 24066; Rhodes: NMW 20.703, 23462–23463, 24065, 27345; Tilos: NMW 32858; Ikaria: NMW 24056; Kiseria (Fourni): NMW 24057; Chios: BMNH 66.6593, 66.6595–66.6598; SMF 73685; Samothrace: NMW 24061.

References

Saseno, or Sazan, in the bay of Valona, half-way between Italy and Albania [556]; Corfu [49, 70–71, 279, 518]; Paxos and Antipaxos [560]; Cephalonia [71, 279, 283, 518]; Zakynthos [50, 71, 518]; Antikythera [350, 352]; Kythera [52–53, 398–399, 518]; Alonissos [289]; Kassidis, Lachanou and Polemika, east of Alonissos [561]; Kyra-Panaghia [289]; Grammiza [289]; Youra [71, 289]; Piperi [289]; Euboea [71, 293, 518]; Antimilos [401]; Milos [51]; Andros [71, 518]; Tinos [518]; Mykonos [71, 518]; Delos [71, 518]; Paros [401, 404]; Naxos [401, 404]; Folegandros [58, 71]; Sikinos [58, 71]; Thera [275, 295]; Astypalaia [58, 61–62, 71, 95, 518]; Crete [71, 95, 276, 279, 297–298, 347–348, 362, 518]; Theodorou [348]; Dhia [71, 348]; Paximada [71, 409]; Ghianissada [409]; Dragonada [409]; Gavdhos [71, 348, 518]; Koufonisi [71, 348]; Armathia [233], Kasos [233, 299]; Karpathos [56, 58, 71, 95, 299, 518]; Rhodes [57–59, 71, 95, 300–301, 518, 562]; Tilos [63, 302]; Gyali [71]; Nissiros [71, 563]; Kos [59, 64, 95, 369, 518]; Patmos [564]; Arkios [453]; Kavolos [453]; Psathonissi [453]; satellite islets of Lipsoi [453]; Agathonissi [453]; Ikaria [58, 71, 95, 518]; Kiseria (Fourni) [58, 71]; Samos [65–66, 71, 95, 305–306, 518]; Chios [64, 67, 69, 95]; Lesbos [64, 70–72, 95]; Bozcaada (Tenedos) [75, 95], Gökçeada [75, 95]; Samothrace [58, 71, 77, 355, 518]; Thasos [71, 356].

Archaeozoological note

According to Vigne [565], commensal rodents were first introduced on the Mediterranean islands between 5500 and 8000 years ago. In fact, the black rat is believed to have arrived along the Mediterranean shores contemporary to other species commensal with humans, such as the house mouse, *Mus musculus* Schwarz and Schwarz, 1943, as well as the house sparrow, *Passer domesticus* Linnaeus, 1758. As far as is presently known, these appearances have been related to prehistoric man's adoption of a sedentary way of life, when a new habitat developed in and around early villages, and to the later development of indirect commensal relationships (compare with [566]). Black rats remains have been described from several Levantine Natufian sites, dated to the 11th–9th millennium BC [567–568]. According to Armitage et al. [569], it was probably in the holds of merchant ships returning from the Middle East to Red Sea ports with cargoes, spices and other exotic products that rats were carried directly to Egypt and the rest of the Eastern Mediterranean in large numbers (compare with [570]). However, as observed by Barrett-Hamilton and Hinton [571], it seems that *Rattus rattus* was unknown to the ancient Greeks. According to Ortalli [572], the species does not appear in European literary sources before the end of the early Middle Ages. In fact, apart from two passages in the pseudo-Aristotle's *On marvellous things heard,* and Aelian's *De natura animalium* (XVII, 17) (compare with [573]) recounting that rats eat iron on the island of Gyaros, located in the northern Cyclades near Andros and Tinos, there is no account of the black rat in the classical and late-antique tradition. Albertus Magnus first mentioned it in his scientific study *De animalibus* (XXII, 2, 1, 123) [572].

The history of *Rattus rattus* is particularly important because of its alleged association with the spread of bubonic and pneumonic plague [574]. Yet these epidemics seem to have been unknown to classical writers [570]. With very few exceptions, possibly the only one being the pestilence described by Rufus of Ephesus (about 200 BC), ancient epidemics since the biblical plague of the Philistines were by no means comparable to the bubonic plague. The outbreak of the bubonic plague perhaps dates to before 541–542 AD, when it was portrayed with remarkable efficacy by Evagrius (*Historia ecclesiastica*, IV, 29), accurately described by Procopius, and then mentioned by an infinite number of sources as a recurrent and ruthless scourge until the middle of the 8th century [572]. The indubitable connections between the bubonic plague and the black rat, on one hand, and the plague and the early Middle Ages, on the other, seem to confirm what historical sources only hint at: *Rattus rattus* had already increased its European distribution by the early Middle Ages. People of the time did not seem to heed the massive presence of this potential bearer of catastrophes, as if it were something negligible [572].

It is noteworthy that, in archaeological sites, the occurrence of *Rattus rattus* among earlier stratigraphic levels could be related to some of its behavioural characteristics [570]. Although black rats are bad burrowers, preferring to live in crevices, holes, and under the eaves of roofs, they are however often able to reach depths of forty to fifty centimeters in a vertical line. Therefore, osteological remains of the species can be found in deeper layers than one might expect, especially when the deposits and sediments are not consolidated. On Crete, for example, black rats have been found associated with remains of endemic faunal elements, such as *Mus minotaurus* Bate, 1942, and *Candiacervus ropalophorus* de Vos, 1984 [106]: both species characterised the late Pleistocene fauna of the island. It cannot be excluded that the murid penetrated the level where it was discovered coming from the surface [570].

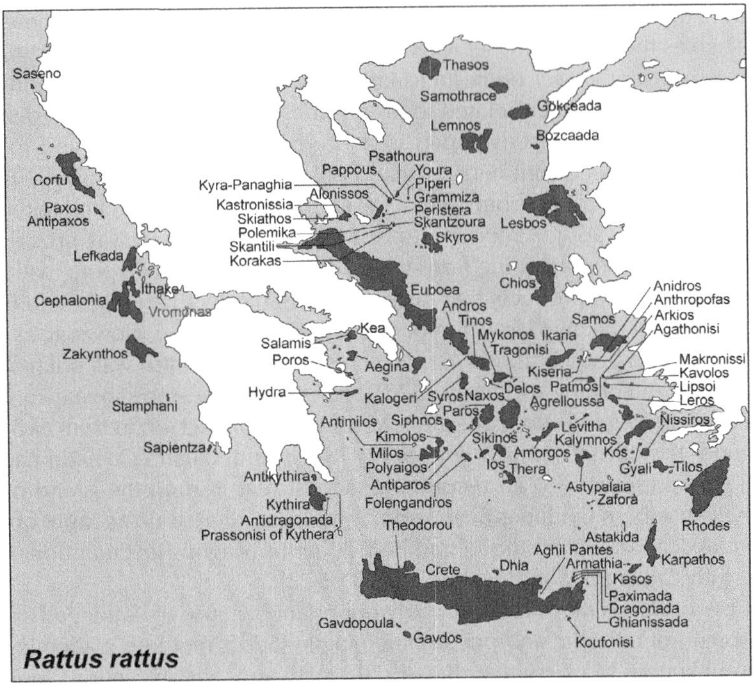

Fig. 60: Distribution of the black rat, *Rattus rattus* Linnaeus, 1758, in the Ionian and Aegean islands.

Spatial and graphic restrictions have made it impossible to include all the islands affected by the diffusion of *Rattus rattus* in the distribution map in ▶Fig. 60.

6.25 House mouse, *Mus musculus* Schwarz and Schwarz, 1943

Greek name: σπιτοπόντικος (*spitopontikos*)　　Turkish name: *ev faresi*

Commonly referred to as the house mouse, *Mus musculus* Schwarz and Schwarz, 1943, together with humans and, perhaps, black rats, this murid is the most widespread mammal on earth and one of the most invasive species in biological terms and most dangerous for human health (compare with [545]) (▶Fig. 61).

The species possibly originated in the steppe of central Asia [575]. Together with the black rat, it is currently the most widespread commensal of humans over most of the planet (with the only exception of Antartica), owing to its strict association with man [373, 472, 576–577]. In fact, although the species can live in the wilderness, it is symbiothic with men (compare with [351]). According to several authors, such as Bonhomme [578] and Auffray et al. [579], the taxonomic group of the so-called "domestic mice" which,

Fig. 61: House mouse, *Mus musculus* Schwarz and Schwarz, 1943. Photo by Sandro Sacchetti.

besides *Mus musculus*, comprises among others also the Balkan short-tailed mouse, *Mus macedonicus* Petrov and Ružić, 1983, is the only group of the genus *Mus* that can originate definite and permanent commensal populations. There are sharp differences between commensal and non-commensal ways of life in mice [580]. Commensal populations form where food is superabundant. They often reach high densities, develop very small home ranges and replace territoriality system with more hierarchically organised societies [581–583]. In contrast, non-commensal populations form where food supply is less available or where rainfall is abundant [584–585]: in this case home ranges/territories turn wide [586]. According to the IUCN (World Conservation Union), the house mouse belongs to the world's hundred worst invasive species, constituting a major threat to biological diversity (disappearance of autochthonous fauna and biodiversity homogenisation) as well as to the activities and health of humans [587]. The house mouse benefits from its interaction with humans: its survival and diffusion depend on the augment of the anthroposphere (the whole of the artificial ecosystems) [543]. Human activities promote its diffusion by eliminating ecological barriers (deforestation, development of agricultural lands and transportation systems) or by increasing the human pressure on the natural ecosystem.

On island ecosystems, in particular, these mice have a variety of negative impacts, which include their predation on reptiles, invertebrates, and the nests of terrestrial birds [212, 588–592]. Evidence of the occurrence of the species was recorded in the majority of the Ionian and Aegean islands visited in the course of the present research. *Mus musculus* was found on the Albanian Saseno, or Sazan [556], Corfu, Paxos, Antipaxos,

Cephalonia, Lefkada, Zakynthos, Arpia (Strofades), Kythera, Antikythera, Milos, Sikinos, Skiathos, Kyra Panaghia (east of Alonissos), Skyros, Euboea, Kea, Andros, Mykonos, Sikinos, Thera, Anaphi, Astypalaia, Crete, Gavdhos, Gavdhopoula, Armathia, Kasos, Karpathos, Rhodes, Tilos, Kos, Leros, Patmos, Ikaria, Psara, Samos, Chios, Lesbos, Bozcaada, Samothrace, and Thasos. The occurrence of the house mouse on Lesbos has been confirmed by A. Legakis (*in verbis*), and on Cephalonia by the finding of osteological remains of the species in barn owl pellets (H. Pieper, *in litteris*). During the present research, the species was also found on Aegina, Hydra, Kymolos and Syros (Y. Ioannidis, *in verbis*), and Bozcaada (Tenedos) (B. Özkan, *in litteris*). It was reported from Itakhe for the first time (C. Papaconstantinou, *in verbis*). Because of the strict dependence of *Mus musculus* on the diffusion of human settlements (compare with [593–594]), it can be presumed to occur on all the 217 Greek islands of the Ionian and Aegean seas permanently inhabited by man.

Systematic data

The systematics of the genus *Mus* Linnaeus, 1758, is rather complex; in fact, it generates wildly contrasting opinions [594–595]. In the last decades, biochemical and genetic studies have shown that the house mouse, which is traditionally regarded as a unique species, is actually a complex taxonomic group of evolutionarily differentiated taxa (compare with [596–607]).

The species *Mus musculus*, in particular, includes two taxa: *Mus musculus musculus* Linnaeus, 1758, and *Mus musculus domesticus* Schwarz and Schwarz, 1943, which are classified to the subspecific, semispecific or even specific level, according to the different opinions of various authors (see Berry and Scriven [608] for their taxonomic review). *Mus musculus domesticus* is widespread in North Africa and ranges in Eurasia from western Europe (and most Mediterranean islands) through southern Eurasia to the Caucasus and eastwards through Iran, Afghanistan and Pakistan to northern India and Nepal. This was the variety also spread accidentally by European colonisation throughout the New World, numerous islands, Australia, and probably Southern Africa ([609]; compare with [610]). However, the systematics of the subspecies is still imperfecly known. *Mus musculus domesticus* has attracted the attention of researchers over the last three decades, from either the genetical [611–612] or a zooarchaeological standpoint [545, 613]. The subspecies has, in fact, demonstrated a remarkable variability in its chromosomal composition [603]. In the Ionian and Aegean islands, *Mus musculus domesticus* has been reported from Corfu [49], Cephalonia [603], Zakynthos [614], Kythera and Andros [603], Astypalaia [61–62], Crete [277], Armathia [233], Kasos [233, 603]; Karpathos [56, 58, 362], Kos [615], Rhodes [55, 57–58, 301, 362, 562] and Lesbos [603].

Museum specimens

Corfu: BMNH 8.10.1.43–8.10.1.48, 8.11.3.14–8.11.3.16; MNHN 1994–2689, 1994–2690, 19331815; Lefkada: NMW 22844–22849; Cephalonia: BMNH 8.11.3.17; NMW

22850; ZMUP 6986–6987, 6993–6994; Zakynthos: ZMUP 6461–6466, 6481–6483, 6486, 65006502, 6504–6509; Arpia (Strofades): ZMUP 6901, 6807; Kythera: ZMUP 7095; Antikythera: ZMUP 67966801; Skiathos: BMNH 66.1472–66.1478; SMF 57334–57336; Kyra Panaghia: SMF 55814–55815; Skyros: SMF 49533–49539; Andros: BMNH 66.5400, 75.1313–75.1319; Mykonos: BMNH 66.5388–66.5392, 66.5395–66.5399; Sikinos: NMW 20.706–707; Anaphi: NMW 20.708–20.711; Astypalaia: NMW 22856–22858; Crete: BMNH 1938.11.25.4–1938.11.25.5, 8.10.24.2–8.10.24.3, 68.612–68.614; NHMC 80.5.42. (1–183); NMW 20.720–20.728, 36175–36179, 36180–36199, 36200–36202; SMF 49526, 49528–49529, 49531–49532, ;66599–66600; ZMUP 6687, 6710, 6731–6733, 6759–6778, 6792–6794; Kassos: ZMUP 6802, 6804; Karpathos: NMW 20.718–20.719, 22859; Rhodes: NMW 20.704, 20.712, 27341–27343, 22851; Tilos: ML139–143, TE933; Kos: NMW 54761; ZMUP 6493–6496; Ikaria: NMW 20.705; Chios: BMNH 66.6585–66.6591; Lesbos: ZMUP 3159; Samothrace: NMW 20.713–20.717.

References

Saseno, or Sazan, in the bay of Valona [556]; Corfu [49, 279]; Paxos and Antipaxos [560]; Cephalonia [70–71, 279, 283, 603]; Zakynthos [50, 71, 614–616]; Kythera [398–399, 603]; Antikythera [350, 352, 603]; Euboea [615]; Kea [449]; Milos [51]; Sikinos [58, 362]; Andros [603]; Astypalaia [61–62, 95]; Thera [70, 275, 295]; Anaphi [58, 70, 362]; Crete [54, 95, 276, 347–348, 362, 615]; Gavdhos [348]; Gavdhopoula [348]; Armathia [233]; Kasos [233, 299, 603]; Karpathos [56, 58, 70, 95, 299, 362]; Tilos [29, 63, 302, 350–352]; Nisiros [563]; Kos [95, 371, 615]; Rhodes [55, 57–58, 95, 301, 362, 562]; Leros [303]; Patmos [303]; Ikaria [58, 95, 362]; Psara [350, 351]; Samos [66, 95, 306]; Chios [67, 69, 95, 617]; Lesbos [72, 329, 603]; Bozcaada [75, 95]; Gökçeada [95]; Samothrace [77, 362]; Thasos [356].

Archaeozoological note

As far as is presently known, the earlier fossil remains of the genus *Mus* have been reported from the site of Ubeidiya, in the Jordan valley, and dated to about 1 million years ago [618]. However, the analyses of series of stratified cave fossils in Israel highlighted the presence of *Mus musculus* on the shores of the Eastern Mediterranean, starting only from 12 000 BC [567]. The diffusion of the murid from south-east Asia throughout Eurasia connects with the last wave of colonisation, during the Holocene [613]. The house mouse is in fact regarded as one of the latest colonisers of the Mediterranean region. Its invasive progression seems than to have stopped or drastically slowed only in the 1st millennium BC. Well-documented archaeozoological records have suggested that the expansion through the northern Mediterranean shores took place in two waves, a first wave reached the eastern Mediterranean, whereas western Europe was colonised more recently during the last millennium BC [613] with a possible separate northern and southern colonisation during the Iron Age [619].

In its expansion towards the western Mediterranean, the house mouse could also colonise Cyprus because this island, even if far from the mainland, was included in the area of pre-ceramic Neolithic diffusion [545, 620]. In contrast, the diffusion of the Neolithic culture in Greece around the 8th–7th millennium BC, from Anatolia through the islands of the Aegean Sea, was not accompanied by the introduction of the house mouse [545]. The species did not reach these areas until after the apogee of the Bronze Age at the time of the Near Eastern empires, the great dynasties of the Pharaohs, and the Aegean thalassocracies. In that period, the maritime trade between empires and city states flourished in the Eastern Mediterranean, as testified by the Late Bronze Age cargo of the Uluburun shipwreck. This rich cargo included raw material, such as copper, tin, ebony, and ostrich eggshells, jewellery, arms and ceramics coming from all over the Eastern basin of the Mediterranean [621]. A jawbone of *Mus musculus domesticus* has also been described as part of the same cargo [622–623], showing that marine navigation was one of the means of passive dispersal of small commensal mammals since protohistoric times. The success this category of mammals can have in colonising depends on the magnitude of the migrant flow as well as on the existence of a favourable ecological niche [545]. The maritime traffic allows small mammals to reach otherwise inaccessible territories, which are far beyond their dispersal capacities. The availability of a favourable ecological niche depends on the intensity of human encroachment as well as on the presence of autochthonous rodent competitors which can constitute an effective ecological barrier. Mice could have hidden in containers with sacks or baskets for seeds, and thus be transported passively towards new territories, exactly as they do today [545].

Given the role of Neolithisation in the renewal of insular Mediterranean fauna [4, 6, 624], the diffusion of the house mouse in the Mediterranean is most probably correlated with the timing of Neolithic migration within the circum-Mediterranean area [545]. A critical appraisal of all available circum-Mediterranean Late Glacial to Iron Age sub-fossil mouse occurrences has shown that house mouse's invasion of Europe was favoured by progress in human evolution [579, 613].

However, although the house mouse rapidly dispersed in the Near East – Cyprus included – in what can be considered the the hub of Neolithisation, it colonised the Eastern Mediterranean in a much longer lapse of time [545]. This diffusion seems to have begun in the 3rd millennium BC, as far as we know today. The species did not appear in Crete before the middle of 2nd millennium BC, by the end of the Middle Minoan period (compare with [347]) and it did not colonise the whole Eastern Mediterranean basin before the 1st millennium BC. According to K. Trantalidou (*in verbis*), few remains of house mice were found, for example, in the Late Bronze Agearchaeological site of Akroitiri, on Santorini (Thera). It is worth mentioning that these murids were also discovered in layers dated from the mid to late 8th century BC (760–700 BC) of the Early Iron age site of the Oropos, in northern Attica, near the seashore [625]. Recent genetic studies [607] confirm the well-documented archaeozoological records that have suggested the above mentioned two-step westward expansion, the early Neolithic one limited to the eastern Mediterranean and the later one, particularly evident in the western Mediterranean, related to the expansion of maritime trade during the first millennium BC and onwards. This last process never really ended, since the most recently populated areas, such as the Atlantic islands, show an unambiguous signal of recent dispersal, while anciently populated islands like Cyprus show the footprints of successive introduction events. Humans accidentally transport founding individuals over long distances and permit this commensal species to increase its distribution and colonise new areas, which would otherwise be inaccessible [545].

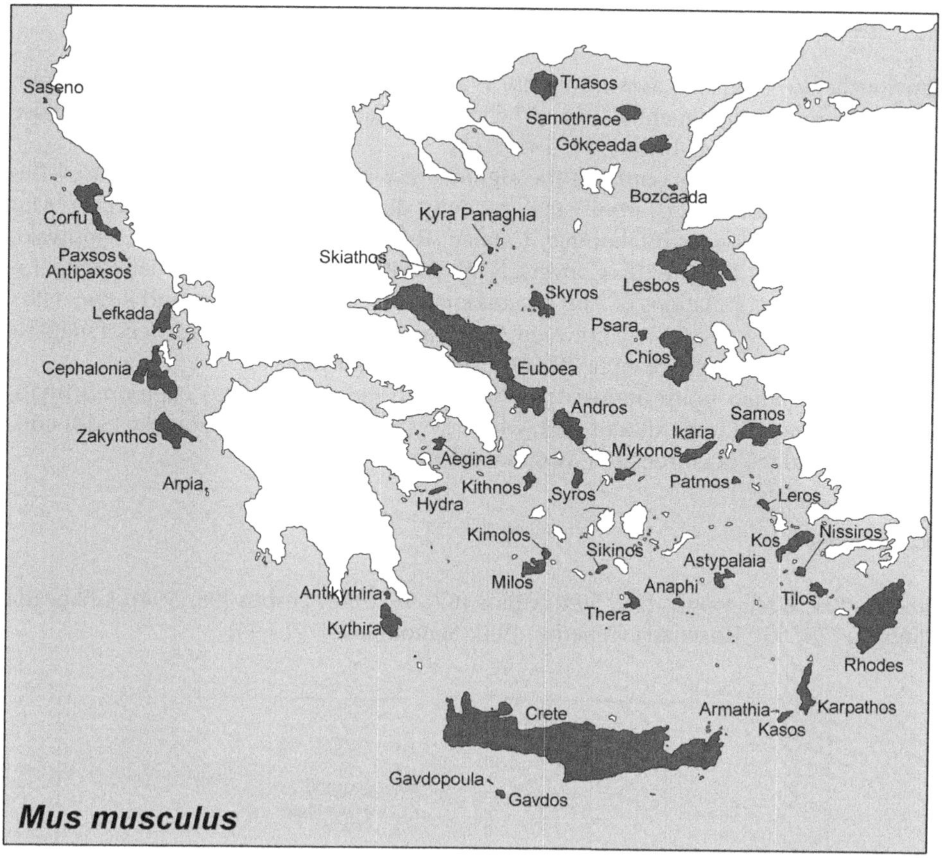

Fig. 62: Distribution of the house mouse, *Mus musculus*, Schwarz and Schwarz, 1943, in the Ionian and Aegean islands.

6.26 Balkan short-tailed mouse, Macedonian mouse or short-tailed mouse, *Mus macedonicus* Petrov and Ružić, 1983

Greek name: Μακεδονικός ποντικός (*Makedonikos pontikos*)

Turkish name: *Makedonya ev faresi*

The Balkan short-tailed mouse, Macedonian house mouse, or short-tailed mouse, *Mus macedonicus* Petrov and Ružić, 1983, occurs in the eastern Mediterranean basin from the western Balkan peninsula to Bulgaria, Asia Minor and the Levant, including Cyprus, and east to Iran [270, 594, 626]. It also occurs on several Eastern Mediterranean islands, such as Rhodes, Samos, Chios, Lesbos, Gökçeada (Imbros), Bozcaada (Tenedos) and Samothrace.

Systematic data

Previously also indicated as *Mus abbotti* Waterhouse, 1837, or *Mus tataricus* Satunin, 1908, the species is frequently confused with the genetically and morphologically close *Mus spicilegus* Petényi, 1882 [594, 596, 627].

Phylogenetic analysis confirms the significant divergence of a mitochondrial lineage of *Mus macedonicus* from Israel, recently described as a new subspecies, *Mus macedonicus spretoides* Bonhomme, Catalan, Britton-Davidian, Chapman, Moriwaki, Nevo, and Thaler, 1984 [628]. Conversely, no genetic hiatus is revealed between European and Asian populations of *Mus macedonicus macedonicus* Petrov and Ružić, 1983 [629]. The mtDNA divergence between *Mus macedonicus* and *Mus spicilegus* suggests a division between them of circa 700 000 and 1 million years ago.

The Macedonian house mouse may have been reported as *Mus bactrianus* Blyth, 1846, by Festa [57] from Rhodes [60], where Kryštufek and Mozetič Francky [524] confirmed instead the occurrence of *Mus macedonicus*.

References

Rhodes [95, 524]; Samos [95, 594]; Chios [67, 95, 594]; Lesbos [95, 594]; Gökçeada (Imbros) [75, 95]; Bozcaada (Tenedos) [95]; Samothrace [77, 594].

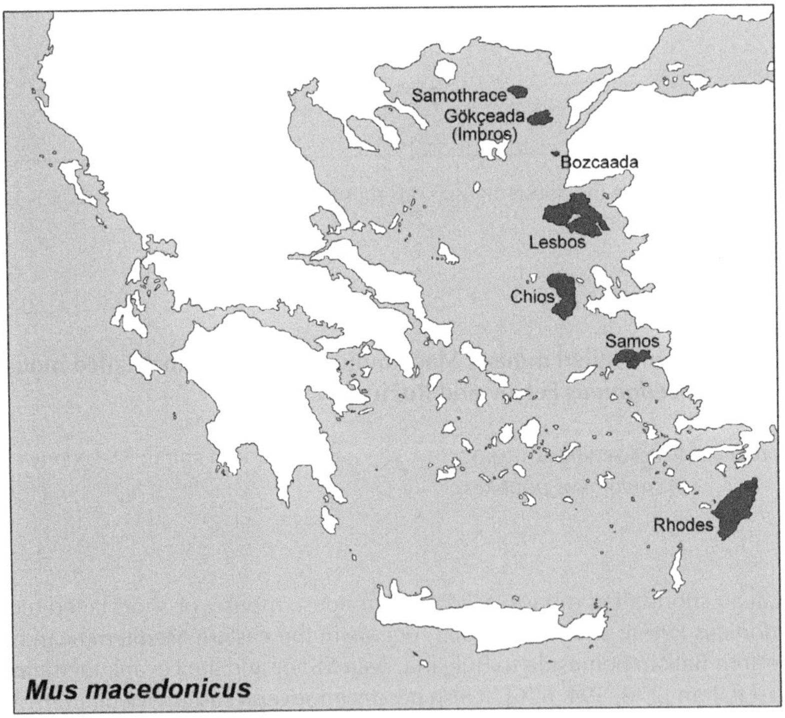

Fig. 63: Distribution of the Balkan short-tailed mouse or Macedonian mouse, *Mus macedonicus*, Petrov and Ružić, 1983, in the Aegean islands.

6.27 Steppe mouse or mound-building mouse, *Mus spicilegus* Petényi, 1882

The steppe mouse or mound-building mouse, *Mus spicilegus* Petényi, 1882, is known from the steppe zone of south-eastern Europe only, where it inhabits grasslands and cultivated land [630]. Its distribution ranges from the lowlands of Austria, southern Slovakia, Hungary, Romania, Serbia, and Montenegro, to Albania, Greece, northern Bulgaria, and the steppes of Moldavia and southern Ukraine [472, 631].

The appearance of *Mus spicilegus* in the southern Balkan peninsula was first confirmed, based on three specimens from Epirus [632], previously reported by Sofianidou and Vohralík [633] as *Mus abbotti* Waterhouse, 1837, while Macholán et al. [629] reported three additional sites for this species, which are located in western Greece, extending as far as Patras, in the north-western Peloponnese. However, the taxonomical certainty of the occurrence of *Mus spicilegus* on the Mediterranean coast is known only since the study of Kryštufek and Macholán [634]. New records have significantly expanded knowledge about the distribution of the mound-building mouse in the southern Balkan peninsula and in the Ionian and Aegean islands (compare with [605]), its present occurrence being reported from Zakynthos, Milos, and Skyathos. Nevertheless, it is very likely that the specimens from Skyathos and Milos in the collection of the Forschungsinstitut and Natural History Museum Senckenberg (Frankfurt am Main), respectively with the number SMF 45196–45203 and SMF 49712, have not yet been revised within the framework of current *Mus* taxonomy. In several cases, the mound-building mouse has been found in Greece in sympatry with *Mus musculus domesticus*, and specimens of the house mouse were trapped inside the mounds, along with steppe mice [605].

In a Holocene fauna provided by the palaeontological exploration of the island of Chios, Kock [67] found the osteological remains of a mouse that he referred to as the *Mus musculus spicilegus* Petényi, 1882.

Systematic data

Mus spicilegus has frequently been confused with *Mus musculus* Linnaeus, 1758, which is morphologically very similar [575, 631]. In fact, despite extensive zoogeographical and taxonomical studies in the subgenus *Mus*, several issues still remain unresolved. In this context, according to Mitsainas et al. [605], the distribution area of *Mus spicilegus* remains undetermined at the southernmost parts of the species' range, possibly due to low population densities and the seasonal character of mound building.

Kryštufek and Macholán [634] proposed two possible routes for the colonisation of Europe by the mound-building mouse. A northern migration route starts from India and Pakistan and proceeds westwards, along the northern shore of the Caspian and Black Seas and then southwards to the Balkans. Another scenario describes a southern route from Asia Minor through the Bosporus landbridge to the Balkans northwards in two colonisation waves; the former wave establishing *Mus spicilegus* in south-eastern Europe from a *macedonicus/spicilegus* ancestor, with later invasion of *Mus macedonicus* (most probably during the Würmian glacial phase), forcing *Mus spicilegus* northwards. The results of a recent genetic study carried out by Mitsainas et al. [605] that place the continental Greece populations of *Mus spicilegus* mouse at the basal clade of all the phylogenetic trees, could reinforce the hypothesis that the invasion of *Mus macedonicus* from the Bosporus forced *Mus spicilegus* not only northwards but also southwards into the southern Balkan peninsula and specifically along the eastern coastal lines. The greater genetic similarity of the western and southern Greek populations with those of the other Balkan and European countries and the positioning of all these populations in a single unresolved clade implies that these Greek populations originated from the ancestral genetic stock, which was initially forced northwards (compare with [605]).

Museum specimens

Skyathos: SMF 45196–45203; Milos: SMF 49712.

References

Zakynthos [71, 605]; †Chios [67].

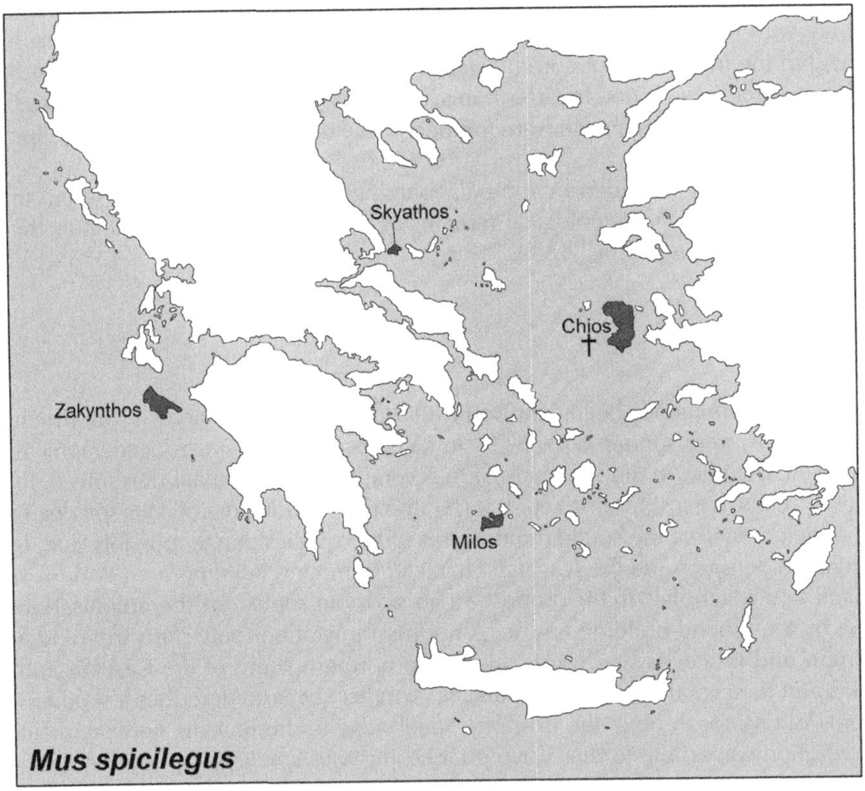

Fig. 64: Former and present distribution of the steppe mouse or mound-building mouse, *Mus spicilegus*, Petényi, 1882, in the Ionian and Aegean archipelagos. Islands and areas that are inhabited by the mouse are shaded and those where they have become extinct are marked with a cross.

6.28 Cretan spiny mouse, *Acomys cahirinus minous* Bate, 1906

Greek name: ακανθοπόντικος (*akanthopontikos*) Turkish name: *Makedonya ev faresi*

Like the Cypriot spiny mouse, *Acomys cahirinus nesiotes* Bate, 1903, the Cretan spiny mouse, *Acomys cahirinus minous* Bate, 1906 (▶Fig. 65) has been described as endemic to the respective islands [475, 635–638]. More in particular, several authors, such as Zima [639] and Musser and Carleton [472], consider it as a separate species, *Acomys minous* Bate, 1906, for its peculiar karyotype and the isolated distribution. Yet the total lack of Pleistocene fossil remains of the genus from Crete [6] rules it out from the island's endemic faunas. Genetic analyses demonstrate the vicinity of the Cretan spiny mouse to the *Acomys cahirinus-dimitiadus* group, which is distributed in Sinai, the Near East, and south-western Asia [638]. Except for *Acomys russatus* Wagner, 1840, which occurs from Egypt to eastern Arabia, the members of this group are the only spiny mice taxa distributed outside the African continent.

Fig. 65: The spiny mouse is regularly found associated with human settlements in the Mediterranean region, where it often behaves as a true commensal with man. Photo by Apostolos Trichas.

Systematic data

The assumed human introduction of *Acomys* on Crete (as on Cyprus) from the nearest mainland has also been supported by the results of cytochrome *b* (non-coding region of mtDNA) analyses [46, 640]. These suggest that the Cretan spiny mouse possibly belongs to the species *Acomys cahirinus* Desmarest, 1819, the Northeast African spiny mouse, *sensu stricto* [640]. In fact, the low divergence between the sequences of *Acomys* from Cairo, Crete, Cyprus and Turkey suggests a possibly recent colonisation of the islands which could be explained as human-derived dispersion [46]. Recently Giagia-Athanasopoulou et al. [47] compared the chromosome banding of the karyotypes of the Cretan spiny mouse with those of the other species in the *cahirinus* group (i.e., *Acomys cahirinus, Acomys cilicicus,* and the Cypriot spiny mouse). These results are supported by the study of the cytochrome *b* sequences, which shows that *Acomys cahirinus minous* possesses the group's ancestral karyotype. This shows that at least two of the karyotypes, that characterise *Acomys cahirinus minous* today, pre-existed in North Africa before the shrew colonised Crete, but it also proves that the specific status of the four members of the *Acomys cahirinus* group needs to be revisited [47].

It is also noteworthy that the spiny mouse is regularly found associated with human settlements in the Mediterranean region, where it often behaves as a true commensal with man [568], generally resulting in a sympatric relationship with the house mouse, *Mus musculus* Linnaeus, 1758, and the black rat, *Rattus rattus* Linnaeus, 1758. In the Levant, according to Tchernov [568, 641], the abrupt increase in the frequency of *Acomys cahirinus* since the Natufian culture layers may be explained as a rapid adaptation of the species to commensal life around human settlements.

Museum specimens

Crete: BMNH 5.12.2.31–5.12.2.33; NHMC 80.5.40.1–80.5.40.18, 80.5.40.20–80.5.40.38, 80.5.40.42, 80.5.40.45–80.5.40.118, 80.5.40.120–80.5.40.122, 80.5.40.124–80.5.40.125; NMW 8999–9003, 36164–36174; SMF 26785–26786, 49549–49554, 66601–66602; ZMUP 6708–6709, 6740–6743, 6758.

References

Crete: [6, 9, 46–47, 276–277, 285, 298, 347–348, 362, 472, 635–636, 638–640].

Archaeozoological note

Several specimens of *Acomys* were present in a sample from the Minoan levels of Kommos, in southern Crete [417]. They closely resemble the present-day Cretan spiny mice. According to Payne [347], the representatives of the genus *Acomys* I. Geoffroy, 1838, reached Crete not before the Iron Age. These rodents must have come from the Levant or North Africa, as they are absent from Europe and western Turkey. Moreover,

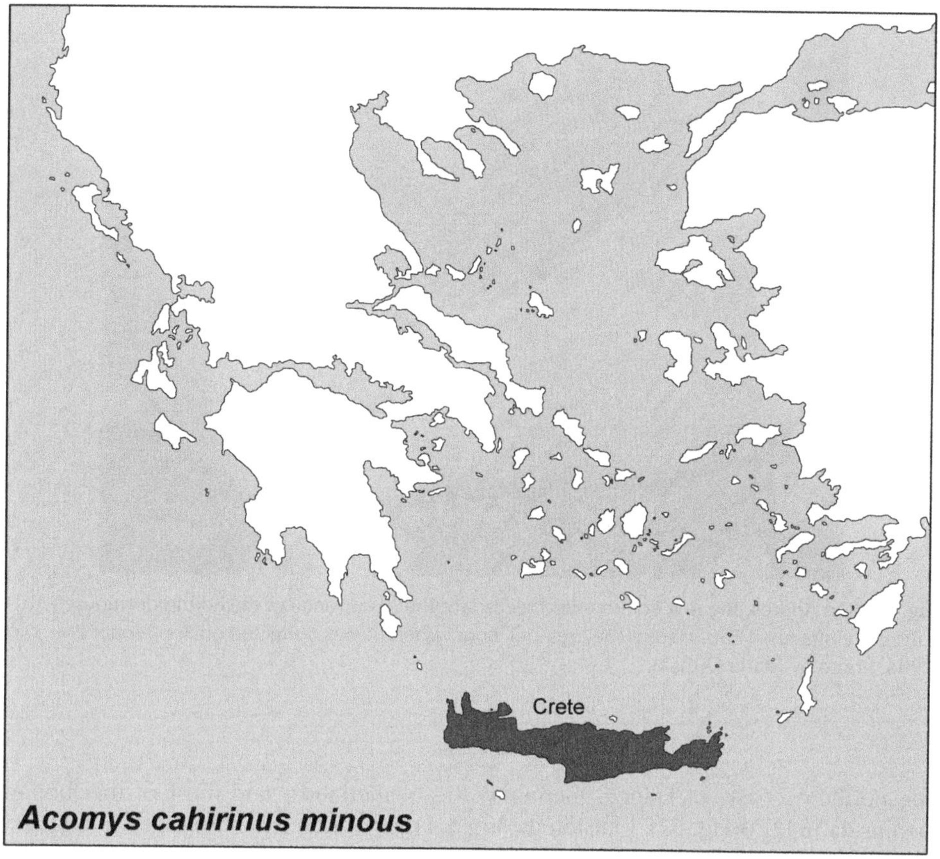

Fig. 66: In the Aegean region, the distribution of *Acomys cahirinus minous* Bate, 1906, is limited to the island of Crete.

it is interesting to note that *Acomys* is present only in the latest period at Kommos [347].

6.29 Fat dormouse or edible dormouse, *Glis glis* Linnaeus, 1766

Greek name: δασομυωξός (*dasomyoxos*)
Greek vernacular names: δενδρογαλή (*dendrogali*)
πέλεκας (*pelekas*, Andros)

Turkish name: *yediuyur*

The edible dormouse, *Glis glis* Linnaeus, 1766, is also sometimes known as the fat dormouse or the squirrel-tailed dormouse. It is dispersed in southern and central Europe, but is absent from most of Scandinavia and the British Isles, as well as from

Fig. 67: On Andros, the first documented record for the occurrence of the edible dormouse, *Glis glis* Linnaeus, 1766, is from the area of Chora, where it was collected on September 2, 1994. Photo by Maria Dimaki.

the northern coasts of France, Germany, the Netherlands, and most of the Iberian peninsula [642] (▶Fig. 67). Outside its original distribution it is introduced in England [643–644]. It is associated with the deciduous forest zones in the western Palaearctic. It is distributed as far north as the Baltic area, and as far south as the island of Crete [645]. In the Mediterranean region, edible dormice may also be found in maquis and shrubland on rocky ground, from sea level up to 2 000 metres. It is not uncommon to see these rodents in orchards and urban margins, and they frequently enter huts and houses [645]. In the southern Balkan peninsula, the species is strictly confined to humid deciduous forests [94]. It occurs also in northern Asia Minor, the Caucasus, north-western Iran [646], and on several Mediterranean islands, where it has often been introduced by man in ancient times.

We have known for some time of the presence of the edible dormouse on islands such as Crete [297–298, 518, 647–648], Euboea [71], Corfu [49, 518] and Cephalonia [282–283, 518]. On the latter island, the occurrence of the species has also been confirmed by Harald Pieper (osteological remains in barn owl pellets, *in litteris* December 4, 2000), and Fausto Barbagli (complete skeleton collected in August 1996, *in litteris*), whereas Dimaki [649] shows that the edible dormouse was also distributed on Andros. Here, and more generally in the whole Cyclades archipelago, the first documented record for the occurrence of the species is from the area of Chora, where it was collected on September 2, 1994 [649]. According to H. Alivizatos and A. Lane (*in verbis*), this rodent is also present on the island of Thasos where they reported the species from the surroundings of the village of Panaghia, on August 30, 2000. Remains of *Glis glis* have been provided by the archaeological exploration of Konispol Cave, on the Albanian coast off the Greek island of Corfu [332].

Systematic data

The subspecies described from continental Greece is *Glis glis pindicus* Ondrias, 1966, which is regarded as synonymous of *Glis glis minutus* Martino, 1930 [650]. The Cretan edible dormouse, *Glis glis argenteus* Zimmermann, 1953, is instead the variety recognised from the namesake island and the Ionian and Aegean archipelagos [70–71].

Museum specimens

Andros: GNHM 9941; Crete: NHMC, 80.5.49.8.

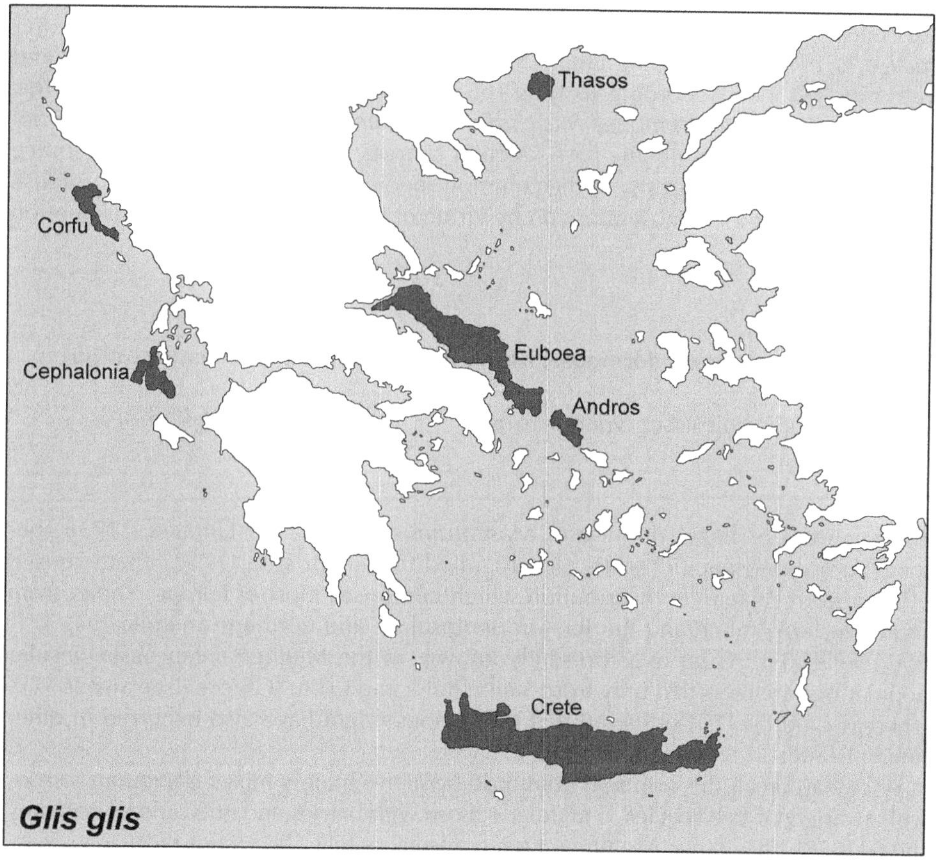

Fig. 68: Distribution of the fat dormouse or edible dormouse, *Glis glis* Linnaeus, 1766, in the Ionian and Aegean islands.

References

Corfu [49, 70, 518, 642]; Cephalonia [282–283, 380, 518, 642]; Euboea [71]; Andros [642, 649]; Crete [70, 285, 297–298, 348, 369, 475, 518, 642, 647–648, 651].

Archaeozoological note

According to Payne [347], the edible dormouse must have come to Crete from Europe or the Levant as it is absent from North Africa.

As observed by Carpaneto and Cristaldi [652] and Colonnelli et al. [653], historical and biogeographical analyses, but also palaeontological and archaeozoological data from the Mediterranean region show that the geographical distribution of edible dormice and their population density have been strongly influenced by humans already since antiquity. Furthermore, ethnozoological enquiries document their utilisation for food or medical treatment, through traditional captive-breeding techniques, up to very recent historical times [471, 654]. The edibility of dormice was one of the major preoccupations of the Classical scholars, such as Varro (*Res rustica*, 3.15. 1–2), Pliny the Elder (*Naturalis historia*, 8.224), Apicius (9.1.1), and Petronius (31.10). According to King [655], it seems however that small glirids, such as the common dormouse *Muscardinus avellanarius* Linnaeus, 1758, were not eaten, unlike *Glis glis*. In any case, ancient sources were imprecise about dormice and could be referring to any of the common species, although the edible dormouse probably attracted the most attention by virtue of its larger size and its consumption as a delicacy.

6.30 Common or hazel dormouse, *Muscardinus avellanarius* Linnaeus, 1758

Greek name: βουνομυωξός (*vunomyoxos*) Turkish name: *findek faresi*

The common or hazel dormouse, *Muscardinus avellanarius* Linnaeus, 1758, belongs to a monospecific genus closely allied to *Glis* Brisson, 1762, characterised by a western Palaearctic distribution which comprises most of Europe – apart from south-western France and the Iberian peninsula – and northern Anatolia [94, 471, 656] (▶Fig. 69). As far as is presently known, in the Mediterranean basin, insular populations are recorded only from Sicily (Italy) and Corfu (Greece) (see also [657]), whereas Ondrias [70] observed that the species might have also occurred in other Ionian islands.

Strictly arboreal, the common dormouse frequents mainly mixed deciduous forests, with a variety of tree species, particularly those with berries and nuts, and understorey shrubs [658]. They range in altitude from sea level up to 1 500 metres [94].

Fig. 69: As far as is presently known, insular populations of the common or hazel dormouse, *Muscardinus avellanarius* Linnaeus, 1758 in the Mediterranean basin are recorded only from Sicily (Italy) and Corfu (Greece). Photo by Fabrizio Darmanin.

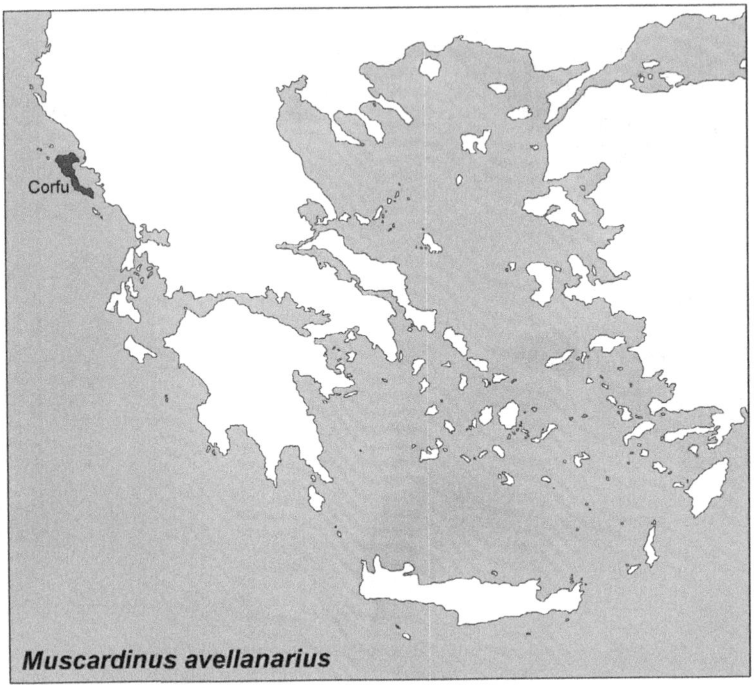

Fig. 70: Distribution of the common or hazel dormouse, *Muscardinus avellanarius* Linnaeus, 1758, in the Ionian islands.

Systematic data

Morris [656] recognises five subspecies, all similar in appearance. *Muscardinus avellanarius zeus* Chaworth-Musters, 1932, is the subspecies typical of Corfu [49], and of the southern Balkan peninsula (compare with [70]).

References

Corfu [49, 70, 344].

6.31 Golden jackal or Asian jackal, *Canis aureus* Linnaeus, 1758

Greek name: τσακάλι (*tsakali*) Turkish name: *çakal*

The golden jackal or Asian jackal, *Canis aureus* Linnaeus, 1758, is one of the most widespread canids of the Western Palaearctic (▶Fig. 71). It is primarily an Asiatic

Fig. 71: Golden jackal or Asian jackal, *Canis aureus* Linnaeus, 1758. Photo by Marco Masseti.

species, ranging from central Asia and India, westwards to Iran, Iraq and Asia Minor, and southwards to most of Africa [373, 659]. According to Hosey [659], it must very likely have entered Europe not earlier than the late Würm, probably via a terrestrial land-bridge across the Bosporus (see also [660]). It is reported as a resident species in the Caucasus and the Balkan peninsula. In continental Greece, it ranges from sea level to 600 metres altitude [661]. The golden jackal is a very versatile carnivore. It mainly lives in the open countryside, but it can frequently be found also around towns and villages. This canid preys extensively upon rodents, but it also feeds on carrion and garbage. It is well adapted to the dry open country settings, but it also spreads into mountainous areas, as well as into forest plantations and riverine thickets. The species is known from several Mediterranean islands: it recently swam at least to some of them according to Kryštufek [661]. Along the shores of the south-eastern Balkan peninsula, it is presently distributed on the eastern Adriatic islands of Korčula, Hvar, Krk, Sipan, Premuda, Vir, Pag, Mljet, Badija and Olipa [662–666]. In contrast, the species appears to be rare or absent in Albania, even in the coastal plains [78].

Until recently, authors disagreed on which Ionian and Aegean islands the jackal had distributed [666]. Although Erhard [401] reported these canids from Syros, Naxos, Tinos and Andros, Ondrias [70] observes that they were unknown in the Cyclades, Dodecanese and Crete. Following Erhard [401], also Heldreich [404] quoted the species among the wild mammals of Andros and Naxos. The ordinary use in Tinian dialect of the popular word τσακαλος (*tsakalos*) to indicate the badger, *Meles meles* Linnaeus, 1758, which still occurs on the island, possibly gave rise to an unproven belief about the presence of jackals on Tinos (Martin Gaethlich, *in verbis*) [478]. It had also been reported from Corfu [49, 666–670], Kythera [398–399], Euboea [401, 404, 646, 670], Skyros [58, 290, 671] and Ikaria [671], but the golden jackal is not present on all these Greek islands any longer. However, while the canid seems to have become extinct on Kythera, Skyros and, possibly, Ikaria, in recent historical times [672], it has disappeared from Corfu not earlier than 1991–1992 (Grémillet, *in verbis*). Individuals from the not too distant Samos might have been introduced in Ikaria. Jackals seem today to occur only on the Ionian island of Lefkada [667, 670] and Cephalonia [670] and on the Aegean island of Samos [65–66, 306, 667–668, 670, 672–675].

Ioannidis and Giannatos [672] found the species on Samos in the same habitats as in the rest of the southern Balkan peninsula, but Giannatos et al. [674] recently reported a drop in the insular population density. Radiotracking conducted by Giannatos and Legakis [675] in eastern Samos revealed that jackals forage in open areas in night hours and use dense thickets as daytime refuges. These canids were already reported in early times in Samos (compare with [676]). On the basis of the accounts on the expedition of the botanist Joseph P. de Tournefort to the Greek archipelago [677], Clarke [678] observed that "*Samos is infested with wolves*". Anyway, this record should refer to jackals rather than wolves. In fact, there is no evidence of the occurrence of *Canis lupus* Linnaeus, 1758, on the Aegean islands during the late Holocene [478]. In the course of the present study, the presence of jackals on Samos could be attested.

Systematic data

Canis aureus moreotica I. Geoffroy Saint-Hilaire, 1835, is the subspecies recognised from mainland Greece, western Asia Minor, and the island of Corfu [49, 70, 279, 659, 667, 679–680]. Lapini et al. [681] consider this form as the subspecies dispersed also in north-eastern Italy, whereas de Beaufort [384] regards it as endemic to Greece. Kryštufek and Tvrtković [665] are, however, of the opinion that the population from Dalmatia, characterised by a broad skull, should retain the name of *Canis aureus dalmatinus* Wagner, 1841, and that the existence of at least two distinct jackal phenotypes in the Balkan peninsula can be ascribed to historic changes in their distribution area.

Museum specimens

Corfu: ZFMK: 61.193, 93.420.

References

Corfu [49, 70, 666–670, 672, 682]; Cephalonia [666, 670]; Lefkada [666–667, 670]; Kythera [398–399, 666, 668]; Skyros [58, 290, 304, 666, 670–671]; Euboea [401, 404, 666, 670]; Andros [401]; Tinos [401]; Syros [401]; Naxos [401]; Ikaria [666, 671]; Samos [65–66, 285, 305–306, 405, 666–670, 672–678, 683]; Thasos [666, 668].

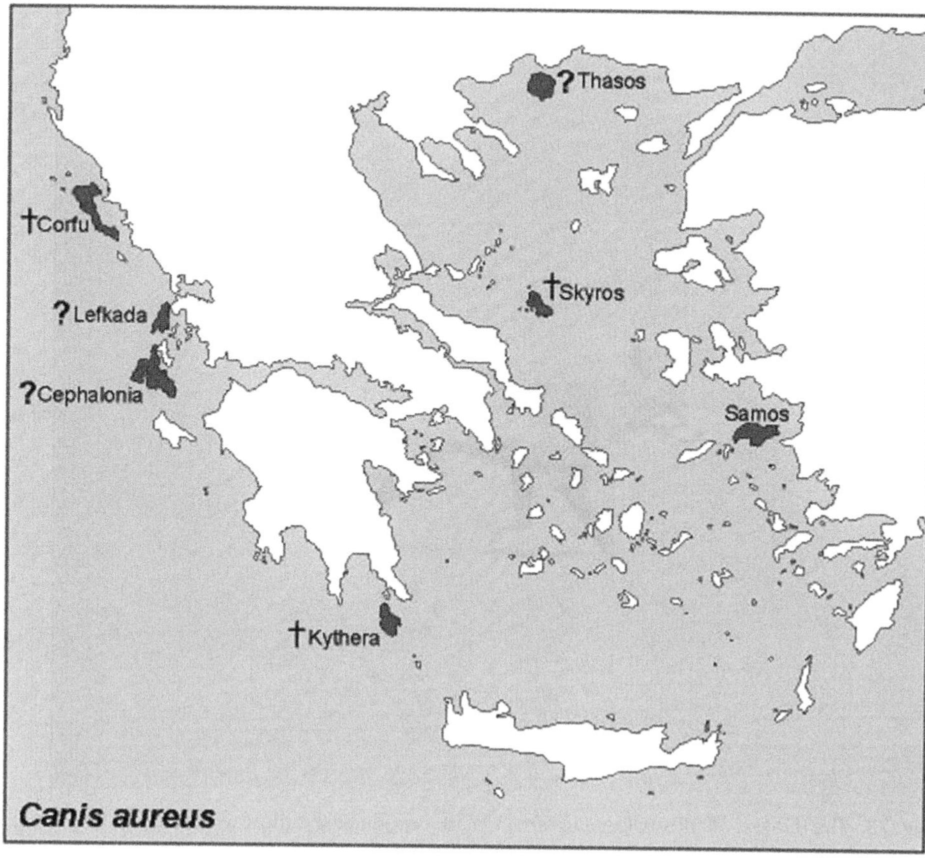

Fig. 72: Former and present distribution of the golden jackal or Asian jackal, *Canis aureus* Linnaeus, 1758, in the Ionian and Aegean archipelagos. Islands and areas that are inhabited by the jackal are shaded and those where they have become extinct are marked with a cross.

6.32 Red fox *Vulpes vulpes* Linnaeus, 1758

Greek name: αλεπού *(alepou)* Turkish name: *tilki*

The red fox, *Vulpes vulpes* Linnaeus, 1758, is perhaps the most widespread carnivore of the whole Holoarctic Zoogeographic Region (►Fig. 73). It has the widest range of any land carnivore, spanning from North America to Eurasia and parts of North Africa. Subspecies exist in Japan. Its distribution spans from Palaearctic Eurasia and north-western Africa as far north as North America, and it distributes in altitude up to 3 000 metres above sea level (compare with [684]). The flexible social organisation of this canid, together with its adaptability, allows it to survive, and often to form flourishing populations, in a variety of environments [685]. The red fox is a species which is found in all

Fig. 73: The red fox, *Vulpes vulpes* Linnaeus, 1758, is perhaps the most widespread carnivore of the whole Holoarctic Zoogeographic Region. Photo by Mauro Pieroni.

types of wooded and open landscapes, and is also adaptable to different feeding conditions [684]. It is often reported in urban and suburban ecosystems, where it behaves as a sort of commensal to man (compare with [686–687]). Red foxes display an opportunistic foraging strategy, which includes the consumption of invertebrates, plants (mainly berries), and vertebrates.

The species occurs on the larger Mediterranean islands, such as Sicily, Sardinia, Corsica and Cyprus (where it is the only carnivore present: see [637, 688]), but is absent from Crete and other smaller Aegean islands.

During the present study the occurrence of the species was confirmed on the Ionian islands of Corfu (Grémillet, *in verbis*), Lefkada, Cephalonia, as well as on the Aegean islands of Euboea, Karpathos, Rhodes, Kos, Chios, Lesbos, where it is the most frequent wild carnivore. No foxes, however, exist on Samos [66, 668, 674]. In 2003, these carnivores have been also reported from the small Turkish islets of Domuz in the Gulf of Fethiye, Gemilè, in the front of the Olüdeniz, and Geykova. Very curious records are also available from Zakynthos [380], Aegina, in the Saronic Gulf [403], and Karpathos [70, 668], where however the red fox seems to be unknown. Moreover, Kemf et al. [689] specify that red foxes do not occur on Zakynthos.

Systematic data

Many subspecies are known, but some need scientific verification [684]. According to de Beaufort [384], the following subspecies have been reported from Greece:

1. *Vulpes vulpes anatolica* Thomas, 1920.
2. *Vulpes vulpes crucigera* Bechstein, 1789.
3. *Vulpes hellenica* Douma-Petridou and Ondrias, 1980, endemic to Greece, but previously classified by Douma-Petridou [667] as *Vulpes vulpes hellenica*, the subspecies commonly reported from the southern Balkan peninsula and regarded as a Greek endemic.
4. *Vulpes vulpes crucigera* is the form reported from Corfu (ZFMK 61.194). Tortonese [562] suggests that the fox of Rhodes is *Vulpes vulpes anatolica*, the same taxon reported from Asia Minor, whereas Festa [57] refers it to the species *Vulpes aegyptiaca* Sonnini, 1816.

Museum specimens

Corfu: ZFMK 61.194.

References

Corfu [49, 70, 304, 668]; Lefkada [668]; Cephalonia [282–285, 380, 668]; Euboea [285, 293, 668, 690]; Rhodes [57, 141, 301, 562, 668]; Kos [141, 285, 668]; Chios [69, 668]; Lesbos [285, 668].

Archaeozoological note

Foxes made their first appearance among Mediterranean insular faunas in the Ancient Neolithic period on Corsica [310, 565] and Sardinia [5, 23], and in the Aceramic Neolithic period on Cyprus [691]. In fact, the species is unknown among the native faunal assemblages of all these islands (compare with [23, 688]). More specifically, Vigne [310] observes that, in most of the Mediterranean area, *Vulpes vulpes* is the object of particular human attention during the Ancient Neolithic period, when the canid might have been important both symbolically and as food. This cultural significance attributed to foxes by early Neolithic man might appear more understandable when compared to the attitude held generally by prehistoric peoples towards natural sources.

Like most of the mammals present today on Crete, *Vulpes vulpes* is an element of continental origin whose extant occurrence on the Greek island is essentially due to human-derived importations since the prehistory. Remains of the species have been identified in post-Minoan sites, such as in Byzantine Gortys and at Prinias, in mixed Late Bronze Age-archaic deposits [692]. As far as is presently known, also the early artificial introduction of foxes onto the Aegean islands dates back to the Neolithic period, when remains of these animals have been unearthed from the exploration of the prehistoric

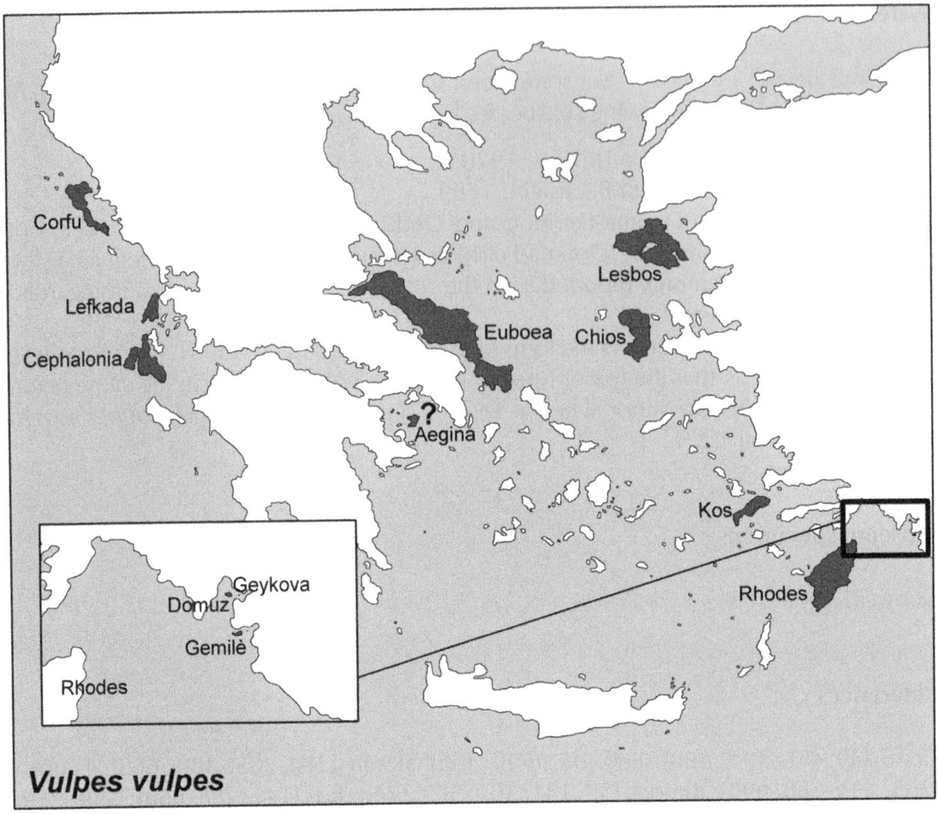

Fig. 74: Distribution of the red fox, *Vulpes vulpes* Linnaeus, 1758, in the Ionian and Aegean islands.

sites of Kalythies (Rhodes) [180], Saliagos, on the namesake islet near Antiparos [693], and Skoteini, on Euboea [694]. It is interesting to note that *Vulpes vulpes* is the only wild mammal present in Late Neolithic Saliagos [693, 695]. Osteological remains of the species have been also provided by the archaeological exploration of the levels referred to as Early Helladic I–Early Helladic II (2900-2600 BC) of Poliochni, on the island of Lemnos [457], and of Bronze Age Chios [696] and Lesbos [697].

6.33 Weasel or least weasel, *Mustela nivalis* Linnaeus, 1766

Greek name: νυφίτσα (*nifitsa*)
Greek vernacular names: καλογιάννου (*kaloiannou*)
σενταχτάρι (*sentakhtari*)

Turkish name: *gelincik*

The weasel or least weasel, *Mustela nivalis* Linnaeus, 1766, is one of the most common mustelids inhabiting the Palaearctic region [698], where it displays a broad circumpolar

Fig. 75: The weasel, *Mustela nivalis* Linnaeus, 1766, is described as a specialised predator of small rodents, especially voles of the genus *Microtus*. Photo by Riccardo Romanelli.

distribution ranging from North America via Asia into Europe and Northern Africa [699–700] (▶Fig. 75). The species accepts a wide range of habitats but with enough food and shelter [700]. The weasel is described as a specialised predator of small rodents, especially voles of the genus *Microtus* [699, 701–703]. Rodents represent 58 to 99% of its diet [704], and consequently the distribution of this mustelid is largely connected with their abundance [705]. The dispersal of the weasel on islands is conditioned by the size and distance of the island from the mainland, but also on its amount of potential rodent prey [706]. The dependence of the weasel's distribution on these insular characters has been verified only on the British Isles [705] and in the Mediterranean archipelagos [707]. As we have already seen, microtinae are, however, very rare in the Mediterranean islands [9] (see also ▶Fig. 137 in this volume), whereas mice of the genus *Mus* Linnaeus, 1758, and rats, *Rattus norvegicus* Berkenhout, 1769 and *Rattus rattus* Linnaeus, 1758, are rather common. Thus, on the Mediterranean islands the weasel switches from vole to murid hunter [707–708].

Several papers and handbooks on Palaearctic mammalian fauna [6, 9, 269, 279, 304, 373, 657, 700, 709] but also species monographs [699, 702, 704], report the presence of weasels throughout the Mediterranean islands. There are also more detailed studies but only on local occurrences, e.g., [707, 710–711]. In the Ionian and Aegean islands, the species has been reported from Corfu, Cefallonia, Zakynthos, Euboea, Samothrace, Lesbos, and Samos [49–50, 66, 282–283, 285–286, 293, 305, 667, 688, 690, 707, 712–713]. Dubious records are instead available for Milos, where an unverified presence of the weasel has been reported to Legakis et al. [51] by local people, Skopelos (Ferdinando Ciani, *in verbis*), and Lefkada. Kock [67] suggested the occurrence of *Mustela* Linnaeus, 1758 – possibly the weasel – on Chios, a hypothesis contested by Spinthakis et al. [69].

At present, this mustelid is not rare on Crete [714–715], where it seems to be widespread, from the sea level, e.g., the coastal palm grove of Vai in eastern Crete, to the mountainous areas. Live individuals and field signs were observed in the area of Xilóskalo (about 1 300 metres), and at the Tripiti gorge (about 800 metres) on the Lefka Ori (White Mountains) (compare with [716]). The species was recorded also from Mount Juktas, and the surroundings of Herakleion. In the course of the present study, weasels could be confirmed on Corfu (Grémillet, *in verbis*), Cephalonia, Zakynthos, Euboea, Crete, Lesbos, Chios, and Samos, but not on the islands of Lefkada, Ithake, Thasos, Ikaria, Fourni, and Rhodes, from where it was previously reported by Douma-Petridou [667] and Ondrias [690]. Its presence is uncertain also on Samothrace were it was reported by Maltezou [413] and Bonetti [285]. In contrast, according to the rangers of the local Forest Department, weasels are still occurring on Lefkada, from where they have been also recorded by Douma-Petridou [667] and Ondrias [690]. The weasel has never been recorded on Rhodes by the majority of the scientists who have carried out field researches on the island, such as Festa [57], Ghigi [394], von Wettstein [58], Tortonese [562], and Masseti [60]. Douma-Petridou [667] and Ondrias [690] recorded this carnivore also from Ithake, Ikaria, Fourni, Chios, and Thasos, and Forsyth Major [717] from Karpathos, but there are no recent or reliable reports.

Systematic data

On the basis of the examination of two skins, Bate [276] described the Cretan weasel, *Mustela nivalis galinthias* Bate, 1906, as a subspecies of large size, showing a very distinct line of demarcation between the colours of the upper and under-surfaces [276, 279]. Douma-Petridou and Ondrias [711] assigned to the Cretan subspecies the population from Peloponnese, continental Greece and Bulgaria. The Mediterranean races of *Mustela nivalis* are all very large in body [699]. Subfossil remains from the western Anatolian site of Norşun-Tepe (south-west of Elâziğ) have also been related by Boessneck [718] to the Aegean weasel, *Mustela nivalis galinthias*.

Museum specimens

Lesbos: ZMUP 3164; Crete: BMNH 45.158; MZUF 12686; NHMC no 80.5.62.(2–9), 80.5.62.11, 80.5.62.15; ZFMK no 65.552, ZMUP 3744.

References

Corfu [49, 667, 688, 690, 707]; Cephalonia [282–283, 285, 380, 667, 690]; Lefkada [667]; Ithake [667]; Zakynthos [50, 285–286, 667, 690]; Euboea [285, 293, 707, 713]; Thera [70, 275, 688]; Crete [58, 276–277, 285, 298, 667, 688, 707, 711, 714–717, 719]; Rhodes [667]; Ikaria [667]; Fourni [667]; Samos [66, 285, 305, 667, 690, 707, 717]; Chios [667]; Lesbos [285, 667, 690, 707]; Samothrace [285, 413]; Thasos [667].

Archaeozoological and ethnozoological note

The presence of the weasel on islands near the mainland, such as Corfu, Euboea, Samos, and Lesbos, could be explained by a natural colonisation via land-bridges or during sea level low-stands in the course of the late Pleistocene (compare with [223, 228, 688]). Its occurrence on Crete [707], as well as on other islands far from the mainland, is instead explained in terms of human introduction (compare with [720]). In fact, as claimed by Miller as early as 1912 [279], the modern distribution of the weasel on the Mediterranean islands has been influenced by man, and its present distribution in these territories reveals its anthropochorous origin. The times and the means of the introduction of the species, however, are still scantily known. Weasels are reported on the Aegean islands at least since the Early Bronze Age [351]. In fact, a hemimandible of mustelid was discovered in the site of Markiani, on Amorgos [416, 419] (▶Fig. 76). These carnivores have also been recorded in a few Late Bronze Age sites of Crete, such as Kommos [417], and Kastro [721].

Several carnivores have been introduced onto islands, already since prehistory, for many purposes. Weasels seem to have been specifically employed to prevent damage produced by small rodents, such as mice and rats [688, 722]. That the species was employed as a house animal in Classical/Antiquity is sufficiently documented. It was, in fact, probably kept in semi-domestic conditions. Eminent witnesses of this custom were the Greek Aristophanes and Aristotle, or the Latin Pliny the Elder and Strabo. In their works, they documented how the *gale* (γαλε), which today might be identified with *Mustela nivalis*, was used to keep cupboards and barns clear of unwanted scavengers such as small rodents, during an age when the use of the cat as a domestic animal was still unknown. With the exception of ancient Egypt, the domestic cat was unknown along the Mediterranean shore before the late Roman period (compare with [723]). On Crete, up to a few decades ago, living individuals of weasel were still released inside the greenhouses where fruits and vegetables were grown to prevent rodent damage to the harvest [688].

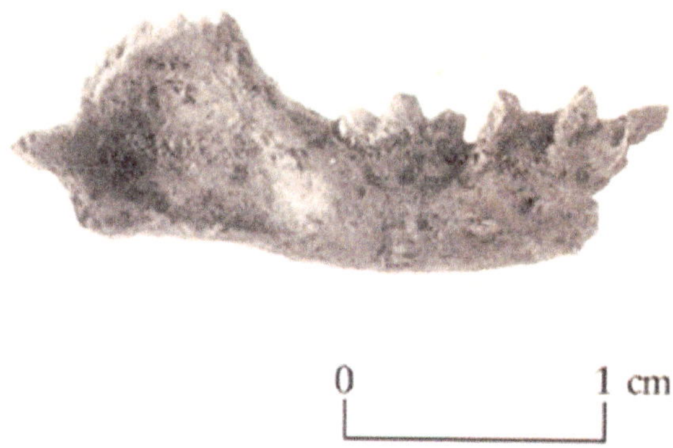

Fig. 76: The lower hemimandible of a weasel found in the Early Bronze Age levels of the site of Markiani, on Amorgos. Photo by Y. Despotidis, from Trantalidou [416].

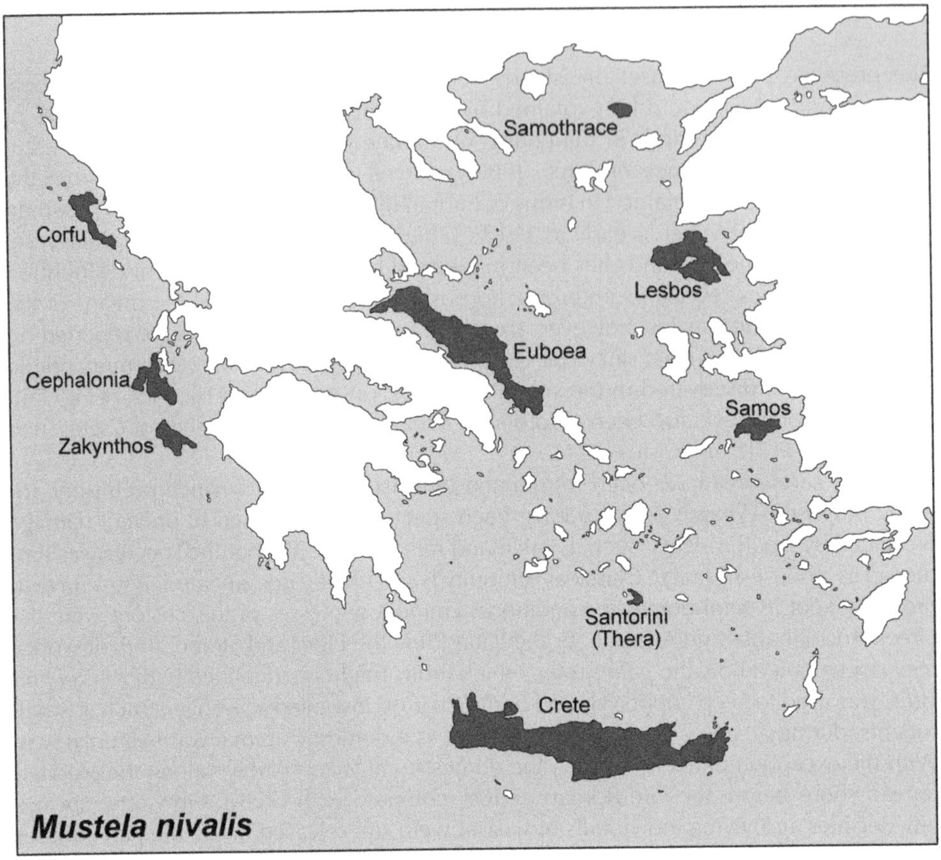

Fig. 77: Distribution of the weasel, *Mustela nivalis* Linnaeus, 1766, in the Ionian and Aegean islands.

6.34 Beech marten or stone marten, *Martes foina* Erxleben, 1777

Greek names: κουνάβι (*kounavi*)
πετροκουνάβι (*petrokounavi*)
Greek vernacular names: ζουρίδα (*zourida*, Crete)
ατσίδα (*atsidha*, Amorgos)
ζουρίδα, ζουρια (*zourida*, *zouria*, Karpathos)

Turkish name: *kaya sansari*

The beech marten, or stone marten, *Martes foina* Erxleben, 1777, is a widely distributed Palaearctic element. It ranges as far west as the Iberian peninsula, and as far east as China, throughout the Caucasus, Kazakhstan, Mongolia and the Himalaya [724].

It is however absent from the majority of northern Europe, the British and the western Mediterranean islands [725–726] with the only exception of Ibiza (Balearic archipelago, Spain) where it became extinct possibly between the second half of the 1970s and the 1980s [727] (►Fig. 78).

Martes foina is however the carnivore most dispersed in the Eastern Mediterranean islands, where it ranges from the Ionian archipelago to the Dodecanese, through most of the Aegean islands, and Crete [715, 728]. According to Erhard [401] and Heldreich [404], the species was very common in the Cyclades. It has been reported from Corfu, Lefkada, Cephalonia, Ithake, Zakynthos, Kythera, Skopelos, Euboea, Kythnos, Serifos, Naxos, Santorini, Crete, Karpathos, Rhodes, Kos, Ikaria, Samos, Chios, Lesbos, Samothrace and Thasos [70, 668, 690, 729]. A. Dimitropoulos (*in verbis*) recorded it also on the island of Syros. The beech marten is the carnivore most dispersed also on the Dodecanese islands, occurring on Karpathos [55, 141, 668, 729–730], Rhodes ([57–58, 141, 668, 729–730]; [301] original data), and Kos [668, 729–730]. According to Cattaneo [303], a carnivore, probably a representative of the Mustelidae family (*Martes foina*? *Mustela putorius furo*?) occurs also on Leros.

On Crete, the stone marten seems to be widespread. Sightings and field signs of it are reported from the sea level, such as the easternmost cliffs of Cape Sideros [688], to the highest mountains, as the White Mountains range (*Lefka Ori*), where it has been reported under the Melindau peak (2 113 metres high) [688, 728]. It has also been reported from the highest plateaux, such as Lassithi, but also from densely populated areas, such as Herakleion, and the Aghios Nicholaos suburbs [728].

Fig. 78: Beech marten, or stone marten, *Martes foina* Erxleben, 1777. Photo by Giulianio Cappelli.

Fig. 79: Cretan marten, *Martes foina bunites* Bate, 1906. Photo by Marco Masseti.

The apparently great concentration of martens currently near human settlements on Crete, as in other islands, suggests that the species may partly behave as a commensal of man. At Amorgos, in the eastern Cyclades, this mustelid was happened to be seen rummaging among the rubbish containers in the very centre of Chora, the capital of the island [60]. During the present research, beech martens were observed for the first time in Andros (Yannis Ioannidis *in verbis*, 2000), Tinos (Martin Gaetlich *in verbis*, 1999), Paros (Martin Gaethlich, Vassili Chondropoulos and Yannis Ioannidis *in verbis*, 1999); Antiparos (Martin Gaethlich *in verbis*, 1999), Hydra, Saronic Gulf (original data, April 3, 1996), and Amorgos (original data, May 4, 1999). The species was also confirmed on Corfu (Xavier Grèmillet, *in verbis* 1993), Cephalonia (Aliki Panou, *in litteris*), Lefkada, Ithake, Euboea, Naxos (Dietrich Ristow, *in litteris* 2000; present work), Karpathos [299]; bones of subfossil and existing martens: Thomas Rathgeber, *in litteris* 2005), Rhodes, Kos, Ikaria, Chios, Lesbos, and Thasos. Martens were also recorded on the islet of Antimilos or Erimomilos (20 km north-west of Milos, western Cyclades) by Koller [731], Ondrias [70] and Cheylan [304], but there is no recent or reliable confirmation.

Systematic data

Three subspecies are recorded on the eastern Mediterranean islands:

1. the European beech marten, *Martes foina foina* Erxleben, 1777, from the Ionian islands (Corfu, Lefkada, Ithake, Cephalonia and Zakynthos) [279, 450, 728–729].
2. the Rhodian marten, *Martes foina milleri*, described by Festa in 1914, presumably confined to Rhodes [57–58, 728–729].

3. the Cretan marten, *Martes foina bunites*, described by Bate in 1905 [276], distributed on Crete, Karpathos and several other Aegean islands, including Naxos, Amorgos, Kythnos, Seriphos, Santorini, Thassos, Samothrace, Lesbos, Chios, Samos, Ikaria, and Kos [55, 60, 70, 269, 290, 728–729].

Martes foina bunites Bate, 1906, was described as a geographic subspecies, characterised by being smaller-sized and having a lighter and more yellowish coat colour than typical *Martes foina* (Erxleben, 1777) [276, 279, 732–733]. The main phenotypic trait is the very much reduced throat patch, which lacks completely in one of the seven skins examined by Bate [276]. Studies on this aspect conducted by Niethammer and Niethammer [48] and Ragni et al. [715], however, failed to record the existence of such a characteristic in the current populations of Crete, suggesting that it should probably be attributed to the appalling conditions of conservation of the specimens described by Bate [276]. Actually, all the Cretan martens examined by Niethammer and Niethammer [48] and Ragni et al. [715] show a pure white or greyish throat bib divided into two narrow, irregularly symmetrical left and right sections by a dark stripe [9] (►Fig. 80). In this regard, the Cretan beech marten shows the phenotypical characters of the Near Eastern representatives, such as the beech martens of the Levant (Syria and Palestine), which are characterised by an irregular horseshoe-shaped throat mark (compare with [9, 270, 715]). It is, however, surprising that the Cretan marten is still described by MacDonald and Barrett [575] as being characterised by "*... a small greyish throat patch*". Furthermore, Corbet and Ovenden [734] have provided a pictorial description of the same taxon, with a small round patch

Fig. 80: It was, at length, believed that the Cretan marten was characterised by the extreme reduction, if not the complete absence, of the throat patch. Photo by Marco Masseti.

Fig. 81: Corbet and Ovenden [734] have provided a pictorial description of the Cretan stone marten, *Martes foina bunites* Bate, 1906, with a small round patch in the middle of the throat, which is scarcely discernible and a far cry from the real throat patches of either the Near Eastern or the European beech martens.

in the middle of the throat, which is scarcely discernible and a far cry from the real throat patches of either the Near Eastern or the European beech martens [9] (►Fig. 81).

Museum specimens

Corfu: ZFMK: 61.196, 93.421; Cephalonia: BMNH: 8.11.3.9–8.11.3.10; Skopelos: NMW: B6320; Euboea: ZFMK: 77.382; Naxos: BMNH: 66.5401; Crete: BMNH: 5.12.2.18, 5.12.2.21–5.12.2.22, 5.12.2.37; NHMC: 80.5.61.(1–29); ZFMK: 66.24, 66.25, 66.26, 93.512).

References

Corfu [49, 70, 668, 688, 728–730, 735]; Paxos and Antipaxos [560]; Lefkada [668, 729–730]; Ithake [668, 688, 728–730]; Cephalonia [282–285, 380, 668, 688, 728–730]; Zakynthos [50, 285–286, 668, 688, 728–730]; Kythera [53, 398–399, 668, 688]; Skopelos [58, 70, 290, 668, 688, 728–731]; Alonissos [289, 730]; Euboea [285, 293, 668, 729–730]; Kithnos [58, 70, 304, 405, 668, 688, 728–730]; Serifos [58, 70, 405, 668, 729–730]; Syros [668]; Naxos [290, 304, 405, 519, 668, 688, 728–730]; Thera [70, 275, 668, 688, 729–730]; Amorgos [60]; Crete [48–49, 70, 269, 277, 285, 290, 298, 373, 405, 647, 668, 688, 714–715, 728–731, 736]; Karpathos [55, 60, 70, 141, 285, 299, 410, 668, 688, 717, 728–730, 737]; Rhodes [57–58, 60, 141, 285, 301,

405, 668, 729–730]; Kos [371, 668, 729–730]; Leros [303]; Ikaria [668, 729–730]; Samos [66, 285, 305–306, 668, 688, 728–730]; Chios [67, 69, 668, 688, 728–730]; Lesbos [285, 668, 688, 728–730]; Samothrace [58, 70, 285, 304, 413, 455, 668, 688, 728–730]; Thasos [668, 688, 728–730].

Archaeozoological note

Though the species is known from Würm deposits in the Levant [738], it does not appear in Europe until postglacial times [148, 644, 738–740]. As observed by Anderson [739], the stone marten probably entered Europe from the Near East at the end of the Pleistocene or in early postglacial times, and spread slowly north-westward; it may have accompanied human dispersal [9, 728].

There is, for example, no palaeontological evidence of the species in the native Quaternary fauna of Crete. Martens are documented on the island only since late

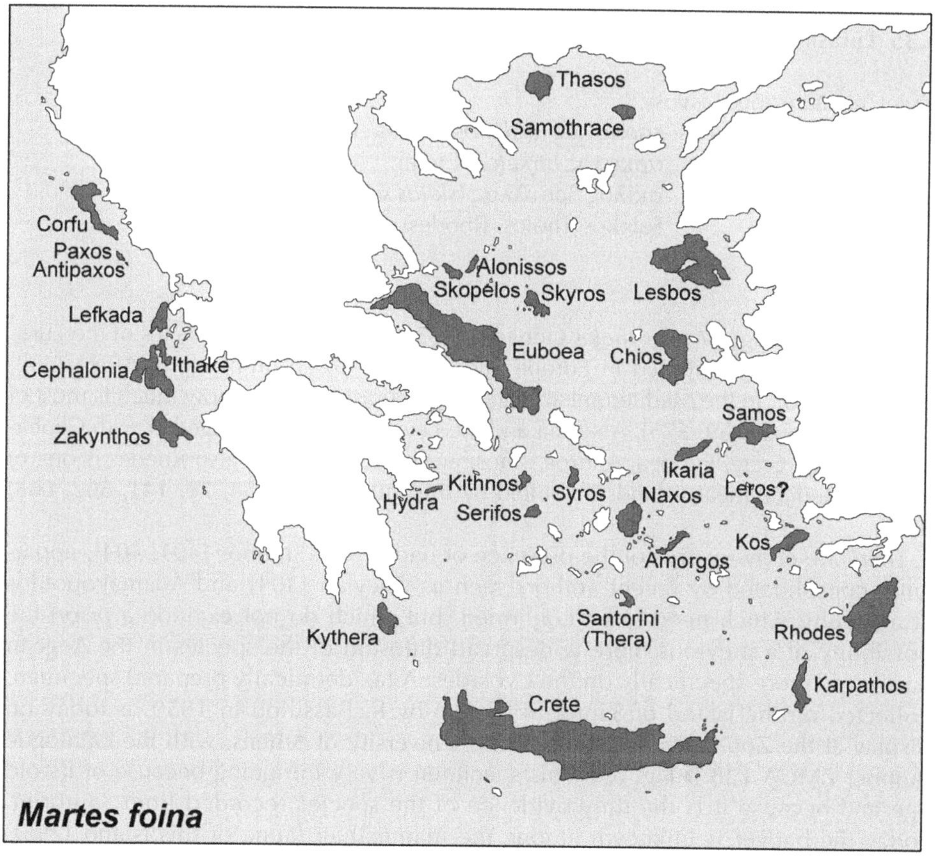

Fig. 82: Distribution of the beech marten or stone marten, *Martes foina*, Erxleben, 1777, in the Ionian and Aegean islands.

prehistoric or early historical times [688] by subfossil osteological remains from four caves, which produced also Pleistocene fauna [106, 728, 733, 741] from Early Neolithic II and Minoan Knossos [418], the Late Minoan levels (about 1550–500 BC) of Hagia Triada [692] and Late Minoan III Kavousi-Vronda [742]. Because Crete was not joined to the mainland during the late Pleistocene, the occurrence of the stone marten in early stratigraphic levels could be related to the animal's fossorial abilities [688, 728], and therefore explained as contamination.

However, the earliest known evidence of the dispersal of stone martens onto the Aegean islands is from Kythnos, in the western Cyclades, where remains of this carnivore have been yielded by the prehistoric site of Maroulas, which is chronologically referred to 8800/88700–8068/7668 BC [421]. Osteological remains of marten have also been found in the Ceramic Neolithic levels of Kalythies, in Rhodes [180], the phases III and IV of the Mycenean shrine at Phylakopi, on Milos [743], and the period V of the Middle Bronze Age settlement of Ayia Irini, on Kea [744]. Small and medium-sized carnivores not only provided food supplies to prehistoric man [688], they might also have been utilised for their fur (compare with [745]).

6.35 Eurasian badger, *Meles meles* Linnaeus, 1758

Greek name: ασβός (*asvos*)

Greek vernacular names: τσακαλος (*tsakalos*, Tinos)
αρκαλος (*arkalos*, Crete)
σκιλος βρουλλας (*skilos vroullas*, Salakos-Tholos, Rhodes)

Turkish name: *porsuk*

The Eurasian badger, *Meles meles* Linnaeus, 1758, is a Palaearctic species of the Euro-Siberian Region, distributed in Europe and Asia but absent from North Africa [432, 497, 698, 746]. In the Mediterranean region, the Eurasian badger now inhabits most of southern Europe [269, 575], Anatolia and the Levant [270, 467]. Together with Cephalonia, Ithake, Crete, Euboea, Andros, Tinos, and possibly Sifnos, also Rhodes is one of the few Mediterranean islands inhabited by this mustelid [57–58, 70, 141, 562, 668, 688].

There are early records of the presence of badgers on Siphnos [401, 404], apparently consolidated by recent authors such as Cheylan [304] and Adamakopoulos et al. [668], which need to be confirmed, but which do not exclude *a priori* the possibility of a previous more widespread diffusion of the species in the Aegean basin, and more specifically on the Cyclades. A taxidermically prepared specimen, collected on the island of Santorini (Thera) by K. Bassiliou in 1859, is today on display at the Zoological Museum of the University of Athens, with the catalogue number ZMUA 128 (►Fig. 83). This specimen is very intriguing because of its old age and because it is the only evidence of the species recorded from Santorini. Today the badger is unknown among the mammalian fauna of this island [295]. If credit is to be given to the alleged origin of the ZMUA specimen, the species must have become extinct around the end of the 19th century, in view of the fact

Fig. 83: Taxodermically prepared specimen of the badger, *Meles meles* Linnaeus, 1758, collected on the island of Santorini (Thera) in 1859. It is part of the collection of the Zoological Museum of the University of Athens (ZMUA 128). Photo by Anastasios Legakis, courtesy of the Zoological Museum of the University of Athens.

that Douglas [275] did not mention *Meles meles* in his list of the insular mammals. Santorini would therefore be included in the list of the Aegean islands inhabited by this mustelid.

Badgers have been recorded on Cephalonia [282], Euboea (Yannis Ioannidis: personal communication; present work), Andros ([668, 688]; M. Gaethlich 1998, personal communication), Tinos ([401, 404, 668]; Martin Gaethlich 1998, personal communication; Y. Ioannidis, personal communication), Rhodes ([57, 405, 562]; [60] original data) and Crete ([70–71, 276–277, 279, 297, 714, 736, 747–748], [715] original data). Mitsainas et al. [284] recorded *Meles meles* for the first time on the Ionian island of Ithake, where in 2007 faecal material had also been observed. According to the rangers of the local Forest Department, badgers were present on the Ionian island of Lefkada in the 1990s, but there is no recent or reliable confirmation. According to M. Gaethlich (*in verbis*), the presence of the species also on Andros cannot be excluded. There is no evidence, however, from Corfu (X. Grémillet, *in litteris*).

Systematic data

The Cretan badger, *Meles meles archalus* Miller, 1907, and the Rhodian badger, *Meles meles rodhius* Festa, 1914, are regarded as the two subspecies respectively endemic to

Crete and Rhodes. On the basis of the examination of four specimens from Crete, the local subspecies was described by Miller [748] as a form of small size and pale colour [70, 276–277, 279, 714, 736, 748]. *Meles meles rodhius* is seriously threatened with extinction.

Museum specimens

Thera: ZMUA 128 (44) (collected by K. Bassiliou in 1859); Crete: BMNH 5.12.2.17, 5.12.2.38; MZUF 361; NHMC 80.5.63.(1–17); NMW B2084, B2087, 18636–18638; ZFMK, no 66.27; Rhodes: NMW 18636.

References

Ithake [284]; Cephalonia [282–283, 285, 380]; Siphnos [304, 401, 404, 668]; Euboea [285]; Andros [668, 688]; Tinos [401, 404, 668, 688]; Crete [54, 70, 276–277, 279, 298, 668, 690, 715, 736, 748]; Rhodes [57–58, 60, 70, 285, 405, 562, 668, 688].

Archaeozoological note

Badgers were introduced by humans on the Ionian and Aegean islands from prehistoric times (compare with [418, 688, 733]). On Crete, the oldest *Meles meles* bones were discovered in the Aceramic Neolithic levels at Knossos, while Ceramic Neolithic and later levels produced numerous remains of the species [418]. Other osteological remains were also found at Aghia Triada, and Kavousi-Vroda which have been referred respectively to the Ancient Minoan period (about 3000–2200 BC) [692] and Late Minoan III C [742]. Why man should have wanted to introduce badgers onto these islands is unclear. Logic wants that these continental carnivores were imported voluntarily by man, because they would have never embarked unobserved on the small boats employed to reach the new territories [310, 544, 688]. Since very ancient times, they may have played an important role in human societies, both symbolically and for food. In fact, the badger is still today caught and eaten, together with other medium-sized mammals, such as wild cats, *Felis silvestris* Schreber, 1775, and porcupines, *Hystrix cristata* Linnaeus, 1758, by local people in several parts of the circum-Mediterranean area (compare with [749]). Moreover, in medieval Europe another use of this mustelid has been documented. Early 16th century wall paintings painted by the Italian artist Giovanni Antonio Bazzi, nicknamed Sodoma, at the monastery of Monte Oliveto Maggiore (Siena, Italy) (►Fig. 84), clearly show badgers used as pets, which very likely represented an authentic status symbol that underscored the affluence and social position of their owner, who was the painter himself [478, 750]. According to Costa [751], this mustelid was easily tamed and kept as a pet at this time. Badgers might also have been utilised for their fur (compare with [688]).

Fig. 84: Detail of the early 16th century wall decoration showing the "Life of St. Benedict" in the Great Cloister of the monastery of Monte Oliveto Maggiore (Siena, Italy), painted by the Italian artist Sodoma (Giovanni Antonio Bazzi).

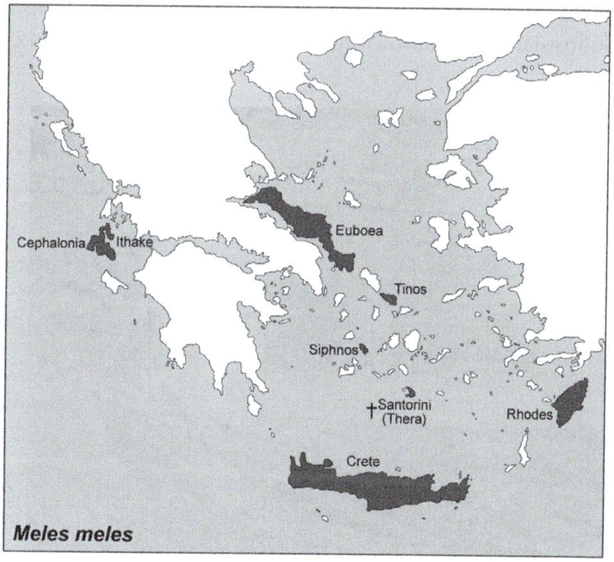

Fig. 85: Former and present distribution of the Eurasian badger, *Meles meles* Linnaeus, 1758, in the Ionian and Aegean archipelagos. Islands and areas that are inhabited by the badger are shaded and those where they have become extinct are marked with a cross.

6.36 Eurasian otter, *Lutra lutra* Linnaeus, 1758

Greek name: βίδρα (*vidra*)
Greek vernacular name: σκυλοποταμο (*skilopotamo*)
(*the dog of the river*)

Turkish name: *sa samuru*

The presence in Europe of the Eurasian otter, *Lutra lutra* Linnaeus, 1758, is only ascertained in early Holocene deposits. In fact, it has never been reported from Pleistocene levels [752] (▶Fig. 86). It represents a Palaearctic species of the Euro-Siberian Region, widely distributed in Eurasia and North Africa [432, 497, 753]. Formerly widespread throughout the western Palaearctic, the species has but declined in central and northern Europe. The Mediterranean diffusion of the otter now occurs throughout most of southern Europe [754], including Albania [81], the southern Balkan peninsula [70, 755] and Turkey [270], with the exception of south-western Anatolia [93].

In the Mediterranean, otters are presently reported only in the Ionian and Aegean islands (compare with [754]), quite close to the mainland coast of Albania, Greece and Turkey, where they mainly live in marshes and deltas. The occurrence of *Lutra lutra* is documented in the Ionian island of Corfu, but also in the Aegean islands of Euboea (Y. Ioannidis, 1999, *in verbis*; Kominos, Galanakis and Alivizatos, 2000: *in verbis*; A. Sfougaris, 2001, *in litteris*), Chios, and Lesbos [49, 69, 71, 81–82, 293, 307, 668, 688, 746, 754, 756–761]. In 1998, the species was also reported from Samos, where a dead specimen was found in the vicinity of Pythagorion and its lagoon (A. Legakis, 25.05.1998, personal communication). According to the rangers of the local Forest Department, otters are still present along the shores of Lefkada. Further investigations on the islands of the Ionian and Aegean seas will probably confirm the occurrence of other populations of otter [688]. All these insular populations might have regular contacts and exchanges

Fig. 86: In the Mediterranean, otters, *Lutra lutra* Linnaeus, 1758, are presently reported only in the Ionian and Aegean islands, quite close to the mainland coast of Albania, Greece and Turkey where they mainly live in marshes and deltas. Photo by Giulianio Cappelli.

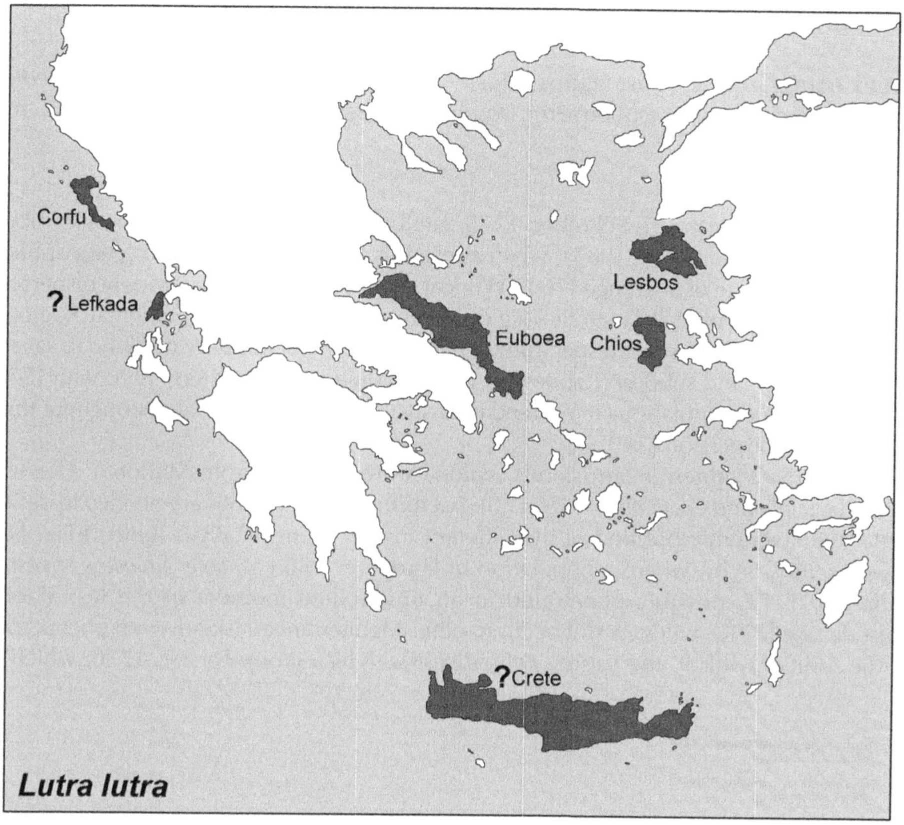

Fig. 87: Distribution of the Eurasian otter, *Lutra lutra* Linnaeus, 1758, in the Ionian and Aegean islands.

with continental groups; some otter populations of the southern Balkan peninsula and western Anatolia inhabit both marine and freshwater settings [688].

Spraints reported from the canyon of Preveli, in Crete, cast doubts on the possible occurrence of the species also on the large South Aegean island [762], despite the remote location of the latter from any continental area that may be naturally colonised by otters.

Museum specimens

Corfu: ZFMK 61.195, 93.422; ZMUA 45.

References

Corfu [49, 81–82, 340, 668, 688, 690, 746, 753–754, 756, 758, 761]; Euboea [285, 293, 340, 668, 690, 746, 754, 759, 761]; Crete [762]; Samos [717]; Chios [69, 668, 688, 746, 757, 761]; Lesbos [307, 746, 760].

6.37 African wildcat, *Felis silvestris, lybica* group Forster, 1770

Greek names: άγριος γάτος (agrios gatos)
Greek vernacular name: φουρογάτος (*fourogatos*, Crete)

Turkish name: *yaban kedisi*

The wildcat, *Felis silvestris* Schreber, 1775, can be considered one of the most biologically successful carnivores [763], when success is measured in terms of geographical and ecological size of the range [764]. Wildcat populations are in fact widely dispersed, from the south-central Palaearctic Region, to the Afro-tropical and Oriental Regions. The species occurs in a broad spectrum of habitats, from deciduous woodland to savannah and steppe and subdesert zones, as well as coniferous forest (compare with [270, 765–768]). Wild populations interbreed intensively with domestic cats throughout their vast range of dispersion [769].

Among the Mediterranean islands, wildcats are reported from Mallorca, Corsica, Sardinia, Sicily and Crete [269, 279, 770], but further investigations are needed to define better the taxonomic position of the different insular representatives [688] (▶Fig. 88). If we exclude Sicily, where a population of European wild cat, *Felis silvestris silvestris* Schreber, 1777, possibly penetrated in an unspecified moment of the last glacial episode (see [771]), wildcats still occur in other Mediterranean islands with phenotypes of the African wildcat, the variety *Felis silvestris, lybica* group Forster, 1770, which is

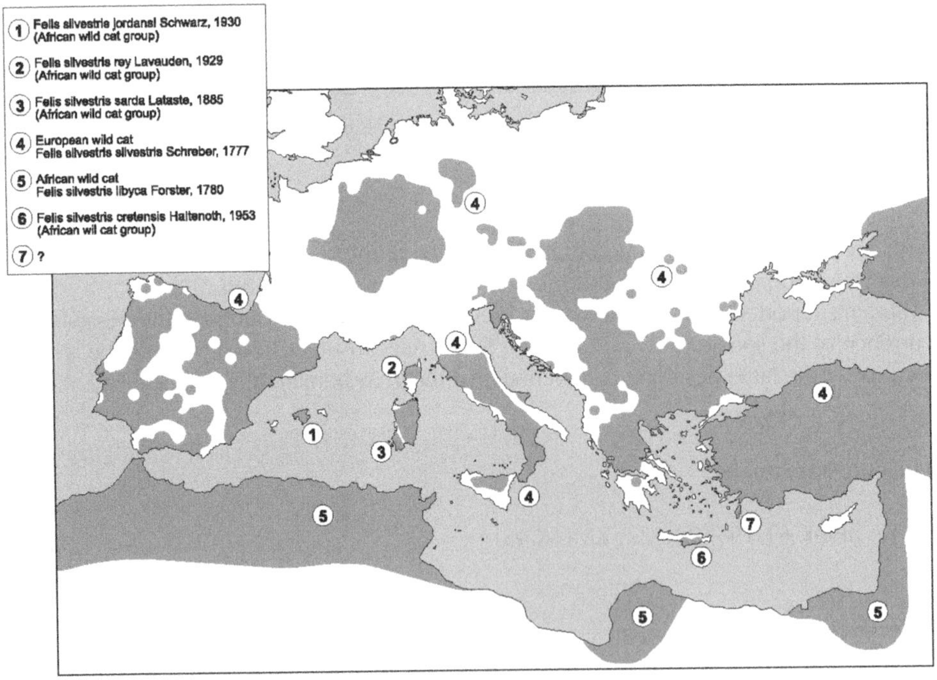

Fig. 88: Among the Mediterranean islands, wildcats are reported from Mallorca, Corsica, Sardinia, Sicily, Crete and, perhaps, Rhodes.

distributed throughout most of non-Saharan Africa and in the Near East (compare with [8, 478]) (►Fig. 89). This wildcat has a very broad habitat tolerance. It is thinly distributed throughout the Nubian, Saharan, and Arabian deserts, where it is generally restricted to mountains and dry watercourses, ranging up to ~3 000 m in the mountains of Kenya, Ethiopia, and Algeria [772]. The African wildcat is generally recognised as the ancestor of the domestic cat [773]. Before being tamed, it very likely already experienced some sort of human cultural control in early Neolithic Cyprus [774].

The wildcat is the only non-mustelid carnivore from Crete. Its presence was first reported by Raulin [747]. According to Ondrias [70], the felid was still present on the Lefka Ori

Fig. 89: Apart from Sicily, the wildcats that are still to be found in the Mediterranean islands present phenotypes that can be referred to that of the African wildcat, *Felis silvestris libyca* Forster, 1780. Photo by Marco Masseti.

(White Mountains) during the 1960s, and there are successive reports of footprints of supposed wildcats were later reported from the gorge of Kalocambos, in the same area [716]. Some individuals were sighted on the Lassithi plateau [775–776]. Catsadorakis [298] instead thought that the wildcat had already become extinct or was an endangered species in the Samaria National Park (Lefka Ori). On April, 10th 1996, a specimen was captured on the south-western slope of Mount Psiloritis, about five kilometres from the village of Platanos [715, 777]. It was a robust male, at least 5 years old, weighting around 5.5 kg, 57 cm long (tail excluded), with a 31 cm long tail, 14 cm high hind foot, and 7 cm long ears. It was in excellent physical condition (▶Fig. 90). Its coat-colour and marking patterns fall in the middle of the range of the *Felis silvestris, lybica* group phenotypes. The trap-site was situated at about 1 000 m in altitude, in a woodland dominated by *Quercus coccifera,* with also *Acer, Platanus, Pyrus,* and phrygana.

Felis silvestris has also been reported from Rhodes (compare with [57, 562, 778]), but this occurrence still needs to be confirmed. Early accounts of the former occurrence of wildcats on other Aegean islands, such as Kythera [778], and Skopelos [290], are also available in literature.

Systematic data

Felis silvestris morea Trouessart, 1904, is the subspecies of European wildcat regarded as endemic to the southern Balkan peninsula [384]. The wildcats from Asia Minor and the Caucasus have been referred to as *Felis silvestris caucasica* Satunin, 1777 [269,

Fig. 90: Adult male of Cretan wild cat, *Felis silvestris cretensis* Haltenorth, 1953, trapped on the south-western slopes of Mount Psiloritis, on April 10th 1996 (photo Apostolos Trichas).

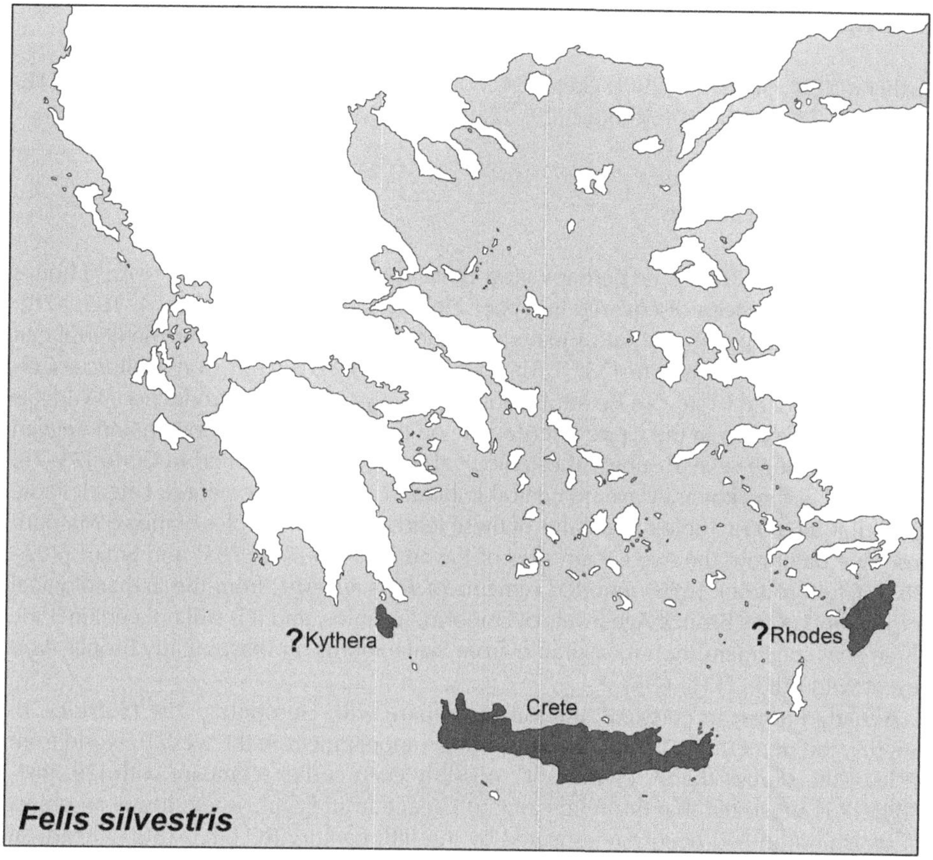

Fig. 91: The wildcat, *Felis silvestris* Schreber, 1775, in the Ionian and Aegean islands.

779]. The Levantine population fall instead in the phenotypes of the African wildcat [270].

As already seen, the Cretan wildcat belongs to the *Felis silvestris, lybica* group, having been recognised as a separate subspecies: *Felis silvestris cretensis* Haltenorth, 1953. This cat was firstly described as *Felis ocreata agrius* by Bate [276], *Felis agrius* by Miller [279], and *Felis silvestris agrius* by Bate [276] and Zimmermann [714]. Later on Haltenorth [778] assigned a specimen caught in the Chania area in 1942 to *Felis silvestris cretensis* and attributed the skins examined by Bate to domestic cats. Some doubts still persist, however, on Haltenorth's classification of the Chania specimen.

Museum specimens

Crete: NHMC 80.5.65.1, 80.5.65.3–80.5.65.4, 80.5.65.7, 80.5.65.9–80.5.65.11; NMW 18635; ZMUP 4027.

References

Kythera [778]; Skopelos [290]; Crete [54, 70, 276–277, 279, 285, 668, 690, 715–716, 747, 775–776, 778, 780]; Rhodes [29, 57, 562, 778].

Archaeozoological note

Apart from Sicily, Corfu, and perhaps a few other continental islands, where fossil bones of wildcats have been discovered in Upper Pleistocene contexts [8, 73–74, 105, 478], felids do not figure in the late Quaternay faunal assemblages of the Mediterranean archipelagos (compare with [137]). The presence of *Felis silvestris* on Mallorca, Corsica, Sardinia and Crete can be attributed exclusively to human introduction. Wildcats were also unknown in the Upper Pleistocene on the majority of the Ionian and Aegean islands, although several subfossil osteological fragments were found in Corfu [73–74, 105], and at the Skoteini Cave, in central Euboea (Late Neolithic period, Late Helladic period) [694–695]. Subfossil remains of these felids figure in samples of mixed Minoan-Iron Age date from the two Cretan sites of Kavousi-Kastro [742, 781] and Smari [782–783]. Clutton-Brock [696] reported remains of *Felis silvestris* from the archaeological exploration of the Bronze Age levels of Emporio, in Chios, and it is still not certain if the three *Felis* specimens (non-measurable) from Skala Sotiros on Thasos (Early Bronze Age) are of wild origin [316, 695].

Already known in classical antiquity (compare with Herodotus, *The Histories*, II: 66–67), the domestic cat made its first sporadic appearances in the western world from at least the start of the 6th century BC, possibly even earlier (compare with [29, 461, 784]). It is suggested that this felid came to Greece from Egypt, where it was an object of worship and had been domesticated by the fifth century BC [785–786]. Already at the end of the Roman Empire it was present more or less everywhere. However, it seems that its more widespread distribution did not take place until the Middle Ages when, concomitant with the establishment of the Arab culture, the domestic cat finally became more extensively diffused, at least in the countries of the northern Mediterranean and its islands [654, 787].

6.38 Leopard, *Panthera pardus* Linnaeus, 1758

Greek name: πάρδαλις (*pardalis*) Turkish name: *pars*
Greek vernacular name: καπλάνη (*kaplani*)

The leopard, *Panthera pardus* Linnaeus, 1758, includes about 30 subspecies, not always clearly defined, distributed in tropical and subtropical Africa as well as in southern Asia, from the Mediterranean, north of Anatolia, as far east as the Caspian Sea and across the southern periphery of Asia to Korea and Java [768]. This felid is found in a wide range of habitats, but preferentially in wooded areas [788]. In recent historical times, it has been extirpated or definitively reduced to tiny scattered populations along the entire circum-Mediterranean region, from the forests of the Maghreb to the deserts of the Near East (►Fig. 92).

A taxidermically prepared adult leopard (►Fig. 93) is on display at the Natural History Museum of the Aegean in Mytelenii, on Samos [478, 769]. This specimen previously belonged to the Town Council (Greek: *Nomarkia*) where it had been exhibited for several decades [66, 306]. The tag on the specimen reads *kaplani*, which is a Samian term for panther. Yet the word derives from the Turkish term *kaplan*, which commonly indicates

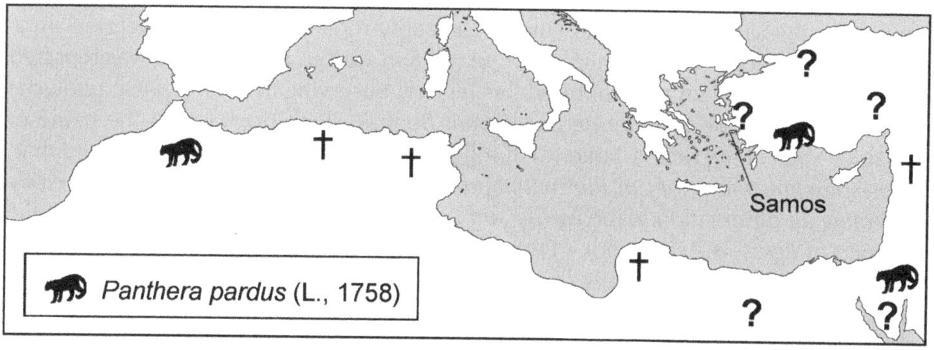

Fig. 92: Distribution of the leopard, *Panthera pardus* Linnaeus, 1758, in the Mediterranean region, with the geographical location of the island of Samos.

Fig. 93: The taxidermic specimen of an Anatolian leopard or Asia Minor leopard, *Panthera pardus tulliana* Valenciennes, 1856, shown at the Natural History Museum of the Aegean, Samos (Greece). Photo by Marco Masseti, courtesy of the Natural History Museum of the Aegean, Mytelenii, Samos.

tiger in Anatolia and erroneously also leopard (see [789]). The information available today is too scanty to ascertain the age and origin of the specimen. The leopard was supposedly killed on Samos between 1870 and 1880. However, Vera Eymorfoutsikos, the Samian museum caretaker in 1997, reports that the animal was killed on the island in 1889 by two local brothers, Nikolaos and Geracimos Gliarmis. The title of one of the most famous novels of the contemporary Samian writer Alki Zei [790], *To kaplani tis vitrinas* (= The *kaplani* of the showcase), better known as *Wildcat under glass*, was inspired by this leopard. Zei described this *kaplani* in speaking of her childhood, and since she was born in 1936, the leopard is presumably older. The animal allegedly arrived swimming across the Samos Strait. In fact there is a deeply-rooted traditional belief on Samos which refers to leopards swimming from Anatolia in various periods. This was reported by de Tournefort [677] who confirmed this legend, observing that: "*Il y passe quelques tigres qui viennent de terre ferme par le Petit Boghas*". Petit Boghas was the name of the Samos Strait at the time. Clarke [678] followed this observation and mentioned that: "*tigers sometimes arrive from the mainland, after crossing the little Boccaze; thereby confirming all observation made by the author in the former section, with regard to the existence of tigers in Asia Minor*". Perhaps de Tournefort [677] was likely inspired by local people and had not actually observed it personally. In any case, leopards are good swimmers, and we cannot exclude that they may have repeatedly reached the island by swimming across the Samos Strait, which is just 1.7 km wide, coming, for instance, from the Samsundag area [769], the last western Anatolian stronghold of the species until the early 1970s [769, 791–793] (▶Fig. 94). Leopards are supposed to have swum from Turkey to Samos after being tossed into the sea by floods of the Menderes (Maiandros) river, or

Fig. 94: A view of the Turkish Samsundag, in the Dilek peninsula, beyond the narrow marine channel that separates it from the Greek island of Samos. Photo by Marco Masseti.

to have fled a fire that burned the wooded shore of Asia Minor [478]. Density-dependent factors, such as competition and predation, may be a plausible cause for venturing into the water in search for new lands ([29, 794] and references therein).

Unfortunately, the Samian specimen looks awkward because of a rather unskilled taxidermic preparation. It is about 235 cm long, with a 90 cm long tail, which makes it a quite large specimen. But the taxidermic procedure might have altered its original size because the skin of felids is extremely elastic. The coat colour is deteriorated and its colours are faded due to its prolonged exposure to daylight. Although the skin's hair is worn in patches, it seems that the colour was originally tawny or buff on the back and paler on the sides, where it could have merged into the white of the belly. Today, the entire coat is uniformly pale, with fairly large (about 3–4 centimetres in diameter), widely spaced and thinly rimmed, dark-brown rosettes on the sides and the back, with central spots slightly darker than the ground tint. The coat's hair is fairly short and thick, that on the nape is long, whereas the tail is decidedly bushy.

Systematic data

Anatolia harbours two varieties of leopards. The Persian leopard, *Panthera pardus saxicolor* Pocock, 1927, is reputed to be still dispersed from the mountains of eastern Anatolia to Baluchistan, throughout northern Persia [373, 795]. This is a large subspecies with thick pelage, generally darker coloured than that of the Arab leopard, *Panthera pardus nimr* Hemprich and Ehrenberg, 1833, still dispersed from the shores of the Dead Sea to the eastern and southern Arabian peninsula [270, 767–768] (▸Fig. 95). The rosettes of the Persian leopard are bigger, more finely rimmed, more annular and deep chocolate

Fig. 95: Adult female of Arab leopard, *Panthera pardus nimr* Hemprich and Ehrenberg, 1833. Photo by Marco Masseti.

brown in colour. According to Harrison and Bates [270], in eastern Turkey, this form would appear to intergrade with the western Anatolian subspecies, the Anatolian leopard or Asia Minor leopard, *Panthera pardus tulliana* Valenciennes, 1856 (▶Fig. 96).

The coat pattern and colour suggest that the Samian specimen belongs to the latter subspecies. It is clearly distinct from other Near Eastern subspecies [769]. In his paper on the intraspecific variations of the leopard and the validity of its subspecies, Leyhausen [798] stated that *Panthera pardus tulliana* was definitely a subspecies, clearly distinct from the geographically adjacent subspecies *Panthera pardus saxicolor*, and *Panthera pardus nimr*. A detailed description of the fur pattern and colour of *Panthera pardus tulliana* was given by Valenciennes [796], who described this subspecies on one single specimen from the type locality of Ninfi, 40 km east of Izmir: "*L'animal aussi grand que nos plus grandes Panthéres africaines, a le pelage cendré ou gris lègérment roussâtre, peu chargé de taches en larges roses ou cercles mal fermés sur les flancs; sur les épatules et sur les cuisses elles sont un peu plus petites; à partir du poignet ou du tarse, les taches deviennent des gros points noirs, que l'on retrouve sur la téte et*

Fig. 96: A rare photograph of the Asia Minor leopard. From Banoğlu, 1958.

un peu sur le cou. Les taches en roses arrondies se continuent sur le dos de la queue. Celle-ci, très-caractéristique, est plus longue que le corps entier de l'animal; le poil fin qui la recouvre s'allonge de plus en plus à mesure qu'il s'approche de l'extrémité, de sorte que le dernier tiers de la queue de cette Panthére esr plus gros ou plus touffu que la racine: c'est précisément le contraire de ce qui existe chez toutes les autres Panthéres indiennes ou africaines dont nous avons parlé. La distance du bou de nez à sa racine ou à la hautcr du nez est aussi un peu plus longue. Cet ensamble de caractères nous parâit suffusant pour bien reconnâitre cette Panthére, trèsdistincte de tuotes celles que nous avans signalées plus haut».

In the course of this research, one specimen of *Panthera pardus tulliana* was examined in BMNH (catalogue number 34.5.14.1), collected in 1931 on Mount Kavajahissar in the mountains south-east of Kuluk, on the south-western coast of Asia Minor. The colour of the middorsal region of this specimen tended to be decidedly tawny, darker than that of *Panthera pardus nimr*, with wide-sized, widely spaced and thinly-rimmed rosettes. Its tail tended to be bushy. The peculiarities of *Panthera pardus tulliana* were also outlined by Pocock [797, 799] and Mendelssohn [800], who observed that the Anatolian taxon was one of the largest leopard subspecies ever known, and perhaps the largest of all. Valenciennes [796], described *Panthera pardus tulliana* in honour of Marcus Tullius Cicero, the Roman administrator of the province of Cilicia who provided the first historical information on the leopards of Asia Minor in 100 BC. Although *Panthera pardus tulliana* was rare in all its range, it was originally spread from the mountains of western Turkey, and as far south as the coastline of the Tauros mountains and of south-eastern Anatolia [270, 373, 765, 767, 789, 791, 793, 795, 801–803]. A few records report its occurrence also in other Anatolian areas, such as the adult male killed near Beypazari (85 km west-noth-west from Ankara), reported by Kumerloeve [804]. To the south, this subspecies formerly reached south-western Galilee, the Golan and Mount Hermon, where it is reputed to have become extinct in 1965 [805].

Museum specimens

There is only one taxidermically prepared specimen, without catalogue number, in the Natural History Museum of the Aegean at Mytelenii (NHMAM), Samos.

References

Samos [66, 306, 654, 769]; (compare with [677–678, 790]).

The survival of the Asia Minor leopard

As reported by several authors, such as Pocock [797], Tortonese [801], Banoğlu [802], Kumerloeve [765, 791] and Borovali [803], even during the 20th century, the Anatolian leopard occurred in the coastal mountainous district of south-western and southern

Turkey (▶Fig. 97). It has been presumed that this felid vanished from its last western Anatolian strongholds before the late 1970s (compare with [795, 805]; S. Güsar, 1997, *in verbis*). But, there are rumours that something has survived.

In fact, the Asia Minor leopard appears to survive in Turkey in spite certain claims that the taxon is extinct [770, 793, 806]. Riffel [807] was the first who supposed that *Panthera pardus tulliana* had not disappeared. During several visits to south-western Anatolia between 1985 and 1992, new information has been obtained by two German biologists, B. Ulrich and M. Riffel. The finding of fresh fecal pellets in the spring of 1992 indicated the survival of the species in the Temessos National Park [793]. The possibility that these faeces might have belonged to some other species can be ruled out by the strong cat-like odour and the size. Their odour and fresh appearance indicated that they

Fig. 97: The 1950s leopard hunter Mantolu Hasan from the town of Kuşadasi, on the Turkish Aegean coast. From Banoğlu, 1958 [802].

had been deposited not long before. Ulrich and Riffel [793] made a cautious estimate that they were no more than one to two weeks old. Unfortunately, no food marks were visible in the surroundings. According to the same authors, recent reports of leopards are not restricted to the Termessos area but have also come from the south of the Lycian peninsula and the vicinity of the town of Alanya: a leopard was shot near Kas in 1989 and a recent sighting of one specimen in the vicinity of Alanya was reported in 1991. In the light of these data, it seems that a scattered population of the Asia Minor leopard exists in eastern Lycia and in parts of western Lycia, between Finike, Antalya and Alanya and other areas to the west of Lycia [793, 806, 808]. According to Kasparek [809], many parts of southern Turkey have remained very remote until recent times, and even now are difficult of access. Thus, it is hardly surprising that leopards have been able to survive in these areas rather than in other parts of Anatolia. Although, for example, local hunters claim that leopards still may exist in the area the Samsundag National Park, in Eastern Anatolia, recent scientific investigations reveal that the last reliable information dated back to the late 1960s [793], or perhaps the early 1970s [792]. This protected area is situated on the Dilek peninsula, south of the town of Kusadasi, within sight of the Greek island of Samos. It covers an area of approximately 11 000 ha consisting mainly of Mediterranean scrubland and pine forests [793]. Until a few decades ago, the Samsundag National Park was famous as one of the better-known strongholds of the Asia Minor leopard, so much so as to justify the choice of the felid's image as its logo [769].

Archaeozoologiocal note

As far as is presently known, the former occurrence of *Panthera pardus* in the southern Balkan peninsula is known only from a few Pleistocene sites, such as Asprochaliko and Kastritsa in Epirus, or Kitsos and Vraon in Attica [416]. No remains of the species have been found so far on any of the Ionian or Aegean islands of protohistoric times (compare with [695]). Leopards in Aegean art are relatively rare ([810]; compare with [811]). However, the presence of bone remains of these felids in Anatolia is reported from several protohistoric sites, such as Neo-Hittite Arslantepe [812]. The occurrence of leopards is even mentioned in the account of classical authors, such as Xenophon (5th century BC), who observed, in his *Kinegeticon* (XI, 1–2), that panthers were fairly common in Asia Minor, beyond the border between the territories of Bithynia and Misia [351]. However, the presence of the species in the Balkan peninsula in historical times has not been confirmed by the findings of osteological remains of leopards (compare with [786]).

6.39 Wild boar, *Sus scrofa* Linnaeus, 1758

Greek name: αγριογούρουνο (agriogourouno)

Turkish names: *domooz, domuz, yaban domuzu, yaban-domooz*

The wild boar, *Sus scrofa* Linnaeus, 1758, is found throughout the steppe and broadleaved forest regions of the Palaearctic (compare with [270]) (▶Fig. 98). Widely distributed

Fig. 98: Wild boar, *Sus scrofa* Linnaeus, 1758. Photo by Giuliano Cappelli.

in central and southern Europe, it is however absent in the southernmost parts of the continental Balkan peninsula [813]. The species is nevertheless considered to be widespread in Anatolia [93]. It is fairly infrequent in most of Albania [78].

As far as is presently known, wild boars have been reported from only very few Ionian and Aegean islands. In relatively recent times, Heldreich [404] reported the historic occurrence of these mammals from Euboea, whereas von Wettstein [58] observed that they occasionally reached the island of Lefkada by swimming from the coast of Epirus opposite. Up to a few years ago, wild boars were said to have also reached Corfu from the Albanian coast (G. Handrinos, personal communication: November 2000). According to Tsachalidis and Hadjisterkotis [131], the species it is not found in Euboea, due to the lack of a suitable habitat and the fact that this island is connected with the rest of Greece by a bridge. Jaia [814], possibly on the basis of island rumours, recorded the occurrence of the species on Rhodes, allegedly introduced there in medieval times by the knights of Saint John of Jerusalem. This occurrence, however, was not confirmed by Festa [57] or by the other Italian zoologists who explored the island in the following years. On Rhodes, wild boars are now kept in a few private enclosures [60]. Field signs and records of the occurrence of wild boars have been reported from the island of Samos since the early 1990s [306]. The species is said to have reached the island by swimming from the nearby Anatolian coast, where boars are still reputed to be the commonest ungulates of the Samsundag range (Dilek Yarimadasi Milli Parki); Samian hunters confirm that they were not deliberately introduced onto the island [478, 769]. Data gathered, however, by Tsachalidis and Hadjisterkotis [131] from the more extensively forested Greek islands, such as Crete, Rhodes, Samos, Lesbos, Lemnos, Thasos,

Corfu, Cephalonia and Zakynthos, did not indicate the occurrence of wild boars on any of them.

In the course of the present research, the species was also seen for the first time on the small Turkish islets of Cleopatra (Şehir Adalari), Değirmen Bükü, the Seven Islands (Yediadalar) in the Gulf of Gökova, the island of Yildiz in the Gulf of Marmaris, the island of Domuz in the Gulf of Fethiye, the island of Gemile opposite the Olüdeniz, and the island of Geykova (see also [267]). All these islands are located very close to the mainland coast and wild boars have often been reported swimming across the narrow marine straits separating the islets from the continental shores. It is interesting to note that, in Turkish, the name of one of these islands is "*Domuz*" or "*Domooz*", which actually means "wild boar". It seems that the few hundred metres which, in most cases, separate these islets from the Turkish mainland areas opposite do not represent any obstacle for the suids which swim across to perpetrate night raids on the island crops. Modern scientific investigation considers wild boars as quite competent swimmers, yet they cannot survive a crossing of more than a few nautical miles of open sea. In this respect, a number of observations are available for the Mediterranean basin. Off the delta of the river Ebro (Spain), for example, adult specimens of *Sus scrofa* are occasionally spotted swimming in the open sea, even at considerable distances from the mainland [267]. It seems, however, unrealistic, or entirely erroneous, that wild boars can swim 20–25 km as affirmed by Schüle [188], without any empirical substantiation.

Systematic data

Sus scrofa scrofa Linnaeus, 1758, is the wild boar of western and central Europe, from France to Belarus, and possibly Albania [815].

Sus scofa libycus Gray, 1868, is the subspecies reputed to be dispersed from south of the Caucasus through the Levant to the Nile delta, and westwards through Anatolia to the Balkans [815]. Also according to Harrison and Bates [270] and Kryštufek and Vohralík [93], this is the taxon occurring in Asia Minor; it was described from a specimen from the locality of Xanthus, near Günek in south-western Turkey.

References

Lefkada [58]; Euboea [404]; Rhodes [58, 60, 394, 814, 816]; Samos [267, 306, 769]; Cleopatra (Şehir Adalari), Değirmen Bükü, the Seven Islands (Yediadalar), Yildiz, Domuz, Gemile and Geykova [267].

Archaeozoological note

Fossilised osteological fragments of wild boars have been found in Upper Pleistocene contexts on Corfu [73–74, 105], whereas the site of Maroulas, on Kythnos, provided Lower Mesolithic remains of the species (9th millennium BC) [421]. On the small island of Youra, in the Northern Sporades, the species is attested since the lowest Mesolithic levels of the cave of Cyclope [420] (►Fig. 99). The oldest remains gave a radiocarbon dating of 7530 cal. BC–7100 cal. BC (8th millennium BC), performed by the Beta Analytic Laboratory of Miami, USA [267, 351]. The particularly ancient age would suggest that these are not domestic animals [267]. Archaeozoological investigation, however, shows that swine were regularly domesticated in the Near East as far back as the Middle Pre-Pottery Neolithic B period (MPPNB) (compare with [111, 817]). Hence, we cannot exclude human cultural control over these pigs, which might have been imported to the island from mainland areas, even very far afield [267].

The earliest introduction of wild boars onto the Eastern Mediterranean islands is nonetheless documented on Cyprus at least as far back as 11 700–11 400 years ago [818]. These Cypriot findings predate the times and means of any known early managing and/or cultural control of *Sus scrofa* by man by about one millennium (11 400 BP) prior to the evidence of pig domestication documented in south-western continental Asia, which dates back to about 10 500 years ago (compare with [819, 820]). We cannot however exclude that wild boars could have lived naturally on Youra, or in its immediate continental surroundings. It is still unclear whether this was in fact already an island of the Northern Sporades during the Late Pleistocene and the Early Holocene

Fig. 99: A *Sus* mandible from the lowest Mesolithic levels of the cave of Cyclope, dated to 7530 cal BC–7100 cal BC (8th millennium BC). Photo by Marco Masseti.

(compare with [166, 228, 420, 422]). At Markiani, on Amorgos, the presence of *Sus scrofa* is recorded in the Early Bronze Age [419], whereas on Crete, its occurrence is documented from the Aceramic Neolithic period (6th millennium BC) archaeological contexts onwards [418].

Feral pigs phenotypically similar to wild boars were reported on Anaphi, Samothrace, and Fourni by von Wettstein [58], on Rhodes by Ghigi [394], Taibel [816] and von Wettstein [58] and on Tilos by Ghigi [394] and Taibel [816].

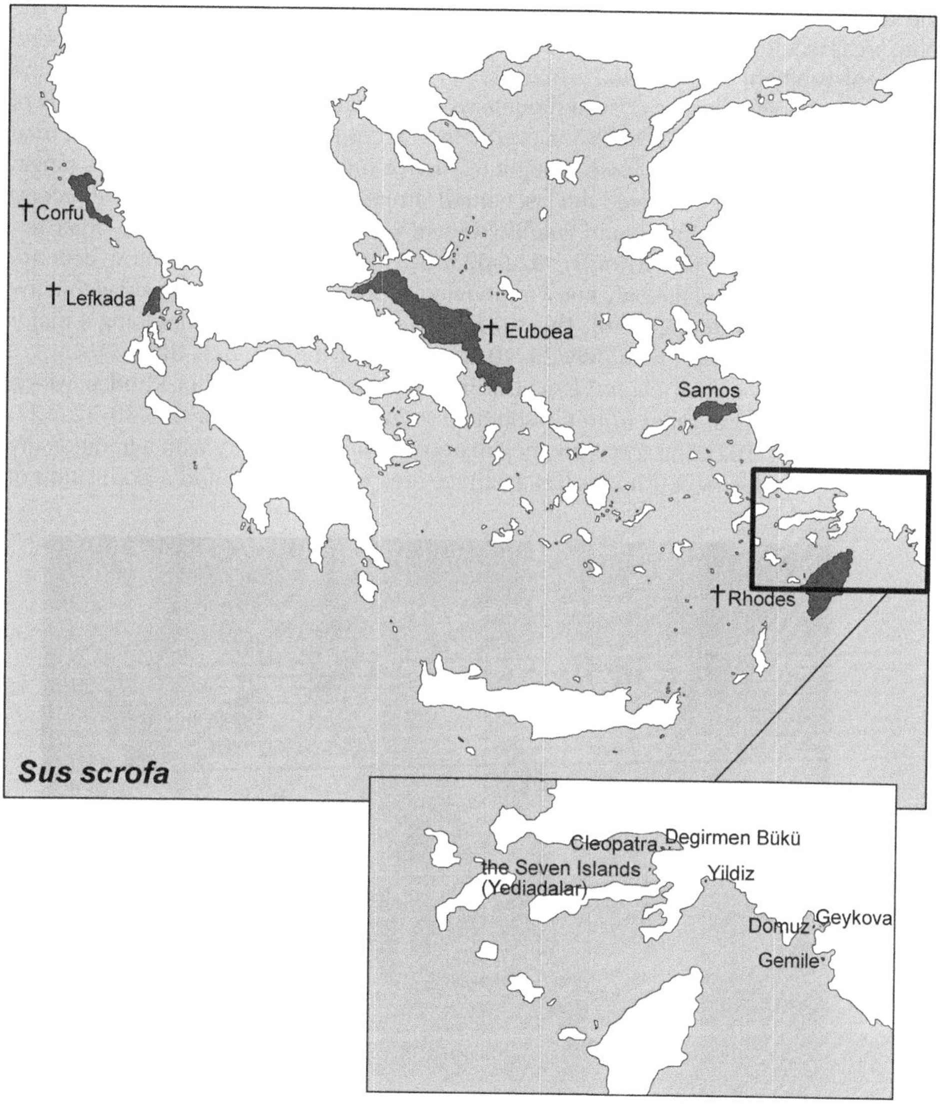

Fig. 100: Former and present distribution of the wild boar, *Sus scrofa* Linnaeus, 1758, in the Ionian and Aegean archipelagos. Islands and areas that are inhabited by the boar are shaded and those where they have become extinct are marked with a cross.

6.40 European fallow deer or common fallow deer, *Dama dama dama* Linnaeus, 1758

Greek name: πλατώνι (*platoni*)
Greek vernacular name: έλαφος (*elafos*)

Turkish name: *alageyik*
Turkish vernacular name: *yamoorcha*

The European or common fallow deer, *Dama dama dama* Linnaeus, 1758, is perhaps the taxon of deer whose current distribution has been most influenced by man. The only areas where it is believed to have persisted as a native form are the coastal wooded plains of southern Anatolia [821–825]. Here, its historical range extended from the Izmir region, following the coastal mountains to the south-east as far as the foothills of southern Cilicia [28, 93, 789, 826]. Nevertheless, according to Chapman and Chapman [826], by 1934 fallow deer were thought to survive only on the western Taurus range. To protect what was considered the last natural stronghold of the species, in the second half of the 1960s the Düzlerçami wildlife reserve was established within the larger Termessos National Park (Antalya) [91, 822–823, 827] (▸Fig. 101). Today fallow deer are very rare in Turkey, and the only known surviving population is that preserved within the Düzlerçami enclosure [32, 828]. This stock, however, is currently undergoing a major reduction in census size, with only 25–30 animals surviving [32, 828–829].

Another population of *Dama dama* still inhabits the Greek island of Rhodes, which is only a few nautical miles from the south-western coast of Anatolia [27, 30–32, 127, 668, 830] (▸Fig. 102). This too is an ancient population in a territory with a traditionally dense human settlement (Rhodes has a surface area of 1 398 km^2 and a population of

Fig. 101: The Düzlerçami wildlife reserve was established within the larger context of the Termessos National Park (Antalya) in the late 1960s to protect the last specimens of the European or common fallow deer, *Dama dama dama* Linnaeus, 1758. Photo by Halil Saribaşak, courtesy of the Forest Department of Antalya.

Fig. 102: Rhodes is the only Mediterranean island still characterised by the occurrence of a population of fallow deer of very ancient origin and with a distinct genetic signature. Photo by Marco Masseti.

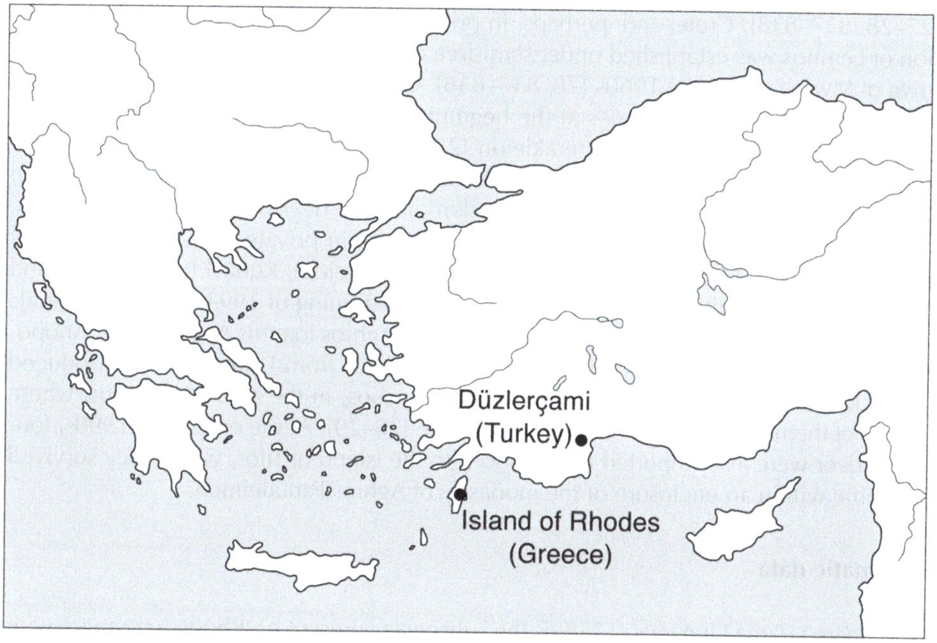

Fig. 103: Population of European fallow deer of very ancient origin still survive is the Turkish Düzlerçami wildlife reserve and the Greek island of Rhodes.

approximately 120 000) (►Fig. 103). In the course of the present study, no evidence was found attesting to the extinction of this species of deer from Rhodes during the period of Turkish rule (1522–1911), nor to its presumed reintroduction by the Italians during the early part of their government (1912–1945), as asserted by Chapman and Chapman

[831], following Dicks [832]. In fact, accounts of this deer's occurrence on the island in the last century of the Turkish dominion have been provided by Wilde [833] and Danford and Alston [789]. In 1880, Charles G. Danford also presented to the Zoological Society of London the complete skeleton of a buck (BMNH: 1880.2.16.2) killed on December 22, 1878, at Laerma, in the central pine-wooded district of Rhodes [789], where fallow deer are still regarded as not uncommon [834].

The latest techniques for investigating population genetics have shown that the Rhodian fallow deer are very special, being of ancient lineage distinct even from the Turkish relic population of Düzlerçami, which was formerly regarded as the likely source of the stock introduced in Neolitic times [31–32, 137, 835]. In fact, according to a recent research carried out by Masseti et al. [31–32], this Rhodian population, presumably founded by humans in Neolithic times, possesses a set of mitochondrial lineages found in no other study populations. All the specimens from Rhodes, except two (haplotypes 5 and 11), have an 80-bp [base pairs] insertion that shows no sequence variation among individuals (►Fig. 104).

Within the Aegean region, the westward transfer of *Dama dama* which began in early Neolithic times and continued throughout antiquity and the Middle Ages, still continues today, as shown by the exports from Rhodes to several islands and various other localities of the Greek mainland, especially in the course of the last century [27–28, 830, 836]. Since the end of the 1960s, fallow deer have been introduced onto Lemnos [27–28, 837–838], Crete, and, perhaps, in private enclosures on Kos [28]. The population of Lemnos was established under semi-free conditions in an area (the castle) of the town of Myrina in the late 1960s [28, 837–838]. On Crete, a number of these ungulates were first imported from Rhodes at the beginning of the 1970s to a private enclosure in the village of Archanes, near Herakleion [28]. A few years later, some of them were transferred to the zoological garden of the Municipality of Herakleion. Around 1990, the Municipality closed the zoological garden, and the deer passed to private hands. Hence, in 1994 Rhodian deer were present on Crete in private enclosures in several localities, such as Gazi (Herakleion), Archanes (Herakleion), Kunavi (Herakleion), and the monastery of Kaliviani (Mires) [28]. Up to the beginning of 1993, a few individuals were also kept in an enclosure of the monastery of Aghios Ioannis Apostolos, at Anopolis (Amnissos). In the winter of 1994, a few individuals from Lemnos were introduced by the local Hunting Society onto the island of Lesbos, in the area of Agiassos, where several of them may have survived for some time [28–29]. At the end of the 1990s, four fallow deer were also imported from Rhodes to the island of Tilos, where they survived for a time within an enclosure of the monastry of Aghios Pantaleimon.

Systematic data

Dama dama dama Linnaeus, 1758, is the subspecies present on Rhodes (compare with [31–32]), where it is characterised by only wild coat colour phenotypes [28, 830].

Museum specimens

Crete: NHMC 80.5.73.2–80.5.73.7; Rhodes: BMNH 67.4.12.240, 1880.2.16.2; NMW 7069/B5445.

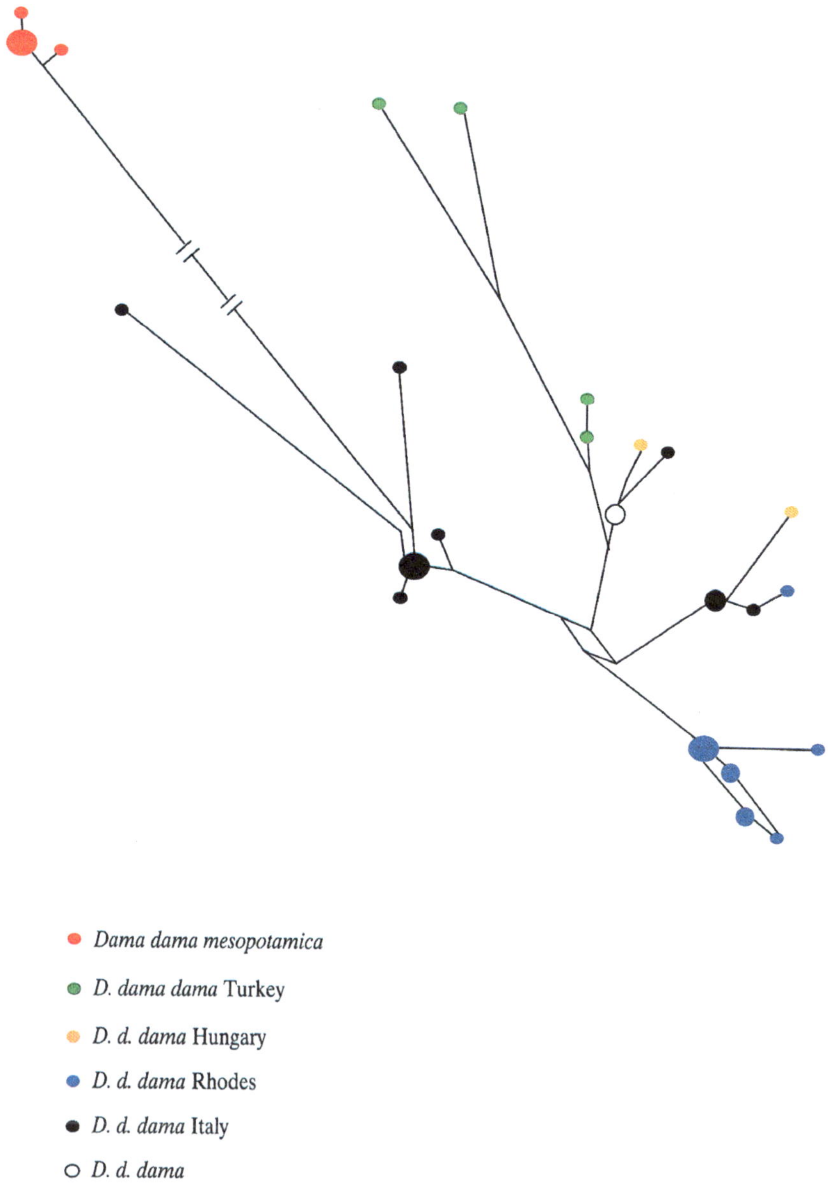

Fig. 104: The Rhodian population, presumably founded by humans in Neolithic times, possess a set of mitochondrial lineages found in no other study populations. From Masseti et al. [835].

References

Lemnos [27–28, 837–838]; Lesbos [28]; Kos [28]; Rhodes [28–29, 31–32, 57–58, 301, 394, 405, 562, 690, 789, 822–823, 826, 828, 830–831, 834, 839–840]; Crete [28, 787].

Palaeontological and archaeozoological data

The common fallow deer is a faunal element characterised by a late Quaternary subtropical distribution [28, 830]. In fact, as far as is presently known, palaeontological and archaeological evidence attests to its occurrence confined practically to the eastern Mediterranean region during the Late Pleistocene and the Early Holocene. Current knowledge indicates southern Europe and Asia Minor as the post-glacial refuges of this deer [28, 31–32, 822–823, 830–831, 841–843].

The palaeontological evidence of the deer's diffusion in this region, however, is extremely fragmentary. As far as is presently known, the only insular fossil evidence comes from Upper Pleistocene Corfu [102–105] and Thasos [203]. In continental Greece, contemporary remains of fallow deer appear to have been reported from Kitson cave, in Attica [844], whereas osteological fragments of the species referred to earlier Pleistocene chronologies are known from Kalamakia, Peloponnese [845], and the Diros cave, Peloponnese [416]. Darlas and Psathi [846] report *Dama dama* from the Upper Palaeolithic levels of the Southern Peloponnesian sites of Kolominitsa, Skoini 4 and Tripsana, but bones of this deer were very likely provided by the exploration of only the deepest strata, around 26 000 BP, of the cave of Skoini, whereas sporadic specimens were found in the other prehistoric sites just mentioned. The species also figures among the main ungulate represented in the fauna of the Upper Palaeolithic sequences at Klissoura Cave 1, in the Peloponnese [847]. The material of the superior strata of Klissoura and Franchthi caves, in the Argolid, Northern Peloponnese, has not yet been published. However, it seems that at Franchthi cave, no *Dama* or cervids of similar size have been found in Late Pleistocene or Holocene deposits. A few fragments of a medium sized cervid were discovered from before 25 000 BP, but their species and genus cannot be identified [28]. According to Sebastian Payne (*in verbis*), it may be possible to refer these fragments to *Dama* on biogeographical grounds, but not on anatomical grounds. K. Trantalidou (*in litteris*), is of the opinion that in Attica the fallow deer was perhaps already extinct in the Late Pleistocene, but osteological remains are very limited. To date it has not been identified in other caves, but most of the material is under re-examination. According to Yannouli and Trantalidou [848], fallow deer are connected with drier conditions, and during the Holocene they lived only in the eastern parts of Greece.

The early history of the species in southern Anatolia is broadly documented by fossil records dating back to the Late Pleistocene [825]. Remains of *Dama dama dama* were found in the Upper Pleistocene levels of the Karain B and Öküzini caves [849–850], and in the Holocene archaeological contexts of Sirkey Höyük (Adana) [851].

There is evidence that man took fallow deer into areas beyond their natural distribution since prehistoric times. Exports of the species are documented in the Aegean islands from Neolithic sites, as attested by the findings from Aghios Petros, an islet in the southern gulf of Kyra-Panagia, Northern Sporades [420, 852–853]. Late Neolithic remains of the species have been produced from the Skoteini cave in central Euboea [694], the islet of Saliagos, near Antiparos (Paros) [854], the sites of Skala Sotiros on Thasos [316], Emporio on Chios [696], and Kalythies on Rhodes [180, 855]. A few bones have also been provided by the excavations of the Rhodian Bronze Age levels at Trianda [856] and Assomatos [416, 857]. Legends recount that fallow deer were introduced onto Rhodes from Asia Minor by the Knights of Saint John of Jerusalem, who conquered the island at the beginning of the fourteenth century [57, 814, 831]. Despite popular beliefs, however, the discovery of osteological remains of this cervid in earlier Neolithic contexts (end of 6th–5th millennium BC) of the prehistoric site of Kalythies suggests that the introduction of fallow deer occurred much earlier [28, 180, 855–856]. This ancient introduction does not signify that all extant deer are descendants of prehistoric importations, although it could explain why the extant population still conserves wild phenotypical characteristics [29].

Furthermore, it was previously believed that the common fallow deer survived on Crete only up to the Roman Imperial and first Byzantine periods, as attested by the discoveries at the site of Eleftherna, near Rethymnon [858–859] and perhaps of the Byzantine age (6th–7th century) (see [692]). However, another finding post-dates the time of the occurrence of the species to later periods. In 1980, the excavations of the early mediaeval settlement at Vori (Timbaki), in southern Crete, yielded one incomplete left antler of *Dama dama*, from a pit dated to between the 8th and 10th century AD [28] (▶Fig. 105). Thus, this antler (total length: 337 mm; brow tine: 154 mm; partial palm

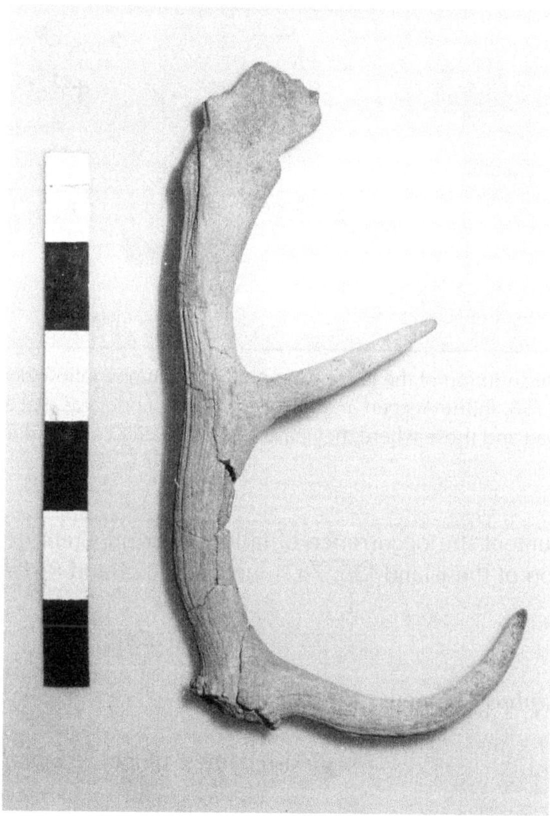

Fig. 105: Incomplete left antler of European or common fallow deer, *Dama dama dama* Linnaeus, 1758, from a pit of the medieval settlement at Vori (Crete, Greece), dated to the time of the Arab dominion of the island between the 8th and 10th century AD. Photo by Christopher Vallianos, courtesy of the Museum of Cretan Ethnology, Vori.

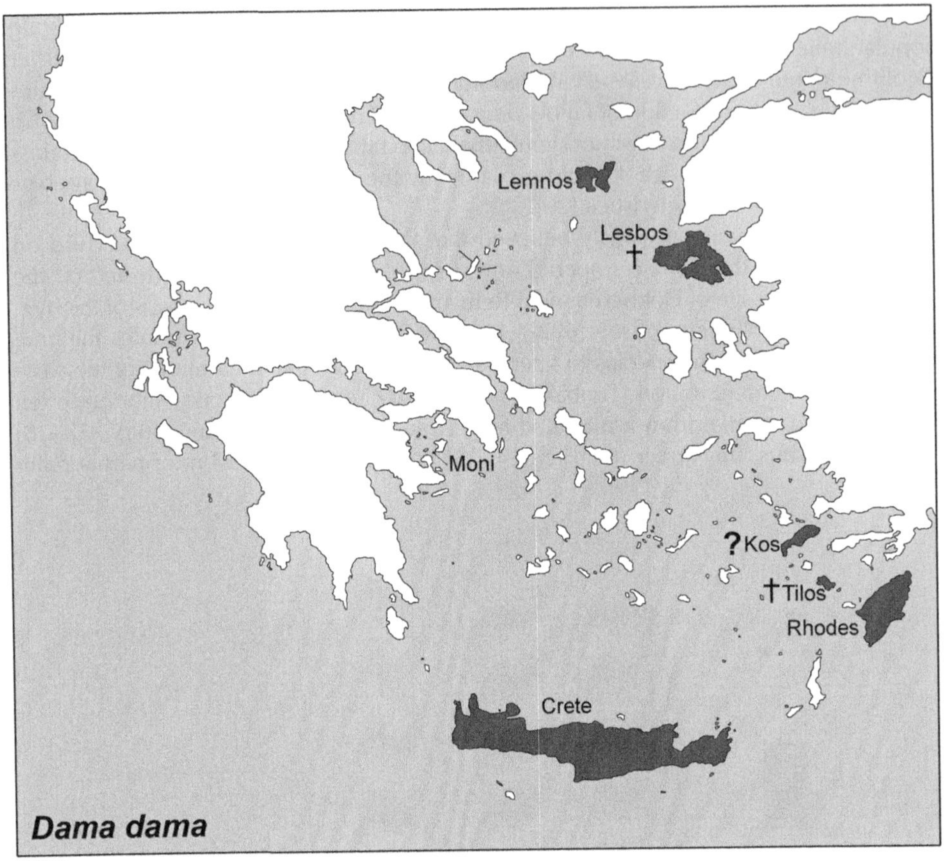

Fig. 106: Former and present distribution of the European fallow or common fallow deer, *Dama dama dama* Linnaeus, 1758, in the Aegean archipelagos. Islands and areas that are inhabited by the deer are shaded and those where they have become extinct are marked with a cross.

width: 69 mm) could document the occurrence of fallow deer on Crete even in the period of the Arab dominion of the island [28, 787], between 824 and 961 AD (compare with [860]).

6.41 Red deer, *Cervus elaphus* Linnaeus, 1758

Greek name: ελάφι (*elafi*) Turkish names: *ulugeyik, maral*[1]

[1] *maral* is the Mongolian name of the red deer in central Asia [861], which is also the name that is sometimes given to the red deer in Turkey and Iran, whose subspecific name is "maral" [862–864].

The red deer, *Cervus elaphus* Linnaeus, 1758, is one of the most widespread and diverse Old World deer species and symbolises deer as such to European cultures [864] (▶Fig. 107). It is a Holartic species spread in North Africa, Europe, part of the Near East (including Anatolia), most of the mountain range in central Asia, southern Siberia, the Far East, and North America [865–866].

Apart from the herd of the Sithonia peninsula, in Chalkidiki, extinct in the 1980s (compare with [867–868]; K. Poirazidis, 2011: *in litteris*), no naturally occurring populations of red deer still exist in Greece. The ones living there today are the result of recent re-introductions in protected areas. They are essentially represented by a few viable populations in several public and private preserves, such as Parnitha in Attika [70, 770, 869] and/or Frakto (Serres) in the Rhodopi mountains at the Bulgarian border. The latter may come from Bulgaria since deer are bred and released there for hunting reasons [870]. Also in Asia Minor, the actual disitribution of the species reflects human manipulations, such as introductions, etc. [93]. Kumerloeve [765] recorded the occurrences all over Anatolia, while Banoğlu [802] reported the presence of red deer in the neighbourhood of Istanbul in the 1930s, and in the Istranca-Bolu forest, in the Taurus mountains, as well as in the eastern Asia Minor provinces in the second half of the 1950s. Turan [871] maps the diffusion of the species from Thrace, Marmara and Black Sea Mountains to eastern Anatolia.

Red deer were still reported to occur in the central-northern part of Euboea in the 19th and the 20th century (A. Sfougaris 2001, *in litteris*). The trophy of a local stag,

Fig. 107: No naturally occurring populations of red deer still exist in Greece. The ones living there today are the result of recent re-introductions in protected areas. Photo by Marco Masseti.

presumably shot in the decade 1930–1940, is still preserved in the village of Prokopi, a few kilometres west of Kirefs, in northern Euboea (▶Fig. 108).

Systematic data

On the basis of morphologic and geographic characters, the red deer of the southern Balkan and Anatolian peninsulas have been attributed to two subspecies:

1. *Cervus elaphus hippelaphus* Erxleben, 1777, is the subspecies indicated by Ondrias [70] and Poirazidis and Paraschi [872] from the southern Balkan territories, possibly including the extinct relic population of peninsula of Sidonia, in Chalkidiki.
2. *Cervus elaphus maral* Gray, 1850, is instead the taxon reputed to occur in Asia Minor, the Caucasus, the Crimea, and the Caspian provinces of Iran [765, 873–874].

Fig. 108: Trophy of an adult male deer shot in the surroundings of the village of Prokopi (Kirefs, northern Euboea) in the decade 1930–1940. The simplified structure of the antlers, with neither a bez-tine nor a crown, is characteristic of the eastern subspecies *Cervus elaphus maral* Gray, 1850. Photo by Athanassis Sfougaris.

The herd bred by the Forest Department at Serres show the phenotypic pattern of this subspecies. The herds of Parnitha, Sidonia and Serres are characterised by simplified structure of the anthers, with no bez tine nor, in many cases, a crown. The Rodopi herd is also characterised by the persistence of spotted coats throughout the life of the animals, even in adult males. The latter pattern falls within the phenotypes of *Cervus elaphus maral*. The extinct population of Sithonia also showed some of the maral phenotypic patterns. Isolated, however, for centuries in a restricted area of low trophic production (compare with [867]), it has probably been at length represented by "maintenance phenotypes", with animals of reduced stature and where the architecture of the stag antlers was very simplified in the old males (about 15 years old), scarcely exceeding the trophy of eight points (compare with [189, 875]).

A recent study carried out on the archaeozoological remains of red deer in Greece and its artistic representation in archaeological contexts, documents the occurrence of both subspecies in the prehistory and protohistory of the southern Balkan peninsula, the Aegean islands of Santorini (Thera) and Euboea [876–877].

References

Euboea [70, 404].

Ethnozoological and archaeozoological note

The former occurrence of vital populations of red deer on certain Mediterranean islands has given rise to legends that emphasised the species' ability to swim [189, 878]. On the basis of deeply-rooted, earlier traditions, several scholars believed that deer were capable of reaching the islands by swimming, even when these were considerably distant from the mainland. Such beliefs could be traced back to Pliny the Elder who, in his *Naturalis historia* (see [314]), mentioned that deer reached the island of Cyprus by swimming from the Anatolian coast, 100 kilometres away. The analogous description given by Aelian (170–235) is so detailed that, according to Held [190], it almost seems to be based on an eye-witness account. The same author further refers to deer swimming from Epirus to Corfu, a crossing of a mere two and a half kilometres: "*And there are deer from other countries too which show this same capacity for swimming. For example, the deer of Epirus swim across to Corcyra: the two countries face each other across a strait*" (see [879]). The belief in the natatory capability of the deer of Cyprus was recalled as late as the 16th century by Sebastian Münster in his *Cosmographei*, published in Basle in 1550 (compare with [880]) (▶Fig. 109). It could be the survival of similar beliefs which led Greek newspapers, some years ago, to give coverage to a story featuring the deer from the private isle of Skorpios (Ionian Sea) considered to be perpetrators of night raids on the crops of the nearby island of Lefkada. Modern scientific investigation considers deer as quite competent swimmers [105, 881–882], yet they cannot survive a crossing of more than a few miles of open sea.

1494 Das siebende Buch

Fig. 109: The belief in the natatory capability of the deer of Cyprus was recalled as late as the 16th century by Sebastian Münster in his *Cosmographei*, published in Basle in 1550.

Fossil osteological remains of the species were discovered in peaty sediments of the Tarovica swamp near Lezhë, in north-western Albania [83], in several Upper Pleistocene contexts of continental Greece, and of the island of Corfu ([102–105, 416, 883]; compare with [876]). Bone fragments of the species have also been found in the Mesolithic layers of the cave of Cyclops, on Youra (see [416, 420] and references therein). However, since no animal bones are reported together with Palaeolithic tools that were collected in the Sporades, we have no indications that *Cervus elaphus* was an indigenous faunal element of the latter island (compare with [420]). Moreover we are not even certain that Youra was not still connected to the mainland by a land-bridge at the time (compare with [228]). According to archaeological evidence, deer began to be imported onto the Mediterranean islands, beyond their natural distribution, in early Neolithic times [6, 28, 118–119]. In the Aegean archipelago, the early occurrence of this ungulate is attested since the Aceramic Neolithic period in Knossos, Crete [418, 884], in the Neolithic period of the Miamou cave [885], in the late Neolithic period of Kephala, on Kea [886], and of the Tharrounia Cave, in Euboea [694]. Osteological remains of the species have also been found in the Bronze Age settlement of Phylla (2nd millennium BC), in the latter island [887], and in Thasos [316], in late Minoan Akrotiri, in Thera [416, 858, 888–889] (▶Fig. 110), in the Middle Bronze Age and later at Ayia Irini, in Kea [886] and in the Middle Bronze Age at Phylakopi II, in Milos [889]. On Crete, the

Fig. 110: Red deer osteological remains and fragments of antlers have also been found in late Minoan (Late Bronze Age) Akrotiri, on Santorini. Courtesy of the Society for the Promotion of Studies on Prehistoric Thera, Athens.

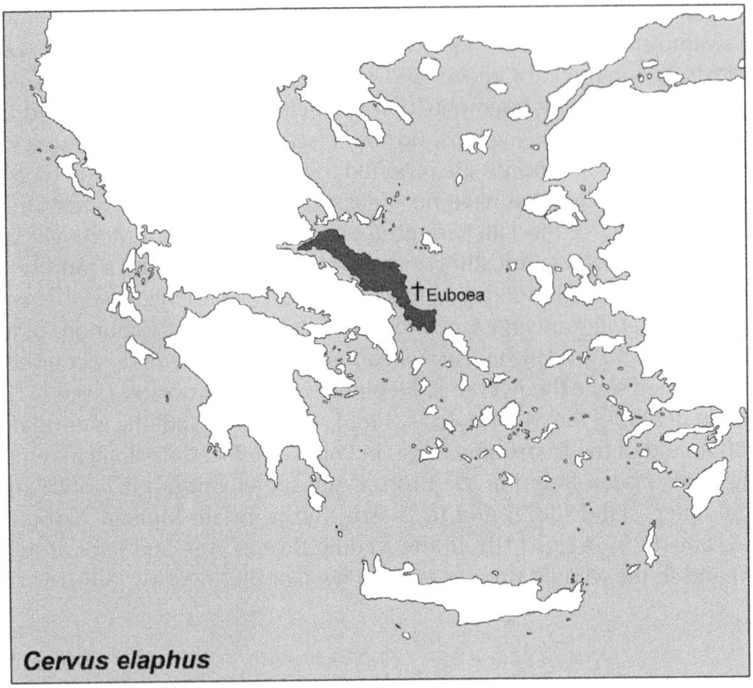

Fig. 111: Euboea is the only Aegean island which was inhabited by a population of red deer up to recent historical times.

red deer is present at Tylissos (Late Minoan I; Late Minoan II) [890], Mallia (Late Minoan) [891], Kommos (Middle Minoan III–Late Minoan III) [417] and Ayia Triada (Late Minoan) [692]. Deer bones have also been provided by the archaeological exploration of the layers of the Geometric period of the hill of Xobourgo, on Tinos [423].

Drandakis [892] reports of an unspecified type of deer from Naxos in the course of the 17th century, when the animal was hunted on the occasion of the visit of European nobles, such as the French traveller Jean Thebenot and/or the Marquis Noadel, French ambassador to Istanbul. Moreover, according to Heldreich [404], red deer were still present in Euboea in the second half of the 19th century. Beyond the peculiar use of islands as natural reserves for fresh meat, in the past centuries the European nobility often regarded these territories, especially the ones closest to the mainland, as true game preserves [189]. The occurrence of wild game on Euboea, or on Naxos, might be regarded as a consequence of the practice of keeping deer in large controlled and/or enclosed areas (deer parks) which had spread throughout Europe since antiquity. This is particularly evident from the late Middle Ages on, when the distribution and popularity of deer was increased enormously by human activities, especially those connected with hunting and the amusement of European kings and their courts (compare with [723, 763, 893–895]). Throughout time, the red deer has been the hunting game *par excellence* for a restricted social class, an exclusive privilege for the sport of kings and nobles [189]. As noted by Cummings [893], in most Western European countries it was the largest animal hunted: it was noble in appearance, and venison was a staple in royal or aristocratic larders. Throughout Europe, and even beyond, a considerable amount of evidence exists to show that, since late medieval times, deer were regularly transferred from one habitat into another, normally for hunting purposes and to create game parks [28, 830, 839, 894, 896–897].

6.42 Roe deer, *Capreolus capreolus* Linnaeus, 1758

Greek name: ζαρκάδι (*zarkadi*) Turkish name: *karaca*

The roe deer, *Capreolus capreolus* Linnaeus, 1758, is the most abundant and widespread deer throughout Europe [882], and possibly also Asia Minor (compare with [93]) (▶Fig. 112).

Fossil and/or subfossil osteological remains of this species are quite infrequent in the Mediterranean islands (compare with [898]). In contrast, they are not so uncommon in the archaeological deposits of the Ionian and Aegean islands, having been discovered in Upper Pleistocene contexts on Corfu [103–104], in Bronze Age levels on Thasos [316] and Euboea [887]. It is noteworthy that all these archaeological sites are located on "continental" islands, not distant from the nearest mainland coasts of the southern Balkan peninsula and/or Anatolia.

Fig. 112: Thanks to its artificial introduction, which occurred in northern Euboea in the course of the 1970s, the roe deer may be today regarded as a full member of the Aegean islands' fauna.

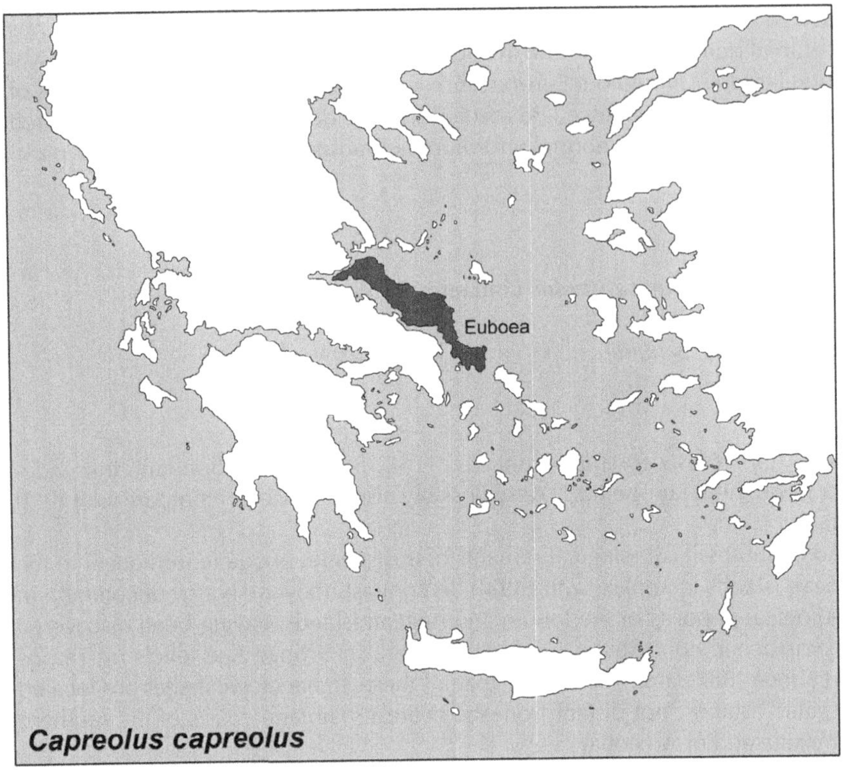

Fig. 113: Distribution of the roe deer, *Capreolus capreolus* Linnaeus, 1758, in the Aegean islands is limited to Euboea.

According to Sfougaris and Chaparis [899], the species is distributed in isolated areas of the southern Balkan peninsula: it disappeared from the Peloponnese (see also [647]), where it has been recently reintroduced in Kalavryta Sanctuary, as well as in in the area of Mount Telethrio, in northern Euboea. Thanks to the latter artificial introduction, which occurred in the course of the 1970s, today the roe deer may be regarded as a full member of the Aegean islands fauna. According to Athanassis Sfougaris (*in litteris*) and Haralambos Alivizatos (*in verbis*), this population still existed at the beginning of the 2000s, although it was declining.

Systematic data

Despite the fact that Danilkin [882] regards the European roe deer as a single, presumably monotypic species, the nominal form, *Capreolus capreolus capreolus* Linnaeus, 1758 seems to be the supspecies occurring in most of the southern Balkan peninsula (compare with [269]), whereas *Capreolus capreolus armenius* Blacker, 1916, is reputed to be dispersed in the Turkish Thrace, Marmara, Black Sea Mountains and Taurus Mountains, and perhaps also eastern Anatolia, whereas *Capreolus capreolus whittalli* Barclay, 1936, has been described on the basis of specimens from a type locality near Alemdagh, 15 miles from Moda, Istanbul [93]. *Capreolus capreolus coxi* Cheesman and

Hinton, 1923, seems to have diffused into southern Turkey, at the border with northern Syria (compare with [900–901]).

References:

Euboea [899].

6.43 Wild goat, Bezoar goat, or Asiatic pasang *Capra aegagrus* Erxleben, 1777

Greek name: αιγαγρός (*egagros*)
Greek vernacular names: αγρίμι (*agrimi*, Antimilos, Crete)
κρι-κρί (*kri-kri*, Crete)
αγριοκάτσικο (*agriokatsiko*)

Turkish name: *yaban keçisi*
Turkish vernacular name: *kayeek*

The wild goat, Bezoar goat, or pasang, *Capra aegagrus* Erxleben, 1777, ranges through inner south-western Asia, from western Sind, in the western Indian subcontinent, Afghanistan and south-western Pakistan to south-western Anatolia Asia Minor and northern Arabia, throughout the eastern Caucasus, Baluchistan, southern Turkmenistan, and Iran [270, 866, 902–905]. It has been introduced in the Czech Republic in the years 1953–1967 [906]. In the Near East, the altitudinal range of the species extends from sea level to the alpine zone. Wild goats are capable of surviving in very arid regions where the vegetation is sparse; their main requirement appears to be precipitous crags, where they can live undisturbed.

It used to be thought that the populations which have been present on several of the Mediterranean islands from time immemmorial, were autochthonous forms (▶Fig. 114).

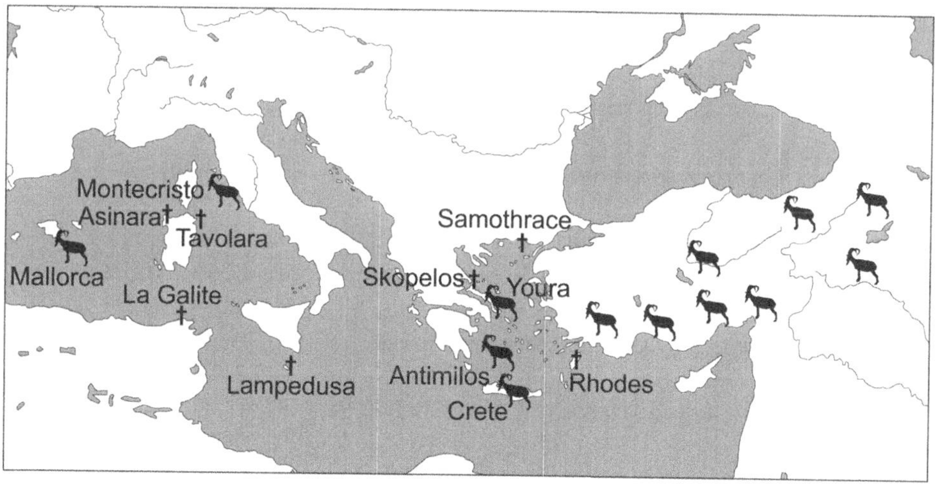

Fig. 114: Former and present distribution of *Capra aegagrus* Erxleben, 1777, in the Mediterranean region. Islands and areas that are inhabited by wild goats are marked with the 'goat' symbol and those where they have become extinct are marked with a cross.

However, in view of the lack of fossil evidence for these ungulates in mainland Europe or on the islands, they are now considered the result of human importations carried out in ancient or more recent times [691, 763, 907–911]. The wild goat is in fact completely absent from the original late Pleistocene fauna of southern Europe [912–913], its original range being documented only from the north-eastern Mediterranean mainland, where wild goat populations still survive in the south-western and eastern Anatolian mountains (compare with [914]). Consequently, all the historical and extant insular populations definitely had a Near Eastern continental anthropochorous origin, being the result of artificial introductions, carried out by man since prehistory [5, 23, 914]. Also according to Kostopoulos [915], *Capra aegagrus* appears in the southern Balkan peninsula only in the course of the Holocene, and today it is the sole living representative of its genus.

At present, ancient anthropochorous populations of *Capra aegagrus* are still present on the Aegean islands of Antimilos, or Erimomilos ("desert Milos"), located five miles off the north-eastern coast of Milos, Youra, in the Northern Sporades, and Crete [45, 70, 401, 912, 914, 916]. According to Hablützel [917], the one remaining "original" habitat where the ungulates survive on Crete is in the White Mountains in the western part of the island. Again possibly introduced in Neolithic times, the population of Samothrace unfortunately became extinct before the end of the 1980s [43, 914, 918]. Reports on this population can also be found in Sfougaris and Lymberakis [45], de Heldreich [404], Couturier [903, 919], Schultze-Westrum [912] and Masseti [904]. See ▶Tab. 2 for a summary of the present occurrence of *Capra aegagrus*.

Tab. 2: Present occurrence of the wild goat, *Capra aegagrus* Erxleben, 1777, on the Ionian and Aegean islands.

Island	Wild Goat Species	Notes	References
Samothrace	*Capra aegagrus pictus*	Extinct	[42, 904, 912, 921, 923]
Youra	*Capra aegagrus dorcas*	Introduced in ancient times	[43, 912, 916, 920] and the present work
Euboea	*Capra aegagrus dorcas*	Introduced during the 20th century	[70, 912, 924]
Atalandi	*Capra aegagrus dorcas*	Introduced 1984–1985	[42–43, 446]
Moni	*Capra aegagrus cretica*	Introduced 1961 and 1983	[42–43]
Psili	*Capra aegagrus dorcas*	Introduced during the 20th century	[70, 912, 924]
Sapientza	*Capra aegagrus cretica*	Introduced 1983	[42–43, 922]
Antimilos	*Capra aegagrus pictus*	Possibly introduced in Neolithic times	[42–43, 904, 912, 916, 921] and the present work
Crete	*Capra aegagrus cretica*	Introduced in early Neolithic times	[42–43, 418, 904, 912, 916, 923] and the present work
Theodorou	*Capra aegagrus cretica*	Introduced 1928–1946	[42–43, 70, 912, 924] and the present work
Dhia	*Capra aegagrus cretica*	Introduced 1957 and now possibly extinct	[42, 912] and the present work
Aghii Pantes	*Capra aegagrus cretica*	Introduced 1951	[42–43, 912] and the present work

In the course of the 20th century, the Cretan wild goat was introduced onto the islands of Theodorou (1928–1946), Aghii Pantes (1951), Dhia (1957), Sapientza (1983), and Moni (1961 and 1983) [34–35, 40, 42–43, 70, 918]. It would appear that only a few animals still survive in the remote and most inaccessible areas of Dhia, as the result of a programme of eradication launched by the Greek government in 1994 (▶Fig. 115). This was because the animals were incorrectly reputed to be not "pure blooded" mainly due to the fact that some individual revealed a melanic colour of the coat or a divergence of the horn tips. Cretan wild goats were also introduced for hunting purposes in 1983 on the Ionian island of Sapientza, off the south-western coast of the Peloponnese [42–45, 914, 918, 922], in 1984–1985 on Atalandi, in the Gulf of Euboea [42–43, 70, 446, 912, 918] and in 1961 and 1983 on Moni, in the Saronic Gulf [42–43, 918].

The occurrence of populations of wild goats in very recent historical times has also been reported on the island of Skopelos, in the Northern Sporades [401, 925] and on Euboea [292]. An old record (May 1942) for the occurrence of *Capra aegagrus hircus* on the small islets of Dragonada (3.5 km^2), off the easternmost coast of Crete, is also available in the collection of the Naturhistorisches Museum of Vienna (NMW 18639–18641). Wild goats of *aegagrus* phenotypes were also reputed to occur in the mountainous areas of Rhodes, in very recent historical times [57–58, 60, 926–927]. In 1888, for example, M.P. Matschie presented to the Zoological Museum of Berlin a specimen from this island, in which the horns were more developed than those of the Cretan wild goat, but less than those of the nominal form, the Asia Minor Pasang [926]. As far back as 1943–1944, Tortonese [801] reported that on Rhodes there was a custom of keeping specimens of wild goat, imported from Anatolia, in conditions of captivity in private enclosures.

Fig. 115: In view of the lack of fossil evidence in mainland Europe or on the islands for wild goats, *Capra aegagrus* Erxleben, 1777, they are now considered the result of human importations carried out since prehistoric times. Island of Dhia. Photo by Marco Masseti.

Systematic data

The populations of wild goat reported from the islands of the Aegean Sea have been described as belonging to three different subspecies:

1. Cretan *agrimi* or Cretan wild goat, *Capra aegagrus cretica* Schinz, 1838 (▶Fig. 116). This is the subspecies described for Crete were it is reported to have occupied most of the mountainous regions as recently as 1850, and/or the beginning of the last century (compare with [276]). However, as a result of habitat loss and hunting, only relic populations remain in the rugged areas of the White Mountains (*Lefka Ori*), in western Crete;
2. Aegean *agrimi* or Aegean wild goat, *Capra aegagrus pictus* Erhard, 1858 (▶Fig. 117). This subspecies is still present on the islet of Antimilos, or Erimomilos ("desert Milos"), off the north-western coast of Milos but, according to Sfougaris [43, 918], is appears to have become extinct on Samothrace before the end of the 1980s. Erhard [401] was the first to distinguish this population as a different taxon, or at least as a distinctly characterised variety, which he described by the name of "*Aegoceros pictus*".
3. Youra wild goat, *Capra aegagrus dorcas* Reichenow, 1888. It inhabits the islet of Youra, in the Northern Sporades. Lorenz-Liburnau [916] called it *Capra hircus dorcas*, since the name *Capra dorcas* had already been applied by Linnaeus (1758) to the dorcas gazelle, *Gazella dorcas* Linnaeus, 1758 (see also [928]) (▶Fig. 118).

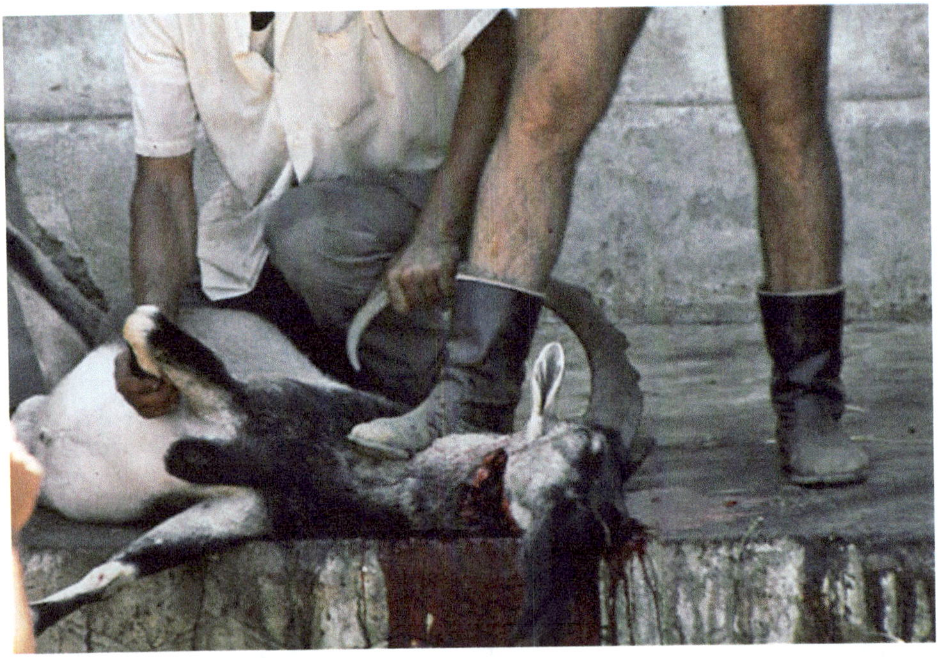

Fig. 116: It would appear that only a few Cretan wild goats still survive in the remote and most inaccessible areas of Dhia, as the result of a programme of eradication launched by the Greek government in 1994. Photo by Marco Masseti.

Fig. 117: Male of the Cretan wild goat or *agrimi*, *Capra aegagrus cretica* Schinz, 1838. Photo by Marco Masseti.

Fig. 118: Male and female of the Aegean *agrimi* or Aegean wild goat, *Capra aegagrus pictus* Erhard, 1858. Island of Antimilos. Photo by Nikos Vyrgiotis.

Schultze-Westrum [912] referred to this goat as *Capra aegagrus hircus* Linnaeus, 1758, indicating it as an animal of domestic origin, whereas other authors, such as Erhard [401], consider it to be a true wild goat identical to those of Crete.

From a morphological point of view, all these insular populations are closely connected with the true wild goat, the Asiatic Bezoar goat, or pasang, featuring the characteristics expressed by what is called the *aegagrus* phenotype, or more precisely, the complex of phenotypes expressed by this species [914].

According to Sfougaris [42–43], Cretan wild goats still occurred on Atalandi, around the mid 1990s. Schultze-Westrum [912] and Ondrias [70] reported that several wild goats of Youra were introduced onto the islets of Atalandi, and Psili, off Nafplion. Today, according to Sfougaris [42–43], only Cretan *agrimi* occur on Atalandi, where they were introduced in 1984–85. Not many years ago, among the domestic goats occurring on both the islands of Alonissos (Northern Sporades) and Tilos (Dodecanese), it was possible to report some individuals characterised by the morpho-phenotype of the extant wild goat of Youra [60, 351, 914].

Museum specimens

Youra: SMF 55791–55796; ZFMK 54.125, 54.155; Antimilos: ZMUA 65 (this specimen, an adult male collected in 1877 by Girtanner, was said to come from "Milos", but might be from Antimilos), ZMUA 74 (this female too, again collected in 1877 by Girtanner, was said to come from "Milos" but might be from Antimilos); Crete: BMNH 5.2.2.36; NHMC 80.5.70.(1–12), NMW 518, 1967, 524/B 3218, 625/B 4386, 626/B 6062, 2066/B 3194, 1552/ST 348, 2068/B 6063, 2069/B 6061, 7068/B 5417, 18642; SMF 16048; ZMUA 63, 71 (collected by Girtanner in 1877), 75; ZSM 1981/0613; Dragonada NMW 18639–18641 (*Capra aegagrus hircus*); Rhodes NMW 7068/B5417 (*Capra aegagrus hircus*).

References

Sapientza [42–44, 918, 922, 929]; Atalandi [41–43, 70, 446, 912, 918, 930]; Moni [42–43, 918]; Psili [70, 912]; Antimilos [42–43, 70, 285, 404, 690, 903–904, 912, 916, 918–919, 921]; Mini, near Aegina [43]; Youra [42–43, 58, 285, 289, 404, 903, 912, 916, 918–920, 931–932]; Crete [8, 29, 42–44, 70, 276–277, 298, 300, 351, 401, 404, 418, 452, 647, 668, 690, 903–904, 908, 911–912, 914, 916, 918–919, 921, 925–926, 933–942]; Theodorou [33–36, 38–44, 285, 903, 912, 918–919, 940, 943]; Dhia [42–44, 452, 903, 912, 918–919]; Aghii Pantes [42–43, 452, 903, 912, 918–919]; Karpathos [410]; Rhodes [57–58, 60, 926–927]; Youra [42–43, 70, 285, 289, 404, 903–904, 911–912, 916, 918–920, 928, 931–932, 944]; Samothrace [42–43, 404, 903, 912, 918–919, 921].

Archaeozoological note

Wild goats are so characteristic of the zoogeography of south-western Asia as to be considered oriental biological and cultural markers [904, 914]. Reports from archaeological sites around the Mediterranean area, yielding the first documentation of the appearance of Near Eastern caprines, reveal that the artificial westward diffusion of the goat in the Mediterranean region seems to be divorced from the cultural context. The goat appears for the first time [904, 914] in the Pre-Pottery Neolithic period in the eastern Mediterranean (end of the 9th–8th millennium BC) and in the Aegean region (8th–7th millennium BC), but in the central and western Mediterranean it appears instead in Early Neolithic archaeological contexts (6th millennium BC), characterised by the production of impressed ware. Owing to its peculiar feeding habits, and the fact that the wild goat is among the species best adapted to the particular environmental conditions found on small and barren Mediterranean islets, it has been preferred for exploiting marginal territories that could not be cultivated and were generally unattractive in human economic terms. Since ancient times, wild goats were brought to the islands and released so that they could breed and provide, at any time, a store of fresh meat that would be readily available along the maritime routes of antiquity [312, 914].

The oldest literary reference for the occurrence of wild goats on one of the Mediterranean islands is Homer's description of the isle of goats 'perpetually empty of men', where Ulysses and his companions landed to hunt before the Polyphemus episode (*The Odyssey*, IX, 116–124; see [945]). Homerian criticism dates the definitive version of the two epic poems – *The Iliad* and *The Odyssey* – to the 7th century BC, but the first occurrence of wild goats on Mediterranean islands is much more ancient. Wild goats made their first appearance on the Mediterranean islands sometime after the end of the 9th millennium BC. The first evidence for this introduction comes from Cyprus, where the earliest osteological remains of the species were provided by the archaeological exploration of the prehistoric site of Shillourokampos, in the south-western part of the island, referred to as the Pre-Pottery Neolithic period [118–119]. Archaeological evidence of the introduction of *Capra aegagrus* on the Aegean islands, such as Crete and Youra (Northern Sporades), also exists, dating respectively to the Pre-Pottery Neolithic and the Mesolithic periods (compare with [418, 923, 946–948]). According to Melas [410], wild goats may have also existed in Karpathos in early times.

Both Crete and Youra are among the few Mediterranean islands still inhabited by populations of wild goats, where the occurrence of the ungulates has been reported since prehistoric times. The archaeological exploration of the Mesolithic levels of the Cave of Cyclops, on Youra, yielded a number of interesting caprine remains [8, 351, 420, 949] (►Fig. 119). More specifically, the date of 7360–7350 BP (Cal BP 8360–8160)

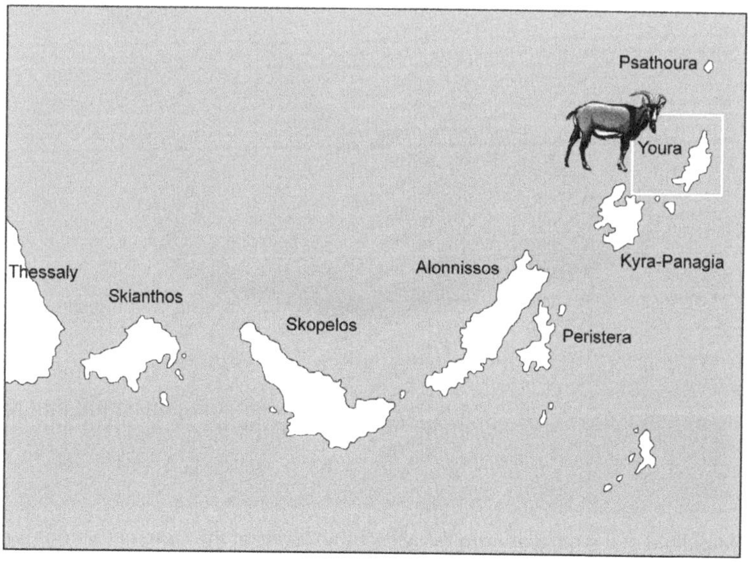

Fig. 119: The island of Youra, in the Northern Sporades, is still inhabited by a population of the namesake wild goat, *Capra aegagrus dorcas* Reichenow, 1888.

was obtained by the Beta Analytic laboratory of Miami (USA), through C-14 dating of goat bones from the earlier Mesolithic levels of Youra [8, 351, 420]. Wild goats, which still display more or less the same phenotypical patterns as *Capra aegagrus aegagrus*, were introduced onto Crete by man in early Neolithic times and generated free-ranging populations that persisted into Minoan periods and beyond [6, 811, 904, 914] (▶Fig. 120). Osteological material recorded from the archaeological site of Festos, however, documents the occurrence on Crete of two distinct forms of goat since the Terminal Neolithic period (about 3000 BC) [692]. One of these is identical to the present Cretan wild goat or *agrimi*, *Capra aegarus cretica* Schinz, 1838, and was often represented pursued by dogs through the rocky landscapes of the island. The other is characterised by the morphological features of a domestic race [6]. Morphologic and genetic analyses of the wild goats of Crete indicate that it very probably experienced some cultural control in prehistoric times before it colonised the mountains of Crete [44, 914, 950].

The maintenance up to the present of the relatively unchanged phenotypical patterns of *Capra aegagrus* in the wild goats of Crete, Samothrace and Antimilos led to the assumption that these populations may have been introduced onto the islands in Neolithic times [914]. For example, as regards the now uninhabited small mass of trachyte of Antimilos, where wrought obsidian has been found, the discovery of several arrowheads further attest to its human occupation even since prehistoric times

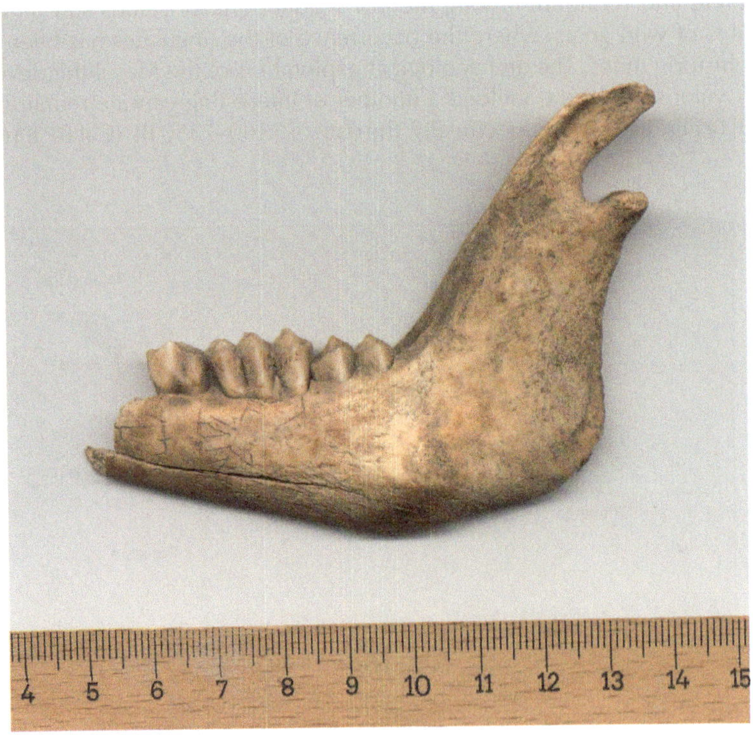

Fig. 120: Mandible of a wild goat from the Mesolithic levels of the Cave of Cyclops, on Youra. Photo by Marco Masseti.

Fig. 121: Libation vase (*rhyton*) (circa 1650–1500 BC) from the Minoan palace at Zakro (eastern Crete). Reliefs show a peak sanctuary on the roof, of which a group of four wild goats is resting, probably ruminating. Photo by Marco Masseti, courtesy of the Archaeological Museum of Herakleion, Crete.

(8000–4000 BC) (Zaphiros Vaos, *in verbis*): the period to which the first importation of the wild goats could date. Nevertheless, the phenotype expressed by the extant Youran goats does not appear to correspond to that of the animals of ancient introduction dating to Mesolithic times. An examination of the morphological patterns displayed by the Mesolithic goats of Youra, compared to the extant specimens, reveals how much they differ in the development of their cranial appendices. The goats of the Upper Mesolithic period reveal small, scimitar-shaped horns, much smaller than those of the extant Cretan agrimi, twisted horns being found only from Neolithic chronologies onwards [415, 420]. Thus, the animals which occur today on Youra are more similar to the Neolithic goats than to the caprines present on the island in Upper Mesolithic times. In this regard it could be interesting to make a comparison between the genetic material of the Mesolithic period goats of Youra and that of the current *Capra aegagrus dorcas* still present on the island [914].

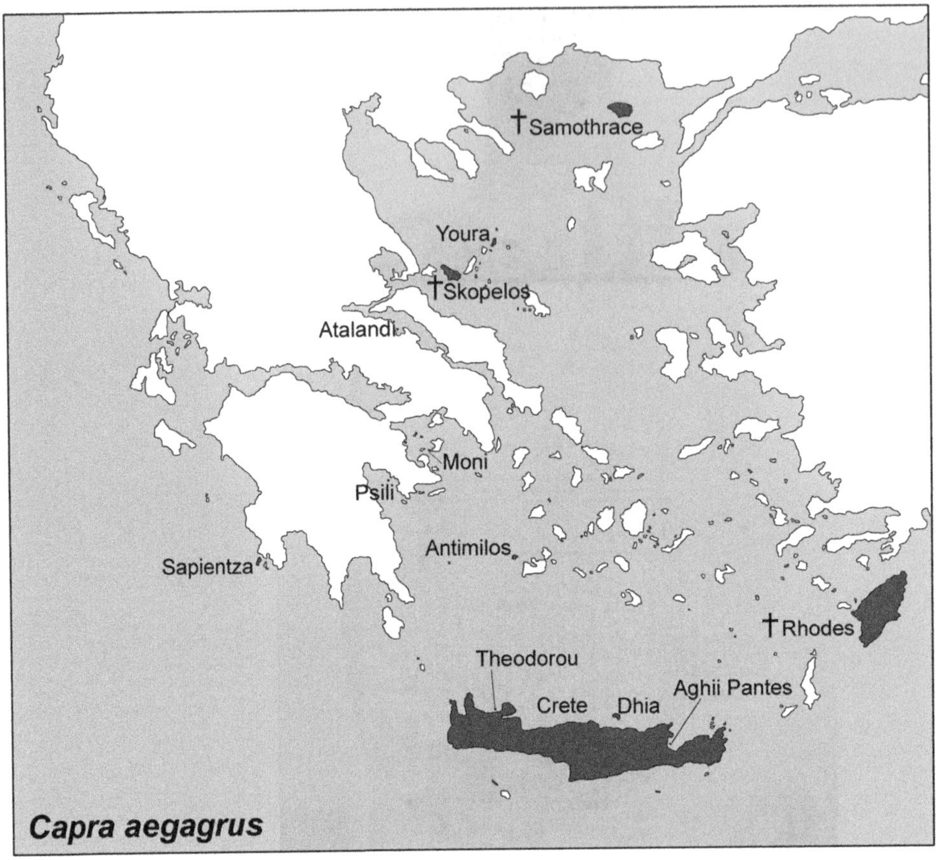

Fig. 122: Former and present distribution of the wild goat, Bezoar goat or Asiatic pasang, *Capra aegagrus* Erxleben, 1777, in the Aegean archipelagos. Islands that are inhabited by the goat are shaded and those where they have become extinct are marked with a cross.

6.44 Asiatic mouflon or wild sheep, *Ovis orientalis* Gmelin, 1774

Greek names: μούφλον (*mouflon*)
αγριοπρόβατο (*agrioprovato* = wild sheep)
Cypriot vernacular name: αγρίνον (*agrinon*)

Turkish name: *yaban koyunu*
Turkish vernacular name: *kotch*

The natural distribution of the genus *Ovis* Linnaeus, 1758, ranges from the eastern mountainous chains of inner Asia westerwards to Anatolia [866, 951–952]. In fact, the mountain range of central Asia is home to wild sheep today. From there they extended westwards into Anatolia and eastwards into America [953] but, as in the case of the wild goat, the Asiatic mouflon or wild sheep, *Ovis orientalis* Gmelin, 1774, is not native to continental and/or insular southern Europe [904, 908] (see ▶Fig. 123). Its natural distribution spans the majority of the Near East, with the exception of most of

Fig. 123: Female of Tyrrhenian mouflon, *Ovis orientalis musimon* Pallas, 1811. Photo by Marco Masseti.

the Arabian peninsula, to southern Iran [763] (compare with [270]). According to van Haaften [954], only one population of the species survived in the whole of Asia Minor, in the Konya-Bozdağ reserve, in the first half of the 1970s. More recently, Kryštufek and Vohralík [93] report the occurrence of *Ovis orientalis* in central Anatolia, Taurus Mountains and eastern Anatolia. According to Mursaloglu [955] and Kaya et al. [952], the main Anatolian population is today confined to the Konya region, in the Bozdağ protected area; rarely a few animals have been observed in the Taurus Mountains and high grassland around Van.

Fossils of *Ovis orientalis* are completely unknown from the Upper Pleistocene levels of the Mediterranean islands, where its extant presence in Corsica, Sardinia and Cyprus can be interpreted as a result of human-derived importations since prehistory [23, 351, 904]. Introduced populations of this ungulate now occur also in several other Mediterranean islands, where they are managed as game species (compare with [770, 956]). From the first half of the 18th century, mouflons from Corsica and Sardinia have been brought to continental and insular Europe [9, 23, 904, 951, 957–960]. In the Greek territory, these ungulates are present on the islands of Atalandi (180 ha), in the northern Euboean Gulf [43], and Sapientza (880 ha), off the south-western Peloponnese (compare with [922]), where they have been imported for hunting purposes in the second half of the 20th century.

Systematic data

The classification of the wild sheep of Anatolia is still uncertain. According to Kryštufek and Vohralík [93], for example, the subspecies of wild sheep that occurs in easternmost

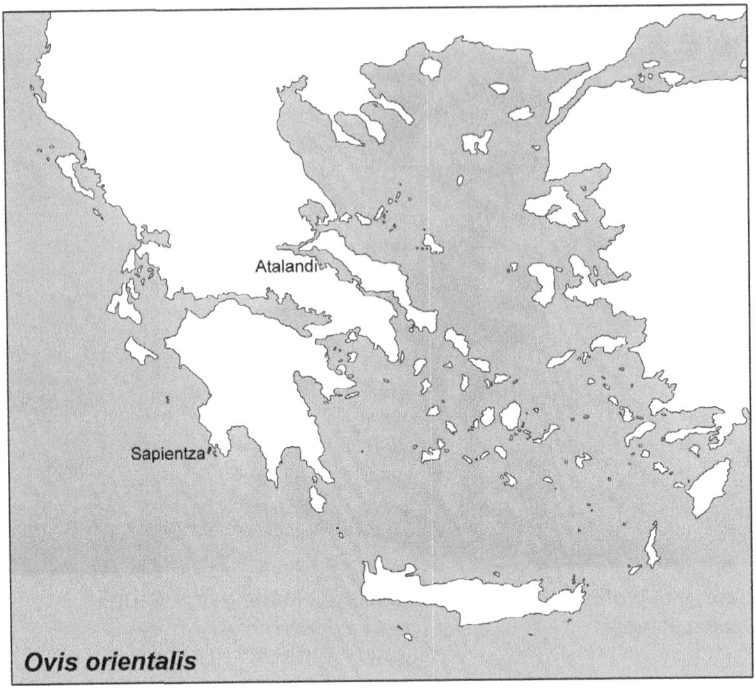

Fig. 124: Distribution of the Asiatic mouflon or wild sheep, *Ovis orientalis* Gmelin, 1774, in the Ionian and Aegean islands.

Anatolia (Ağri Daği, near Doğubayazit), has been classified as *Ovis orientalis armeniana* Nasonov, 1919, whereas the so-called Konya wild sheep was described by Valenciennes in 1856 [796] as *Ovis orientalis anatolica*. Phenotypically quite similar to this latter sheep, apart from the smaller size, the Cypriot mouflon, which occurs only on the island of Cyprus has been described as *Ovis orientalis ophion* Blyth, 1841, while the Tyrrhenian mouflon, which is still dispersed on Corsica and Sardinia, is regarded as *Ovis orientalis musimon* Pallas, 1811. Apart from their geographic distribution, the main morphological differences between the two representatives are in the shape and development of the male horns. Both are twisted: those of *Ovis orientalis musimon* are homogymous with divergent tips, while those of *Ovis orientalis ophion* are heteronymous with convergent tips (compare with [954]). The Greek island of Atalandi is today inhabited by individuals of the Tyrrhenian wild sheep [446]. Mouflons of the same subspecies are also present on Sapientza.

References

Sapientza [43, 922, 929]; Atalandi [43, 446].

7 Species dubiously and/or erroneously recorded from the Ionian and Aegean islands

As observed by Razzetti and Sindaco [961], ruling out the presence of a species in a specific geographical area is a more complex operation than demonstrating its presence. In fact, not finding a species does not automatically prove that it is not present and, while the scientific debate is undoubtedly useful, it is unfortunately not always exempt from personal aspects which are extraneous to matters scientific [962]. Nevertheless, erroneous evaluations or rather inattentive reading of the publications of early authors and travellers of the past have sustained cultural models which are still difficult to eradicate, e.g., the consideration of the diffusion of various mammalian species in the Ionian and Aegean islands (compare with [7, 654]). Several European natural history museums conserve material collected on these archipelagos that pose problems in the attempt to arrive at their origins [654]. In other cases, specimens represent species which are completely unknown on the islands they are reported to originate from [654]. Detailed below are reported the species of alleged and/or erroneous insular occurrence.

7.1 Southern white-breasted hedgehog, *Erinaceus concolor* Martin, 1838

Greek name: ακανθόχοιρος (*akanthohiros*)
Greek vernacular name: σκανζόχιρος (*skanzohiros*)
Turkish names: *kirpi, doğu avrupa kirpisi*

The southern white-breasted hedgehog, *Erinaceus concolor* Martin, 1838, occurs in most of the Near East, from Asia Minor to Israel, Syria, Lebanon, Iraq and Iran, including the southern Caucasus [268, 270], with the exception of extreme desert, very cold and tropical areas [269]. According to Kryštufek and Vohralík [93], it is widespread in Turkey, but of marginal occurrence in south-eastern Anatolia. It was formerly regarded as the species occurring throughout most of Eurasia and widely dispersed in south-eastern Europe, west to about 14°00′E, north to 60°00′N in Russia, east to about 80°00′E (River Ob) and south to the Levant and Iran [268, 270, 274]. *Erinaceus concolor* prefers lowlands and hills up to 300–800 m above sea level, where it occurs in urban and suburban habitats and agricultural lands [274]. It was also regarded as dispersed on the islands off the western coasts of Anatolia, such as Gökçeada, Lesbos, Chios, Kos, and Rhodes [93, 269]. Today, in all these territories, the representative of the genus *Erinaceus* Linnaeus, 1758, is however reputed to be the northern white-breasted hedgehog, *Erinaceus roumanicus* Barrett-Hamilton, 1900 [268] (see the entry on *Erinaceus roumanicus*, Section 6.1).

7.2 Blind mole, *Talpa caeca* Savi, 1822

Greek name: τυφλασπαλακας (*tiflaspalakas*) Turkish name: *Akdeniz köstebeği*

According to Lindermayer [292], the blind mole *Talpa caeca* Savi, 1822, was dispersed on Euboea. However this 19th century report strikes a false note, since the species is limited in its south-eastern European distributional range to the continental Balkan peninsula [478]. No moles have ever been reported from the eastern Mediterranean islands, with the only exception – as already noted – of the *Talpa stankovici* on the Ionian islands of Corfu [49, 372, 379] and Cephalonia ([282–283]; C. Stamatopoulos, *in verbis*).

Remains of a mole taxonomically referred to as the common mole, *Talpa europaea* Linnaeus, 1758, have been provided by the archaeological exploration of Konispol Cave, on the Albanian coast opposite the Greek island of Corfu [332].

7.3 Red squirrel, *Sciurus vulgaris* Linnaeus, 1758

Greek name: σκιόυρος (*skiouros*) Turkish name: *Avrupa sincab*

The red squirrel, *Sciurus vulgaris* Linnaeus, 1758, is a medium-sized arboreal squirrel, essentially characterised by a Palaearctic distribution, ranging from the British Isles in the west, south to the Mediterranean and the Caucasus – where it has been introduced – the southern Urals and the Altai mountains in central Mongolia, to north-east China [963]. Red squirrels are also found on the island of Sakhalin off the eastern coast of Russia, and on the most northerly Japanese island of Hokkaido. The natural occurrence of the species in Mediterranean insular environments is practically unknown, and their presence on some of these islands, such as Veli Brijuni (Croatia) ([470]; personal observation), is essentially regarded as a consequence of recent human intervention.

Werner, as far back as 1928 [290], quoted the occurrence of a kind of squirrel on the island of Skyros, in the archipelago of the Northern Sporades (north-western Aegean sea, Greece), where he collected a specimen between the villages of Skyros and Linaria which he described as "*Sciurus lilaeus*". According to Ellerman and Morrison-Scott [373], this taxon is used to define a Greek subspecies of the red squirrel, *Sciurus vulgaris lilaeus* Miller, 1907, characteristic of the region of Mount Parnassus in continental Greece. Nevertheless, the occurrence of the same species on Skyros was subsequently also recorded by other authors such as von Wettstein [58], followed in very recent times by Cheylan [304]. On the basis of the authority of Werner [290], and to an even greater extent that of von Wettstein [58], it is very difficult to refute the truth of these reports, even if red squirrels are today completely unknown on Skyros and the other islands of the Aegean and Ionian basin. According to Marinos and Symeonidis [104], for example, red squirrels never reached and/or colonised the Aegean islands. At the same time, however, there is no evidence to exclude the former diffusion of red squirrels on Skyros, where a population of the species could have existed up to the first half of the 20th century, later becoming extinct. Red squirrels could have been imported by man onto the island from Euboea, where their presence was reported by Lindermayer [292] in the first half of the 19th century. It cannot be excluded that red squirrels occur today on the latter island, although there is no recent or reliable evidence (A. Sfougaris, *in litteris*). In the light of modern ethnozoological enquiry, it would also appear that red squirrels figure among those species of mammals which have been the subject of particular human attention for a variety of cultural purposes (see the entry on *Sciurus anomalus*, Section 6.11).

7.4 Common pine vole, *Microtus subterraneus* De Sélys-Longchamps, 1836

Greek name: σκαπτοπόντικος (*skaptopontikos*) Turkish name: *küçük oyucu fare*

The systematic position of some *Microtus* populations in the Balkan peninsula is still not clear [71, 77, 356, 964] and specimens of this area previously ascribed to *Microtus majori* Thomas, 1906, are today recognised as belonging to *Microtus subterraneus* De Sélys-Longchamps, 1836 [485].

The distribution of the common vole ranges from western Europe to western Anatolia and Ukraine. It is a highly tolerant species as regards habitat, occurring from dry or marshy meadows in the lowlands to alpine pastures [485]. In Macedonia, Vohralík and Sofianidou [356] collected the species in the mountain only, in altitudes between 1 250 and 1 550 metres, where its typical habitats were mostly represented by small stone walls in either neglected or abandoned gardens, together with meadows and clearings with a thick layer of humus covered with a high and dense herbaceous vegetation. The subspecies *Microtus subterraneus dacius* Miller, 1908 was accepted by Ondrias [71] as occurring in Bulgaria.

Although Cheylan [304] reported the common pine vole from Euboea, Niethammer [965] and Kryštufek [485] mentioned it as absent from the entire Mediterranean coast and islands (see also [9, 478]).

7.5 Forest dormouse, *Dryomis nitedula* Pallas, 1778

Greek name: δενδρομυωξός (*dendromyoxos*) Turkish names: *ağaç faresi, karagöz*

The forest dormouse, *Dryomis nitedula* Pallas, 1778, occurs in a wide range of habitats, including deciduous and coniferous woodlands, Mediterranean evergreen shrubs, and alpine meadows, from the sea coast up to 2 300 metres above sea level [270, 495, 966]. This species is distributed from eastern Switzerland and north-eastern Italy, across eastern Europe and Asia Minor to the east as far as Mongolia and the Tien Shan mountains in China [71, 270, 967]. There are some isolated populations in southern Italy (Basilicata and Calabria) [710, 968–970]. Ondrias [71] reported the occurrence of the subspecies *Dryomis nitedula wingei* Nehring, 1902, from the Peloponnese and through the Greek mainland to the south-eastern Balkan peninsula and probably Bulgaria, while *Dryomis nitedula phrygius* Thomas, 1907, is widespread in Anatolia, and possibly further south along the eastern Mediterranean coast [94]. Fossil remains of the species are known within the late Middle Pleistocene fauna from the island of Chios, a few miles off the western Anatolian coast [106, 204].

Although Ondrias [71] and Cheylan [304] include this species among the non-flying terrestrial mammals of Euboea, Kryštufek [495] recorded it as not occurring on European islands. The occurrence of the rodent on the Cycladic island of Andros, recorded by Chondropoulos and Frageudakis-Tsolis (*in verbis*), is still to be confirmed [471, 478]. It was perhaps the observation of one specimen of *Glis glis* that led the latter authors to mistakenly infer that a population of forest dormouse occurred on Andros, where Dimaki [649], as already observed, recorded the edible dormouse for the first time on the island.

In considering the diffusion of several of the species of glirids in the Mediterranean region, and in particular on the islands, it is useful to cast further light upon the way in which the erroneous evaluation, or rather the inattentive reading, of the pages of authors of the past has given rise to cultural models which are still difficult to eradicate. Erhard [401] reported, for example, the diffusion of a species of glirid, similar in name to the forest dormouse – and which he called *Myoxus nitela* Schreber, 1782 – on Andros, Naxos and Siphnos, where it occurred in orchards and orange groves; it was in fact believed that it fed on citrus fruit. This report gave rise to the belief in the diffusion of this rodent on these islands, but in reality the taxonomic classification does not correspond to that of the forest dormouse. In fact, in Ellerman and Morrison-Scott [373], *Myoxus nitela* Schreber, 1782, is indicated as one of the synonyms of *Eliomys quercinus* Linnaeus, 1766, that is, of the garden dormouse, a species currently unknown in the Aegean area, but widespread, as already observed, in the central-western Mediterranean basin, where it is not found further east than Dalmatia and the north-western Balkan peninsula. Another report of the occurrence of a sort of dormouse, possibly the forest dormouse, is recorded by von Wettstein [58] from the island of Rhodes where, however, according to other authors, the species was, and still is, unknown (compare with [54–55, 57]).

7.6 Crested porcupine, *Hystrix cristata* Linnaeus, 1758

Turkish name: *oklu kirpi*

The European distribution of the crested porcupine, is limited to peninsular Italy, Sicily and the island of Elba [312, 749, 914, 971] (▶Fig. 125).

Despite what has been suggested by several authors [58, 269, 304, 972–974], the species has never occurred on the Ionian and Aegean islands, the Balkans or the Iberian Peninsula in the Holocene, even in the very recent past [749]. This erroneous suggestion has, however, become embedded in the literature to the point that even recently Gisbert et al. [975] and Woods and Kilpatrick [976] have suggested a European distribution of the crested porcupine that extends beyond Italy, to Albania and northern Greece. Von Wettstein [58] observed that local people referred to the presence of porcupines also on

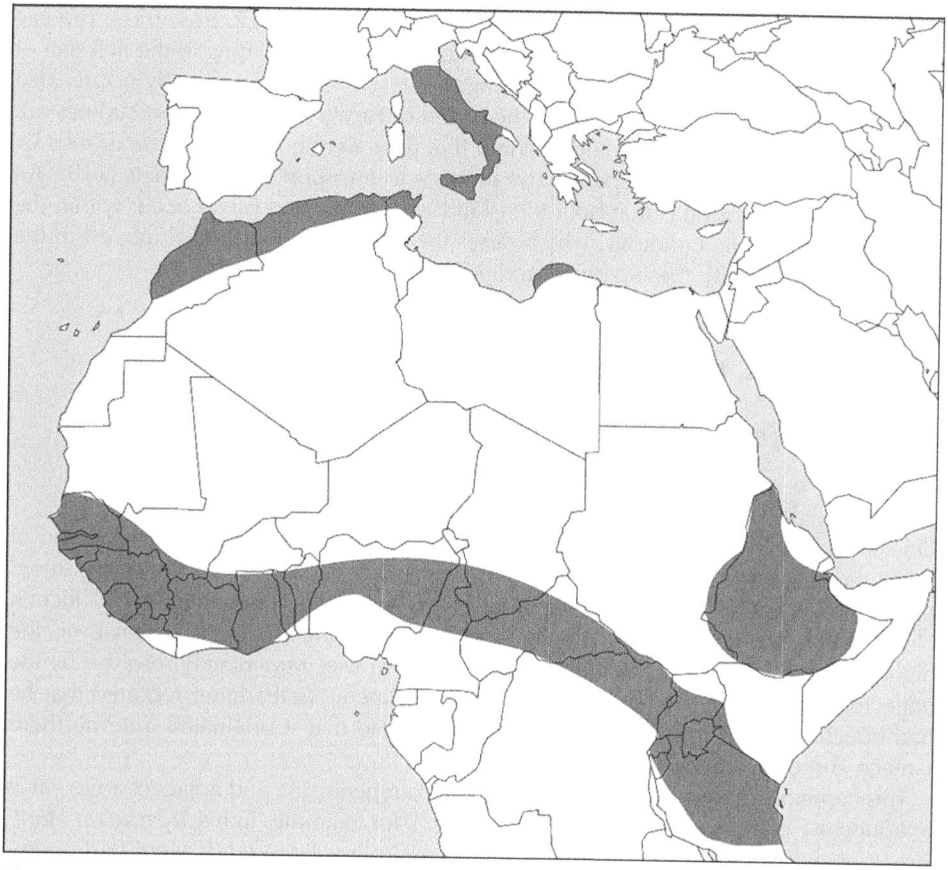

Fig. 125: Current distribution of the crested porcupine *Hystrix cristata* Linnaeus, 1758. From Masseti et al. [749].

the Eastern Aegean islands of Ikaria and Lesbos, but he added that this might have been the result of confusion with the hedgehog. The Greek term used to indicate this latter mammal is *skanzohiros*, which means 'spiny pig', which might easily have given rise to the confusion with the English 'porcupine' (the Italian *porcospino* and/or the French *porc-epic*). All these reports have given rise to several unproven beliefs which still persist. Cheylan [304], for example, still erroneously quotes the occurrence of *Hystrix cristata* on the Eastern Aegean islands of Lesbos, Ikaria and Rhodes, and Sarà [657] that of the same species on Samos and Rhodes. It seems, however, that the Indian crested porcupine, *Hystrix indica* Kerr, 1792, is found on the Turkish islet of Domuz, in the gulf of Fethiye, as part of a fauna almost entirely comprised of anthropochorous species [749]. The latter is the species of porcupine occurring in Asia Minor and the Middle East (compare with [976–977]), being distributed from Anatolia and Arabia, to Kashmir and Nepal, through the Indian subcontinent and Sri Lanka (compare with [270]).

Apart from the Indian porcupine of Domuz, the insular dispersion of a representative of the genus *Hystrix* in the Mediterranean basin is limited *to Hystrix cristata* which – as already mentioned – is found in Sicily and the island of Elba [312, 914, 971]. The last occurrence is the result of very recent introductions, probably dating to the first half of the 1980s [978]. It is also worth mentioning that *Hystrix cristata* currently occurs also off the north-western Libyan coast, on the island of Farwa [979]. Since porcupines are not competent swimmers, it is fairly certain that they were brought to these islands by people, which indicates a propensity by humans to transport these animals [749]. An additional consideration is that both the islands of Domuz and Farwa occur within the boundaries of Islamic countries, which raises the possibility of a special interest in the species by people with these cultural and religious backgrounds.

7.7 Eurasian lynx, *Lynx lynx* Linnaeus, 1758

Greek name: ρήσος λύγκας (*risos lygas*) Turkish name: *vaşak*

One specimen of Eurasian lynx, *Lynx lynx* Linnaeus, 1758, collected on the island of Corfu, is part of the collection of the Museum Alexander Koenig in Bonn, registered under the collection number ZFMK 93423. The specimen was purchased by Jochen Niethammer during the mammalogical exploration of the island [49]. However, the highly questionable occurrence of the lynx on Corfu was immediately resolved by the collector himself. On the label of the ZFMK specimen, Niethammer reported that he had bought it at the market, where he had been told that it originated from northern Greece, more specifically from Macedonia.

This carnivore is recorded in the southern Balkan peninsula and adjacent areas since antiquity (▶Fig. 126). Xenophon (5th century BC), for example, in his *Treatise on Hunting* (*Kinegeticon*: XI, 1–2), stated that lynxes and other large felids were fairly common in Asia Minor, beyond the border between the territories of Bithynia and Misia [769, 980]. The occurrence of Eurasian lynxes in Greece is attested by the capture of an adult female in Mount Parnitha onthe 18th of March, 1862. This taxidermically prepared specimen is still preserved in the collection of the Zoological Museum of the University of Athens with the registration number ZMUA 114. There are no official reports on the lynx in Greece, at least since the end of the Second World War [981]. Today, the species seems however to be restricted to the north-western areas and along the Albanian borders, where the felid is considered "critically endangered" and there are doubts as to whether it maintains a reproductive population [982]. According to Huş [766] and Aymerich [983], the Eurasian lynx is still relatively widespread within the Turkish borders, being recorded from several areas in northern, southern and eastern Anatolia. For the former diffusion of this felid see also Kumerloeve [765], Harrison [767] and Harrison and Bates [270]. It is probable that individuals from Turkey should be referred to as the North Caucasian lynx, *Lynx lynx dinniki* Satunin, 1915, which differs from *Lynx lynx lynx* Linnaeus, 1758, in its slightly larger size [270, 984].

Mediterranean Turkey still represents one of the last Near Eastern strongholds of another two medium-sized felids, the jungle cat, *Felis chaus* Gueldenstaedt, 1776, and

Fig. 126: The Eurasian lynx, *Lynx lynx* Linnaeus, 1758, has been recorded in the southern Balkan peninsula and adjacent areas since antiquity. Photo by Giuliano Cappelli.

the carcacal, *Caracal caracal* Gray, 1843 [765, 769]. Less rare than previously believed, the latter carnivore is recorded from the Aegean and eastern Anatolia [89, 91, 93, 829, 985–986]. Neither caracals nor jungle cats have ever been recorded on any of the Aegean islands.

8 Archaeozoological evidence of the diffusion of mammalian species of different and/or exotic origin or durable parts of them

...sailing the wine-dark sea to men of alien language...

(Homer, *The Odyssey*, I: 183)

The discovery in insular archaeological contexts of mammalian species extraneous to the native Quaternary faunas is not surprising.

In the course of the 2nd millennium BC, the evidence of trade between Eastern Africa and the Eastern Mediterranean basin is attested by the importation through Egypt of Ethiopian faunal elements. In fact, excavations of Minoan sites, both on Crete and other islands, have often yielded exotic cultural and biological elements. Archaeological evidence suggests that they were possibly imported from the Greek mainland, North Africa, or the Near East [811]. Plants and animals, together with artistic ideas and fashions, were spread by trade and especially by the exchange of costly gifts between rulers [987]. The excavations of the Minoan palace of Zakros, on Crete, yielded copper ingots from Cyprus and elephant tusks, possibly imported from Syria [988]. The discovery of exotic species or durable parts of them attests to the importation into the Aegean Bronze Age world of exotic species and raw materials that were precious even for the countries of origin. All these materials must be considered luxuries which the flourishing Cretan palace economy could afford [811]. On the other hand, the keeping of strange living animals by the elite was a widespread phenomenon in contemporary Egypt and the Near East [941, 989].

From the so-called "House of the Frescoes", west of the palace of Knossos (Crete), comes a remarkable series of paintings showing blue monkeys and blue birds in a rocky landscape among streams and waterfalls with luxuriant vegetation (Late Minoan 1A, about 1550 BC, *Herakleion Museum*). Blue monkeys are a recurrent motif in the production of Minoan artists. Their painted images have also been discovered in Theran frescoes (Late Minoan IA, about 1550 BC *Athens, National Museum*) (►Fig. 127), and in the paintings from the Pillar Crypt area on Milos [990–991] (►Fig. 128). The illustrations of these monkeys feature the unmistakable morphological patterns of the green monkey or grivet, *Chlolocebus aethiops* Linnaeus, 1758 (►Fig. 129) [941, 991–992]. The representatives of this taxonomic group are distributed in the African savannahs and steppes south of the Sahara from about 18°–15°N [993–996]. Together with the Barbary macaque, *Macaca sylvanus* Linnaeus, 1758, and the baboon, *Papio* sp., this was possibly the species of monkey best known in the ancient Mediterranean world. Grivets have been kept as pets since time immemorial and numerous exemplars have been exported out of Africa since antiquity. Vague accounts of an East African green monkey have circulated since the days of the ancient Greeks, who adopted the

Fig. 127: "Blue monkeys" are a recurrent motif in the production of the Aegean Late Bronze Age artists. Detail of the "blue monkeys wall-painting" from Akrotiri, Thera (Santorini).

Fig. 128: Archaeological location of the Late Bronze Age Aegean sites that provided painted images of "blue monkeys". On the upper left, the stone rashly claimed by Poulianos [1231] as the fallacious discovery of a *Chlorocebus* monkey skull, was apparently found on the Aegean island of Santorini (Thera) in the summer of 1966. From Masseti and Bruner [991].

Fig. 129: Although the grivet, *Chlorocebus aethiops* Linnaeus, 1758, is not (and perhaps never has been) a *taxon* naturally dispersed within the boundaries of the Western Palaearctic, there exists considerable archaeological evidence to show that it regularly occurred (even, perhaps, as a naturalised species) in the Aegean region in the course of the 2nd millennium BC. Olduvai Gorge, Tanzania. Photo by Marco Masseti.

term *callithrix* ("beautiful hair") [997–998], originally introduced by Homer [993]. *Callitriches* were also mentioned by Pliny the Elder who observed that: "*Hoc animal negatur vivere in alio quam Aethiopiae quo gignitur caelo*" (= "It's said that this animal can survive only in the climate of Ethiopia, where it comes from") (*Naturalis historia*, VIII, 216). There exists considerable archaeological evidence to show that it regularly occurred – even, perhaps, as a naturalised species – in the Aegean region (eastern Mediterranean) in the course of the 2nd millennium BC. As far as is presently known, *Chlorocebus aethiops* was imported there from Egypt as a precious gift [811, 941, 991]. The value of the monkeys may have been related more to their curiosity appeal than to their effective economic worth, or possibly to both (compare with [239]). These animals were deemed so exotic and important by the Aegean Bronze Age culture that they were not thought of simply as animals, but as creatures invested with the role of intermediary between the human and divine worlds [989, 999–1000].

On the walls of Building Beta (room Beta 1) at Akrotiri (Thera), six "*white antelopes*" (▶Fig. 130) and the human figures of two boxing boys were discovered. The ungulates

Fig. 130: Detail of the wall-painting of Building Beta (room Beta I) at Aktoriri (Santorini), showing two "white antelopes". Their morphology has been referred to that of East-African gazelles, such as the Grant's gazelle, *Nanger granti* (Brook, 1872), or perhaps the Soemmering's gazelle, *Nanger soemmeringi* (Cretzschmar, 1826). Both of these species show long, lyre-shaped horns and dark markings in the coat colour.

have been identified as oryxes by Marinatos [1001] and Sapouna-Sakellaraki [1002] or gazelles by Masseti [992, 1003] and Säflund [1004]. In a recent revision of the Theran frescoes, Porter [1005] suggested interpreting them as a liberal elaboration of the morphology of the Cretan *agrimi*, on the basis of arguments which are not wholly convincing. Masseti [992, 1003, 1006] referred their morphology to that of East African gazelles, such as the Grant's gazelle, *Nanger granti* (Brook, 1872), or perhaps the Soemmering's gazelle, *Nanger soemmeringi* (Cretzschmar, 1826) (▸Fig. 131). Both of these species feature long, lyre-shaped horns and dark markings in the coat colour. They belong to the taxonomic group of the large-sized gazelles, whose distribution spans most of East Africa [994, 1007]. The *"white antelopes"* of Akrotiri are portrayed in antagonistic attitudes. In fact, as indicated by their long horns, both are bucks [1006]. Effectively, they are not depicted fighting but in a phase of intimidation, known by the specialists as "lateral or broadside display" [193, 1008] (▸Fig. 132). The manner in which their fluid movements have been so well captured suggests that the artist who portrayed these East African ungulates was familiar with them by sight and not just by reputation [1006]. The practice of breeding exotic mammals as hunting game or for cult purposes is also documented since at least early protohistorical times. In Minoan Crete, the local *agrimi* was regarded as an authentic wild animal [811, 904, 941], whereas a special type of bull was bred to be employed in the ritual bull sports [478, 1009].

Fig. 131: Adult male of Soemmering's gazelle, *Nanger soemmeringi* Cretzschmar, 1826 at Dahlak Kebir, Eritrea. Photo by Giuseppe De Marchi.

Fig. 132: Intimidation display of Grant's gazelle, *Nanger granti* Brooke, 1872, showing a combination of lateral display and facing away. From Walter [1010].

The importation into the Ionian and the Aegean world not only of species of different and exotic origin, but also of durable parts of them, such as ostrich egg shells, hippopotamus ivory and/or the aforementioned elephant tusks, was merely the repetition of a practice which had been going on for centuries – if not for millennia – whenever political and economic conditions were favourable. This tradition had found its oldest roots in the trade of faunal and botanical elements that took place between North Africa, the Near East, the eastern Mediterranean basin and the Middle East since prehistorical times (compare with [239, 1006, 1011]). Fox osteological fragments and teeth, for example, made their first appearance in the Mediterranean insular archaeological contexts in the early Neolithic period [5, 23, 310, 565, 691]. More specifically, Vigne [310] observes that, in most of the Mediterranean area, *Vulpes vulpes* was the subject of particular human attention during the Ancient Neolithic period, when the canid could have been important both symbolically and as food (compare with [1012]). This cultural significance attributed to foxes by early Neolithic man may appear more comprehensible when compared to the attitude held generally by prehistoric peoples towards natural sources. In the archaeological contexts of the Aegean islands, it is still possible find remains of exotic or dangerous carnivores, such as brown bears, *Ursus arctos* Linnaeus, 1758, and lions, *Panthera leo* Linnaeus, 1758.

8.1 Brown bear, *Ursus arctos* Linnaeus, 1758

Greek name: καστανή αρκούδα (*kastani arkouda*) Turkish names: *ayi, boz ayi*

Populations of brown bear, *Ursus arctos* Linnaeus, 1758, have today withdrawn to remote areas of the southern Balkan peninsula, in northern Greece, but there are reports of their presence in the southern parts of continental Greece as late as the 19th century [340, 1013] (►Fig. 133). The extant range of the brown bear in Greece consists of two geographically distinct nuclei, located in the Pindos mountain range (NW Greece) and the Rodopi mountain complex (NE Greece) [1014]. According to Turan [871], the distribution of the species in continental Turkey ranges from the area of Marmara and the Black Sea mountains to the Taurus chain and eastern Anatolia.

In prehistoric times their range extended further south, but the genus *Ursus* was never a member of the Cretan fauna and it is somewhat unlikely that it inhabited the larger islands of the Aegean, at least not for a long time [695]. Its presence in the Final Neolithic/Chalcolithic period of Chios [696] and in the Bronze Age of Samos [1015] is based on the evidence of single specimens, which very probably represent imported finds

Fig. 133: Tamed bear in the town of Korçe, south-eastern Albania. Photo by Luigi Forte.

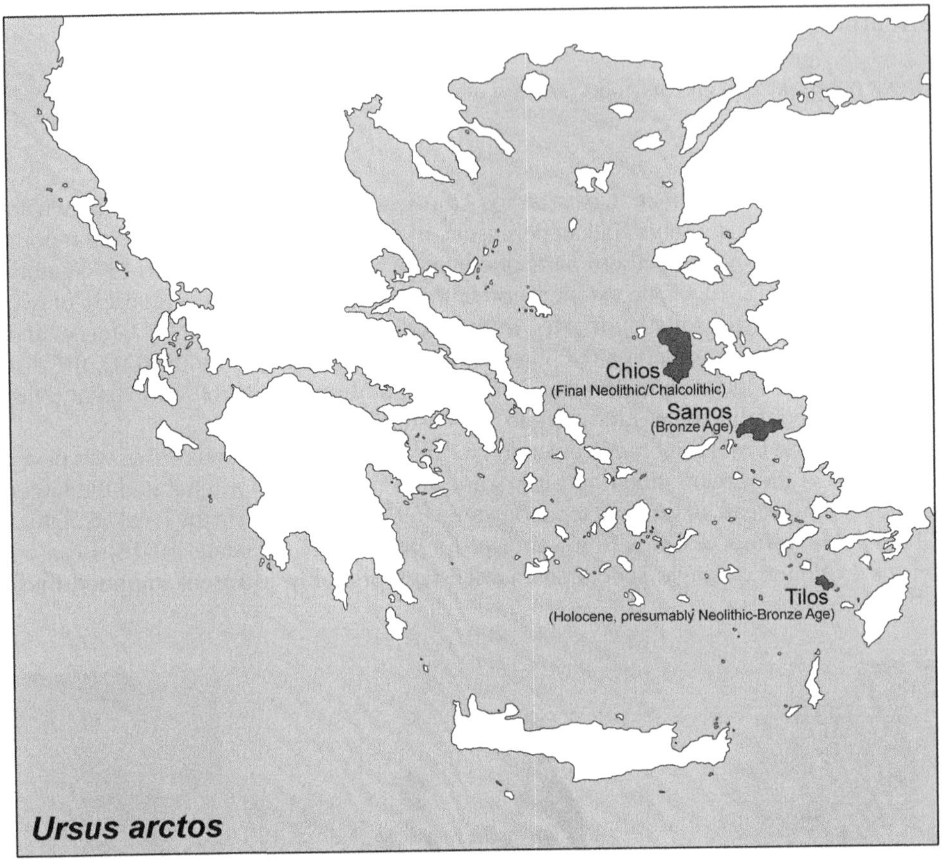

Fig. 134: Geographical location of the insular archaeological sites of the Aegean that provided remains of brown bears.

(▶Fig. 134). In particular, according to Clutton-Brock [696], the example from Emporio (Chios), a slightly polished radius, may also suggest a talismanic function of the find. Together with dwarf elephants, the Holocene stratigraphy of the Charkadio cave, on the island of Tilos (Dodecanese), also yielded remains of *Ursus arctos*, which have been assumed to be a hunter trophy imported from the nearest mainland [101, 106, 150].

In the course of the 1980s one specimen, at least, of brown bear was kept in a free ranging state on the small island of Tersane in the Gulf of Fethiye (Turkey) by its owner, the king of the gipsies of the islands, to prevent the visit of unwanted guests. According to Kasparek and Kasparek [795], a small population of the species still survives not far from there, in the higher parts of Mount Latmos (Beşparmak Dağlari) on the northern shore of the lake of Bafa.

Systematic data

At present, the taxonomic situation of this polymorphic species is still not fully determined. Seven to eleven subspecies are recognised with large differences in body

Fig. 135: The Syrian bear, *Ursus arctos syriacus* Hemprich and Ehrenberg, 1828, is distinguished from the European brown bear, *Ursus arctos arctos* Linnaeus, 1758, by its average paler pelage colour, generally a uniform yellowish or greyish white. Photo by Paolo Cavicchio.

size and fur colour [1016]. Specimens from southern Turkey are referred to as the subspecies *Ursus arctos syriacus* Hemprich and Ehrenberg, 1828, which is distinguished from the European brown bear, *Ursus arctos arctos* Linnaeus, 1758, by its average paler pelage colour, generally a uniform yellowish or greyish white (▶Fig. 135). According to Harrison [767] and Harrison and Bates [270], it seems clear that this subspecies extends to Asia Minor, Iraq, Transcaucasia and northern Persia.

References

Tilos [101, 106, 150]; Samos [1015]; Chios [696].

8.2 Lion *Panthera leo* Linnaeus, 1758

Greek name: λεοντάρι, λιοντάρι (*leondari, liondari*) Turkish name: *aslan*

Africa is the homeland of lions, *Panthera leo* Linnaeus, 1758 [1017], where they are dispersed from the sub-Saharan Sahel to the Cape region (compare with [697]). In recent historical times, this species was also distributed in the south-western Palaearctic from western India to almost all the Near East, from Persia and Asia Minor to North Africa [1018]. Until 1942 the Barbary lion, *Panthera leo leo* Linnaeus, 1758, lived on the western Atlas, in Morocco [1019–1020], while the Asian or Indian lions, *Panthera leo persica* Meyer, 1826, survived in Iran up to the second half of the 20th century. A group of these big cats also survived in the gallery forests along the Euphrates until at least the 12th century AD [90, 1021]. Remnant populations were reported from the region of Fethiye, southern Turkey [1022], the southern bank of the river Esen (Koca Çay, south-western Anatolia) [795, 809] and the Upper Euphrates valley [765, 769, 789, 1023]. Kasparek [809] reported that a small population of lions certainly survived in southern Turkey at least until the mid 19th century.

Scholars from the 8th century BC to the 2nd century AD, such as Homer (*The Iliad* and *The Odyssey*), Herodotus (*Histories*), Xenophon (*Treatise on Hunting*), Aristotle (*Historia animalium*), Pliny the Elder (*Naturalis historia*), Dion Chrysostom, Pausanias (*Description of Greece*), and Helianus (*On the characteristics of animals*), warned about the treacherous presence of lions in eastern Europe as well as in the southern Balkan regions between the 8th and 1st centuries BC [695, 980, 1024–1025]. In *The Iliad,* for example, Homer (circa 8th century BC) uses over two dozen similes in which lions attack domestic animals and man may be described attempting to drive the lion away [980]. These similes have been traditionally interpreted as being based on Homer's personal observation, presumably in Ionia [1026]. Three hundred years later Herodotus (5th century BC) described an attack by lions on the camels of King Xerxes' caravan, in the course of the Persian wars (*Histories*, VII, 125) [769]. He also noted: "*The boundary of the lions' country is the river Nessos that flows through Abdera and the river Achelous that flows through Acarnania. Neither to the east of the Nessos anywhere in the nearer part of Europe, nor to the west of the Achelous in the rest of the mainland, is any lion to be seen; but they are found in the country between those rivers*" (*Histories,* VII: 125–126). The river Nessos (230 kilometres) rises in the Rila Mountains (Bulgaria) and flows into the Aegean Sea. It forms today's boundary between Greek Macedonia and Thrace [980]. Nearly a century later, Xenophon, in his *Treatise on Hunting* or *Kinegeticon* (XI: 1–2), observed that lions were fairly common in Asia Minor, beyond the border between the territories of Bithynia and Misia [769]. He also wrote that "*lions ... are caught in foreign countries in the neighbourhood of Mount Pangaeum...*" (*Treatise on Hunting,* XI: 1). Mount Pangaeum is located near Thessaloniki. The next author to mention lions some 50 years later was Aristotle (384–322 BC), who already referred to the lion as: "*a scarce animal ... not found in many places. In the whole of Europe it occurs only in the tract of country between the rivers Achelous and Nessos*" (*Historia animalium,* VI 579b). Aelian (2nd century AD) also reported the occurrence of lions "*... high up on Mount Pangaeus, in Thrace*" (*On the characteristics of animals,* III) (compare with [573]). The latest archaeozoological reports from Greece, i.e., one specimen from Kastanas [1027] and a claw from Delphi (identified by Poplin [1028] which, however, could be of Geometric or Mycenean date), date to the archaic period [695]. A lion's tooth of MMIII–LMIB date was identified at Aghia Irini on the small island of Kea (Keos); two more, from different animals, were found in House M at the same site in an LMIB–LHII context [416, 695, 980, 1029–1030] (▶Fig. 136). These teeth and a third metatarsal from Early

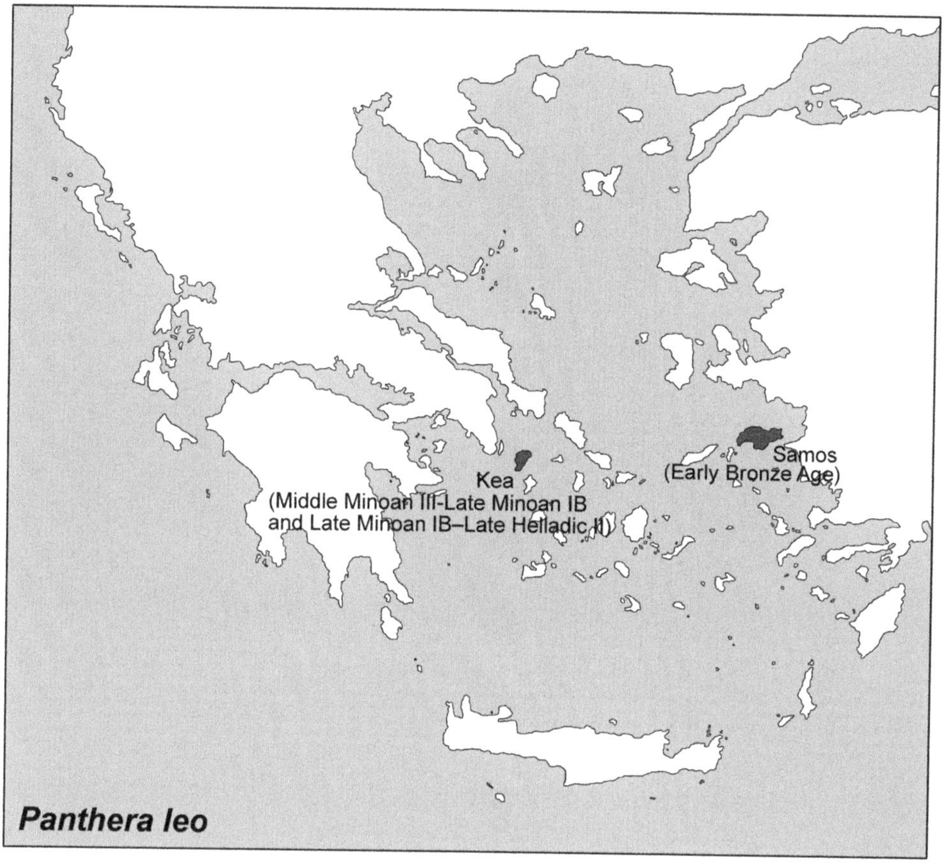

Fig. 136: The Aegean islands that provided ancient lion remains.

Bronze Age Samos [1015] may be regarded as imports, since the islands are unlikely to have sustained lions [416, 695]. Moreover, in the case of Aghia Irini, where the lion is represented only by loose teeth, these would have served as items of prestige or amulets [695].

When lions actually became extinct in the southern Balkan peninsula is still uncertain. It is generally agreed that this occurred in the second half of the 1st millennium BC, or at the beginning of the Christian era [695, 1031]. They are believed to have died out within the borders of present-day Greece around 80–100 AD [1032].

References

Kea [416, 695, 744, 980, 1030]; Samos [695, 1015].

9 Biogeographical remarks

Small hares are common in most of the islands, both deserted and inhabited; they are more numerous in the islands than on the mainland because here there are neither foxes, which attack and kill the hares and their young, nor eagles, birds of prey which prefer the higher peaks, while those of the islands are fairly low.

(Xenophon, *Kynegeticon*, 24–26)

44 extant species of non-volant mammals have been recorded so far from the Ionian and Aegean islands (▶Tab. 3), the sole endemic being the Cretan white-toothed shrew,

Tab. 3: Distribution types of the non-flying terrestrial mammals reported from the Ionian and Aegean islands.

Species	Endemic species	Palaearctic pluriregional species	Western European species	Eastern European, near Eastern and North African species	Human commensal species
Erinaceus roumanicus				•	
Sorex minutus		•			
Neomys anomalus			•		
Crocidura leucodon		•			
Crocidura suaveolens		•			
Crocidura zimmermanni	•				
Suncus etruscus		•			
Talpa stankovici				•	
Lepus europaeus		•			
Oryctolagus cuniculus			•		
Sciurus anomalus				•	
Cricetulus migratorius				•	
Microtus arvalis		•			
Microtus guentheri				•	
Microtus thomasi				•	
Chionomys nivalis		•			

(*Continued*)

(*Continued*)

Species	**Endemic species**	**Palaearctic pluriregional species**	**Western European species**	**Eastern European, near Eastern and North African species**	**Human commensal species**
Meriones tristrami				•	
Nannospalax leucodon				•	
Apodemus flavicollis		•			
Apodemus mystacinus				•	
Apodemus sylvaticus		•			
Apodemus witherbyi				•	
Rattus norvegicus					•
Rattus rattus					•
Mus musculus					•
Mus macedonicus				•	
Mus spicilegus				•	
Acomys cahirinus minous					•
Glis glis		•			
Muscardinus avellanarius		•			
Canis aureus				•	
Vulpes vulpes		•			
Mustela nivalis		•			
Martes foina		•			
Meles meles		•			
Lutra lutra		•			
Felis silvestris		•			
Panthera pardus				•	
Sus scrofa		•			
Dama dama				•	
Cervus elaphus		•			
Capreolus capreolus		•			
Capra aegagrus				•	
Ovis orientalis				•	
Total = 44	**1**	**20**	**2**	**17**	**4**

regarded as a true relic of the island Pleistocene faunas, which has survived up to now. It is also the rarest mammal reported from the area.

The number of mammals recorded on the Ionian islands is 22, with 40 recorded on the Aegean islands. Two species – *Talpa stankovici* and *Muscardinus avellanarius* – are present only in the Ionian islands, while no less than 42 taxa have been formerly reported from the Aegean islands. At present, mammals occurring on the Ionian archipelago include: 1 representative of the Family Erinaceidae; 2 of the Family *Soricidae*, 1 of the Family *Talpidae*, 2 lagomorphs, 9 rodents, 5 carnivores and 2 ungulates. Currently occurring mammals on the Aegean islands include: 6 insectivores, 2 lagomorphs, 19 rodents, 7 carnivores, and 4 ungulates. Rodents represent the most widespread taxa, occurring on almost all of the islands, with 20 species, including *Rattus rattus*, which is the most dispersed element being reported from 137 islands of the total considered in the present study, followed by *Mus musculus*, which has been found on 48 islands. Lagomorphs are represented only by hares and rabbits. *Oryctolagus cuniculus* has been reported from 88 islands. Erinaceids have been found on 54 islands, but with only one species, the northern white-breasted hedgehog, while the soricids are represented by 6 taxa (*Sorex minutus, Neomys anomalus, Crocidura leucodon, Crocidura suaveolens, Crocidura zimmermanni, Suncus etruscus*). Carnivores comprise 7 taxa (*Canis aureus, Vulpes vulpes, Mustela nivalis, Martes foina, Meles meles, Lutra lutra, Felis silvestris*). Ungulates occur at present on only a few islands with 5 species, including the Rhodian fallow deer and the Cretan wild goat.

Several of these species originated populations that presumably colonised the islands naturally. This may be the case, for example, of the extant mammal fauna of Euboea, the second largest Greek island with a total surface area of 3 658 km^2, the highest mountain being Mount Dirphys at 1 745 metres above sea level. The narrow Epirus Strait which separates Euboea from Boeotia, in mainland Greece, has always favoured the natural colonisation of the island by continental biological elements. However, Euboea is still smaller than the largest of the islands in the area in question, namely Crete. Despite this, it is the Aegean island with the largest number of mammalian species: 25, in all, including a population of roe deer introduced several decades ago. Euboea is also the only island of the Ionian and Aegean Seas from where the recent occurrence of red deer is reported. There are also some accounts on the presence of the wolf, *Canis lupus* Linnaeus, 1758, in the 19th century, although these need to be confirmed (A. Sfougaris, *in litteris*). Moreover, the presence on Euboea of the grey hamster and three microtines, such as *Microtus arvalis, Microtus thomasi* and *Chionomys nivalis*, is quite surprising since very few Mediterranean islands are known to be currently inhabited by voles [914]. Widespread in the continental areas and islands of more northerly Europe, microtines are considerably less diffuse in the Mediterranean insular ambits (compare with [88, 265, 269]), even though, for example, the biogeography of islands such as Corsica, Sardinia and Malta, was at length characterised by the presence of several endemic representatives of this taxonomic group during the late Pleistocene/early Holocene [310, 565, 1033–1037]. As already seen, the common vole is also known on Lesbos [64, 72], the only Mediterranean island from where the Guenther's vole, *Microtus guentheri*, has been reported [64, 72, 372]. A subfossil variety of the same species, characterised by very large size, has been documented among the Holocene fauna of the island of Chios (Eastern Aegean Sea) [67, 336]. Snow voles, have been found only in Euboea, where they have been reported from Mount Dirphys, at an altitude of 1 700 metres above sea level [70–71]. Cheylan [304] also reported from Euboea the occurrence of the common pine vole, *Microtus subterraneus* De Sélys-Longchamps, 1836, whereas Niethammer and Krapp [518] and Kryštufek [479] observe that this species is absent from the entire Mediterranean coast and islands.

Beyond those just mentioned, very few other Mediterranean islands are known to be presently inhabited by voles (▶Fig. 137). These are essentially Krk, in the Quarnero Gulf (Croatia), from where Tvrtković et al. [1038] dubiously record the presence of the bank vole, *Myodes glareolus* Schreber, 1780, and the common vole, and Elba from where the Savi's pine vole, *Microtus savii* De Sélys-Longchamps, 1838, was reported only once by Vesmanis and Hutterer [1039] and has never since been found in any of the subsequent studies conducted on the island [1040–1043]. The Guenther's vole is also dispersed in Cyrenaica (north-eastern Libya), which can be regarded as an ecological island,

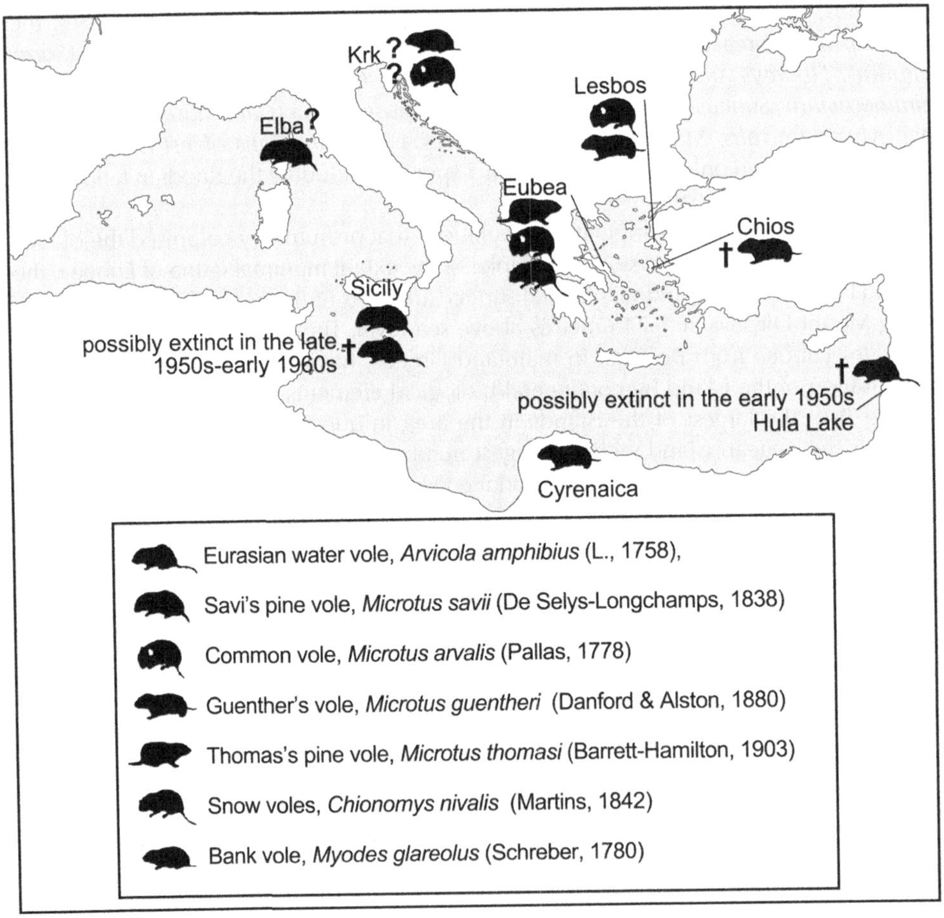

Fig. 137: Present distribution of the representatives of the taxonomic group of *Arvicolinae* on the Mediterranean islands. The occurrence of the Guenther's vole, *Microtus guentheri* Danford and Alston, 1880, in continental North Africa (Cyrenaica, Libya) and the location of the Hula Lake (Israel) are also indicated.

being the only territory of the entire African continent where a representative of the *Microtinae* family occurs [480]. The Eurasian water vole, *Arvicola amphibius* Linnaeus, 1758, appears to have survived in Sicily through the Neolithic chronologies [1044–1045] up to relatively recent times [1046]. The present paucity of voles in the Mediterranean islands may even be related to the particular xerothermic conditions of these territories [914]. In Israel, for instance, evidence of the occurrence of the Eurasian water vole was found only in barn owls' pellets on the banks of the lake of Hula before this was drained during the early 1950s [1047], and it would appear to have been extirpated as a consequence of this event [467, 1048]. On the Mediterranean islands several carnivores, such as the weasel, change their feeding ecology from specialised microtine hunter to murid hunter, due to the lack of voles, their favourite continental prey (compare with [707–708]). As far as is presently known, for example, representatives of the Microtinae subfamily have been reported from only three Mediterranean islands inhabited by *Mustela nivalis*, that is to say Euboea, Lesbos and Sicily [0009].

With a surface area more than twice that of Euboea and the highest summit at 2 453 m above sea level on Mount Pachnes, in the Lefka Ori, Crete is instead inhabited by 18 mammalian species, including the endemic shrew and the Cretan spiny mouse. The presence on the island of the otter and of wild populations of rabbit still needs to be verified. Apart from *Crocidura zimmermanni*, the current mammalian fauna of Crete – like that of many other Aegean islands – is the result of human-induced introduction carried out since prehistoric times ([1049]; compare with [1050]). Crete reached its present configuration at the end of the Pliocene or in the Early Pleistocene, when it could be colonised only by overseas sweepstake dispersal [184]. In any case the Upper Pleistocene mammalian fauna was already completely extinct at the Aceramic Neolithic level of Knossos, and replaced by newcomers that accompanied or were brought by the humans [226, 418, 923]. The new mammalian fauna included the ancestors of the current species, such as the beech marten, the badger, and the wild goat (▶Fig. 138). Moreover, the Holocene osteological remains of the extant mustelids are intermediate between the recent local varieties and their Near Eastern relatives, as pointed out by Steensma and Reese [733], Ragni et al. [715], and Masseti [914]. This indicates an eastern origin of the human settlers and the accompanying fauna, as previously suggested by Evans [1051].

9.1 Invasive alien species

From the large Aegean islands to the smaller ones, just a few hectares in size, the mammalian composition changes drastically, becoming strictly oligotypical and featuring the presence of one or only a very few species. What is particular striking is the spread of 4 mammals which with human aid managed to reach the islands farthest from the mainland and/or from other island complexes. These are the hedgehog, the rabbit, the black rat and the house mouse. As we have already observed, in the first two cases these were voluntary introductions, designed primarily to ensure a nutritional resource that would always be available on the more barren islands along the maritime routes since ancient times. By contrast, in the case of the rat and the mouse, the importations were carried out by man involuntarily, and are essentially connected with the ability to colonise new environments expressed by the biology and ecology of these two murids. It is interesting to

Fig. 138: Diorama of the Natural History Museum of Crete, with an environmental reconstruction of an aspect of the region of Sfakià, south-eastern Crete. We can admire some of the most characteristic mammals, including (from right to left) a brown hare (*lagos*), a domestic goat, a badger (*asvos*), and a stone marten (*zourida*). Photo by Apostolos Trichas, courtesy of the Natural History Museum of Crete, Herakleion.

note that on certain islands, even of not too reduced dimensions and of a carrying capacity that can sustain populations of wild goats, such as Antimilos (8 km^2) and Youra (11 km^2), or together with a stock of Tyrrhenian mouflons, Sapientza (9 km^2), the only other species of non-volant mammal reported is *Rattus rattus*. On Dhia (11.909 km^2), wild goats and black rats are present along with rabbits, while wild goats and rabbits are reported from Atalandi. Regardless of whether the islands were close to or distant from the continental coast, they were frequently represented essentially by the black rat and the rabbit, the two species that adapted most easily to the difficult environmental conditions of the Mediterranean islands. In this regard, Flux [431] observed that rabbits can tolerate temperatures up to 50°C provided they have access to a minimal supply of water or vegetation. The black rat's capacity for environmental adaptation is well known, as is its intelligence that has made the rat the most widespread mammalian species in the world (compare with [570, 574, 1052]). To mention only the islets around Crete, for example, *Rattus rattus* has been reported from Theodorou (0.68 km^2), Dhia, Paximada (0.3 km^2), Ghianissada (2.08 km^2), Dragonada (2.84 km^2), Gavdhos (29.585 km^2), and Koufonisi (4.15 km^2), whereas populations of *Oryctolugus cuniculus* are known from Theodorou, Dhia, Aghii Pantes (0.32 km^2), Gaidouronisi (Chrysi) (4.75 km^2), and Koufonisi. The islands permanently occupied by a human community can instead also be inhabited by a species strictly commensal to man, such as the house mouse (compare with [1053]). As already seen, the brown rat on the other hand requires damper habitat conditions than those tolerated by the rabbit and the aforementioned murids. Some of the islands farthest from the mainland can even be entirely devoid of mammalian fauna (compare with [227, 1054]).

On the other hand, the current presence of various continental species in several Ionian and Aegean islands located just off the mainland coast can be traced to a kind of natural colonisation. This may be the case, for example, of the Balkan mole on Corfu and Cephalonia and of the lesser mole rat on the islands of the north-eastern Aegean Sea. The insular distributional range of both these micromammals cannot be explained as the result of a human-induced introduction, since they are fossorial animals adapted to digging and life underground. As already noted above, at present *Nannospalax leucodon* is to be found on Euboea, Lesbos, Bozcaada, Lemnos, Gökçeada, and Samothrace. The occurrence of other continental mammals on islands off the Thracian and the western Anatolian coast, such as Thasos, Gökçeada, Samothrace, Bozcaada, Lesbos, Chios, Samos, and even Kos can again only be explained through a direct, natural colonisation at an unspecified moment of the Last Glacial episode, when the islands were connected to the nearest mainland (▶Fig. 139). The occurrence of the jackal on Samos, the Persian squirrel on Lesbos and Gökçeada, the common vole and the Guenther's vole on Lesbos, the Tristram's jird on Kos, the steppe field mouse on Bozkaada, and Kos, and the Balkan short-tailed mouse on Gökçeada, Samothrace, Bozkaada, Lesbos, Chios, and Samos can also be referred to a similar natural colonisation. It is a different story for Rhodes, since the island was linked to the nearest mainland only up to the end of the Middle Pleistocene, permitting the entry of the zoological varieties which, again in the course of the Middle and Upper Pleistocene, gave rise to faunas of a moderate endemic character, such as *Elephas (Palaeoloxodon)* sp. of medium size, *Sus* sp., *Hippopotamus* sp. and *Cervus* sp. [101].

One specimen of leopard is reputed to have been killed on Samos about a century ago and it has been taxonomically referred to *Panthera pardus tulliana*, the variety distributed in western Anatolia up to few decades ago. As noted above, it is believed that the felid passed to the island from the opposite coast of Turkey, swimming across the narrow channel, the *Dar Bogaz*, (no more than 1 650 metres long) that separates the island from the Samsundag range which, as already observed, up until the early 1970s was reputed to be the last Western Anatolian stronghold of the species (compare with [791, 793]). The island of Samos is characterised by very peculiar zoogeographical features linked to its particular geographical location. It is, for example, the only island of the Eastern Aegean still inhabited by a well developed population of jackals, *Canis aureus* Linnaeus, 1758 [672, 683], whereas the Mediterranean chameleon, *Chamaleo chamaleon* Linnaeus, 1758, is also reported from Chios [1055–1057]. Nevertheless, it would appear that the populations of chameleon in Samos are healthier and more numerous than in any other Greek area in which they occur [306]. Furthermore, field signs and records of the occurrence of wild boars have been reported from the island since the early 1990s. It is also said that this species reached Samos by swimming from the opposite Anatolian coast, where boars are still reported as the commonest ungulates of the Samsundag range [769]. Moreover, Samian hunters confirm that this type of wild game was not deliberately introduced onto the island.

9.2 Characters of the extant insular mammals: generalist colonisers of pre-deserts

The non-flying terrestrial mammals which today inhabit the Ionian and Aegean islands are mainly represented by generalist species, animals which are able to thrive in a wide variety of environmental conditions and can exploit a range of different resources.

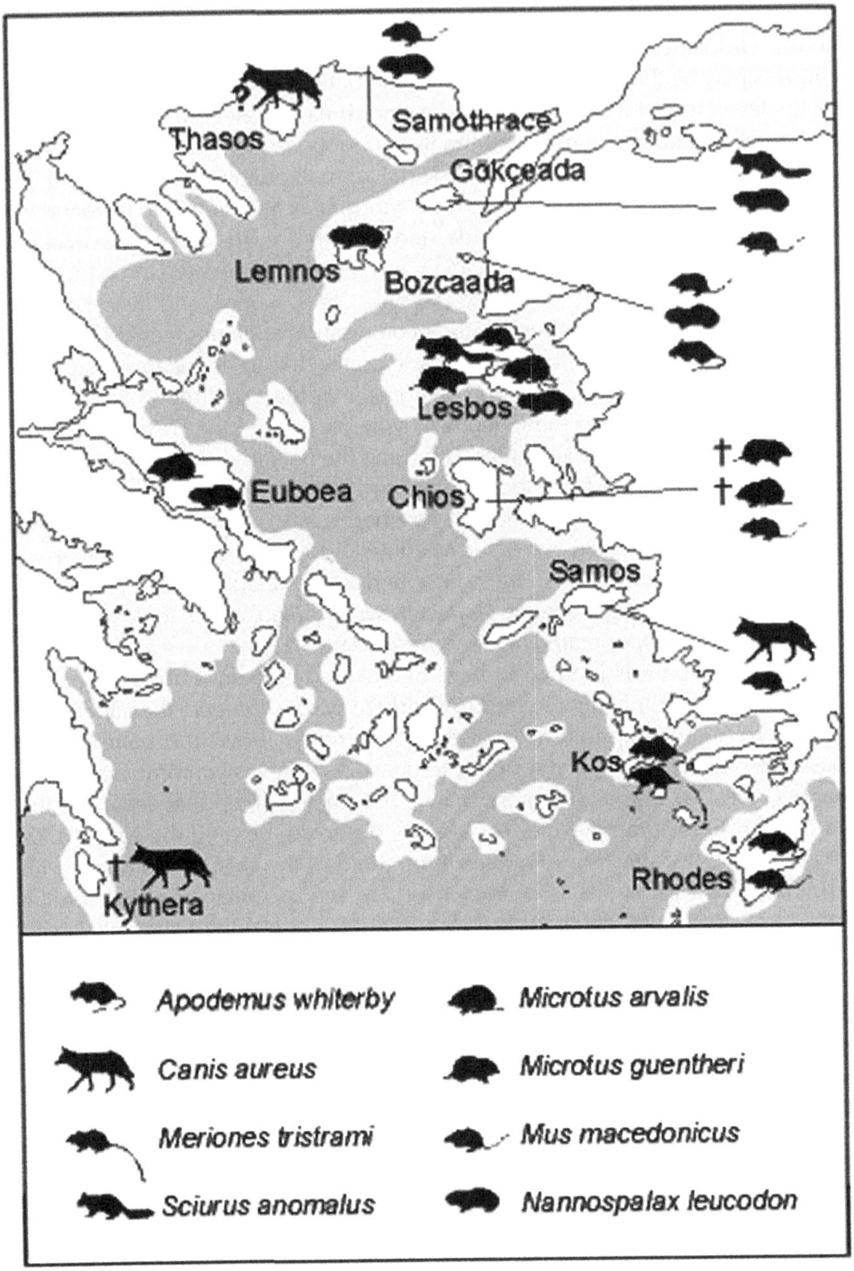

Fig. 139: The occurrence of several continental mammals on islands off the Thracian and the western Anatolian coast, such as Thasos, Gökçeada, Samothrace, Bozcaada, Gökçeada, Lesbos, Chios, Samos, and even Kos can only be explained through a direct, natural colonisation at an unspecified moment of the Last Glacial episode, when the islands were connected to the nearest mainland. It is a different story for Rhodes, since the island was linked to the nearest mainland only up to the end of the Middle Pleistocene. The pale grey area represents land lost during Holocene sea level rises.

When environmental conditions change, generalists are better able to adapt, while specialists tend to fall victim to local extinction much more easily [1058]. Since these are almost exclusively species of continental origin, they are also those that manage to react better to the conditioning imposed by human cultural control and even to the effects of domestication. It is scarcely surprising that ungulates such as wild boars, mouflons, and wild goats, and the red deer, which were imported as tamed and/or semi-domestic livestock onto the islands since Neolithic times, occasionally escaped from their guardians' control, giving origin to the free-ranging population, the descendants of which persist up to the present. Back in the wild, they maintained the morphological patterns of their Near Eastern ancestors. In fact, as observed by Ryder [953], the recognition of the origin of domestication is complicated by the fact that the first domestic animals did not differ greatly from their wild counterparts.

There are even phytophages of small and medium size that can survive in the most hostile of insular habitats, comparable in terms of both environmental and trophic characteristics to veritable pre-deserts. As we have already seen, these are wild goats, hares, and rabbits [8, 388]. Apart from Crete, all the extant populations of wild goats of the Mediterranean islands survive in territories characterised by very low trophic conditions, and often without any artificial food supply. And yet, if we consider some of the biogeographical parameters of the islands, such as area and altitude, as indirect indices of habitat heterogeneity (*iihh*) (▶Tab. 4), we note that wild goat populations can still prosper in small rocky islands with a mean altitude of less than 600 (571) metres above sea level, and an average surface area of less than 10 km^2 (9 km^2) [8, 914] (▶Fig. 140). Survival in such peculiar insular environments may also be rendered possible by the integration into the diet of mineral salts, which are assumed both by licking the deposits that form on the foreshore and by consuming the excreta of birds. For example, we can cite the case of the wild goats of Montecristo, in the northern Tyrrhenian sea (Italy), which occasionally feed on the excreta of yellow-legged gulls, *Larus michahellis michahellis* Naumann, 1840 [914].

The only analogous feeding habit reported to date is that by Hobbs [1059] of the Nubian ibex, *Capra ibex nubiana* F. Cuvier, 1825, in Egypt's Eastern Desert, which eats the excreta of another species of bird, the crowned sandgrouse, *Pterocles coronatus*

Tab. 4: Former and present occurrence of populations of *Capra aegagrus* Erxleben, 1777, on some of the Mediterranean islands without artificial food supply (iihh = indirect index of habitat heterogeneity).

Island	Subspecies	Altitude (metres)	Area (km^2)	iihh	Notes
Montecristo (Italy)	*Capra aegagrus hircus*	645	10.39	0.016	
Tavolara (Italy)	*Capra aegagrus*	555	6	0.010	Extinct
La Galite (Tunisia)	*Capra aegagrus*	400	10	0.025	Extinct
Samothrace (Greece)	*Capra aegagrus pictus*	1 840	178	0.096	Extinct
Youra (Greece)	*Capra aegagrus dorcas*	570	11	0.019	
Antimilos (Greece)	*Capra aegagrus pictus*	686	8	0.011	

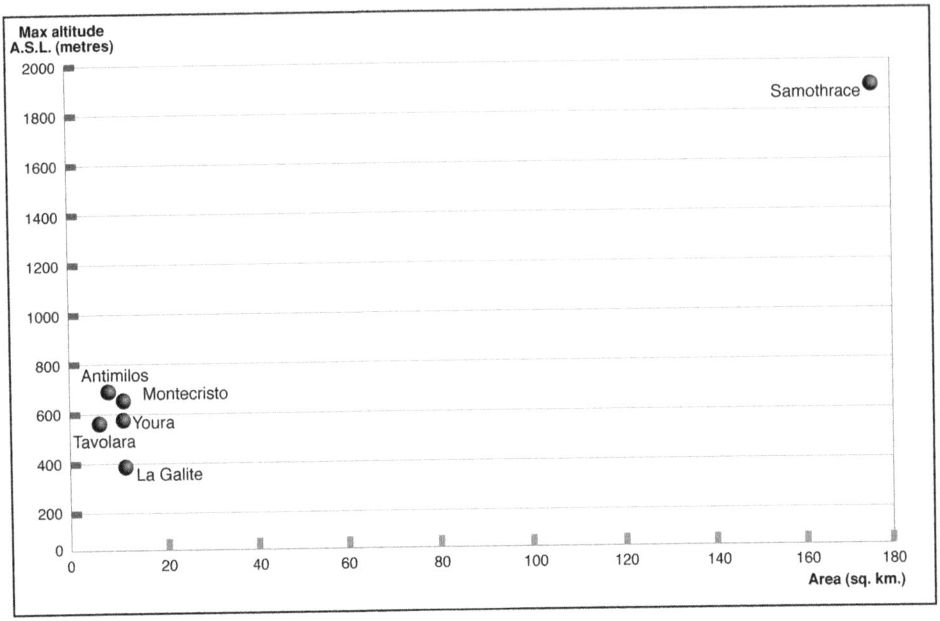

Fig. 140: Former and present occurrence of populations of wild goat, *Capra aegagrus* Erxleben, 1777, on Mediterranean islands <200 km^2, without artificial food supply.

Lichtenstein, 1823. In this way, the wild goats of the Mediterranean islands ingest mineral salts otherwise not available in their insular habitat.

Hares can survive on even smaller islands than wild goats, with an area of 3–3.5 km^2, as shown by the populations that still inhabit several of the islets around Crete ([8]; compare with [388]). And we have already seen that rabbits too can now be found on many Mediterranean islands of even much smaller dimensions [388]. Occasionally, wild boars also are able to survive on small islands very near to the mainland, such as the populations that frequent several of the the small Turkish islets of the Gulf of Gökova, the gulf of Marmaris, or the gulf of Fethiye, all located very close to the south-western coast of Anatolia [30]. The weasel and the stone marten figure among the extant carnivores most widespread on the Ionian and Aegean islands [688, 707]. This is probably due to the fact that they were imported into the insular environment to prevent the damage produced by unwanted rats and mice in an age when the domestic cat was still unknown along the shores of the Mediterranean basin, apart from Egypt [654, 787]. From this point of view, the full commensal behaviour performed by *Martes foina* within the human settlements of several Ionian and Aegean islands may appear more comprehensible (compare with [9, 60]). Both, the weasel and the stone marten can be regarded as generalist predators, since they are able to vary their diets according to local and seasonal food availability (compare with [708, 1060–1061]). They are carnivores that humans have accepted as neighbours, but are also among those most adept at avoiding contact with man [29, 688, 715].

10 A weighty legacy: concluding remarks

As pithily expressed by Xavier de Planhol [1062] in his book *Le paysage animal. L'homme et la grande faune: une zoogéographie historique*: *"... man, a superior predator, was destined to devastate and destroy. And he did."* However, expounding an analysis rich in quantitative clarifications, the author goes on to argue that this phase is by now over and behind us: *"For at least the last fifty years, humanity has shifted from the destruction stage to that of protection, even of resurrection, already recording brilliant successes, the number of which is bound to increase."* Nevertheless, the legacy of the global redefinition of the original ecological equilibrium of the Ionian and Aegean islands – like that of all the remaining Mediterranean islands – raises not inconsiderable problems of conservation and management. First of all, in the vast majority of cases, it is impossible to reconstruct the natural ecosystems of the past which have been irretrievably destroyed, lost thousands of years ago [29, 878]. In fact, if we pause to reflect on the general appearance of these islands in the Upper Pleistocene, or even only in the Early Holocene, we realise how incommensurably different it was from that of today. In the Late Quaternary, instead, when the existing botanical and faunal assemblages were already defined, the majority of the islands must still have been cloaked in the densely tangled vegetation that covered most of these environments before any human explorer set foot on them. Oak jungles consisted of downy oaks, *Quercus pubescens* Willd.; evergreen oaks, *Quercus ilex* Linnaeus, 1753; Kermes oaks, *Quercus coccifera* Linnaeus, 1753; and of endemic species, such as the golden oak of Cyprus, *Quercus alnifolia* Poech, 1842 (compare with [9, 128, 1063–1070]), not to mention the dense undergrowth (see ►Fig. 141). These practically impenetrable vegetable formations must have been home to an entire fauna that was particularly adapted to living, moving, feeding, resting and hunting in such an environment. Few other regions of our planet, have been maltreated by man more than the Mediterranean, to the extent that we can now have only the vaguest impression of what the ancient natural riches, vegetable lushness and primeval faunal luxuriance of the region may have been. The result is that much of the environment of the Mediterranean islands is by now reduced to little more than a mineral skeleton (►Fig. 142).

What now appears evident is that, starting from the Early Holocene, there began a human colonisation of the Mediterranean islands that entailed the massive introduction of continental fauna species, accompanied by the gradual disappearance of the endemic elements. This effectively took place in a diachronic and differentiated form in relation to the various species and the different insular complexes. The Neolithic settlement of the islands cannot be explained as the result of a merely casual maritime dispersal of the hunter-gatherers, but appears much more plausibly a movement of intentional and planned colonisation [1071]. The Neolithic settlement of islands such as Crete support this hypothesis. Here, the colonisation was accompanied by the transfer of the continental ecological appurtenances, also comprising domestic animals, the wild progenitors of which did not exist on the islands [6, 1049, 1072–1073].

Fig. 141: Dense jungles of evergreen oaks, such as the endemic golden oak, *Quercus alnifolia* Poech, 1842, of Cyprus, must have occurred on the Mediterranean islands before any explorer set foot on them. Photo by Marco Masseti.

Fig. 142: Much of the environment of the Mediterranean islands is by now reduced to little more than a mineral skeleton. Island of Tilos, Dodecanese (Greece). Photo by Marco Masseti.

Therefore, it is not simply a question of the transfer of breeding knowledge, but also of the physical transportation of the animals themselves. The human group and the related domestic animals that were moved by sea had to be sufficiently important and balanced to establish a vital population of men and beasts so that the agrarian way of life could be reproduced in a new territory. This ecological and cultural transplantation could not have been the result of casual maritime prospecting, but rather the outcome of an expedition, or a series of expeditions, planned and prepared with a specific objective: the colonisation of an island.

Previous authors classified many of the modern insular mammals, the majority of which could be regarded as free-ranging or even feral, as subspecific geographic forms, almost entirely on the basis of arbitrary criteria and the examination of scattered materials [878]. Based on the data given in literature, the various subspecies are distinguished by the coat patterns and by the size of body and skull. As is consequently understandable, this led to a multiplication of forms which now call for better taxonomic and genetic definition. Throughout most of the 19th and the 20th centuries, it was common practice among scientific explorers to bring home an excessive number of subspecies from their explorations of the Ionian and Aegean islands. This is the light in which we must consider the taxonomic treatment, on the part of the international scientific community, of the insular lagomorphs, mustelids or wild goats (compare with [127]).

As we have already noted, the majority of the mammalian populations described in this volume are the result of introductions performed by man since prehistoric times, and continued possibly without interruption throughout historical chronologies [6]. As a consequence of this, they can be defined as populations of anthropochorous origin. In view of the vulnerability of the insular ecosystems, however, it is critically important to prevent further introductions. From both a scientific and a cultural point of view, it would be advisable to take seriously into consideration the eradication of the alien species of very recent invasion. But this leaves the question of how to treat the allochthonous populations of certified ancient anthropochorous origin [29, 878, 914]. Each case has to be considered individually. Therefore, while the eradication of rats, rabbits, and feral goats imported in recent historical times is in many cases fundamental for the recovery of island ecosystems, several of the other anthropochorous populations represent images which remain fixed in time and history. For instance, this is true for the hares of the Dodecanese, the pine martens of the Ionian and Aegean archipelagos, the wildcats of Crete, the fallow deer of Rhodes, the wild goats of Antimilos, Crete, or even Youra, and many other extant insular mammalian populations (compare with [1074]) (►Fig. 143).

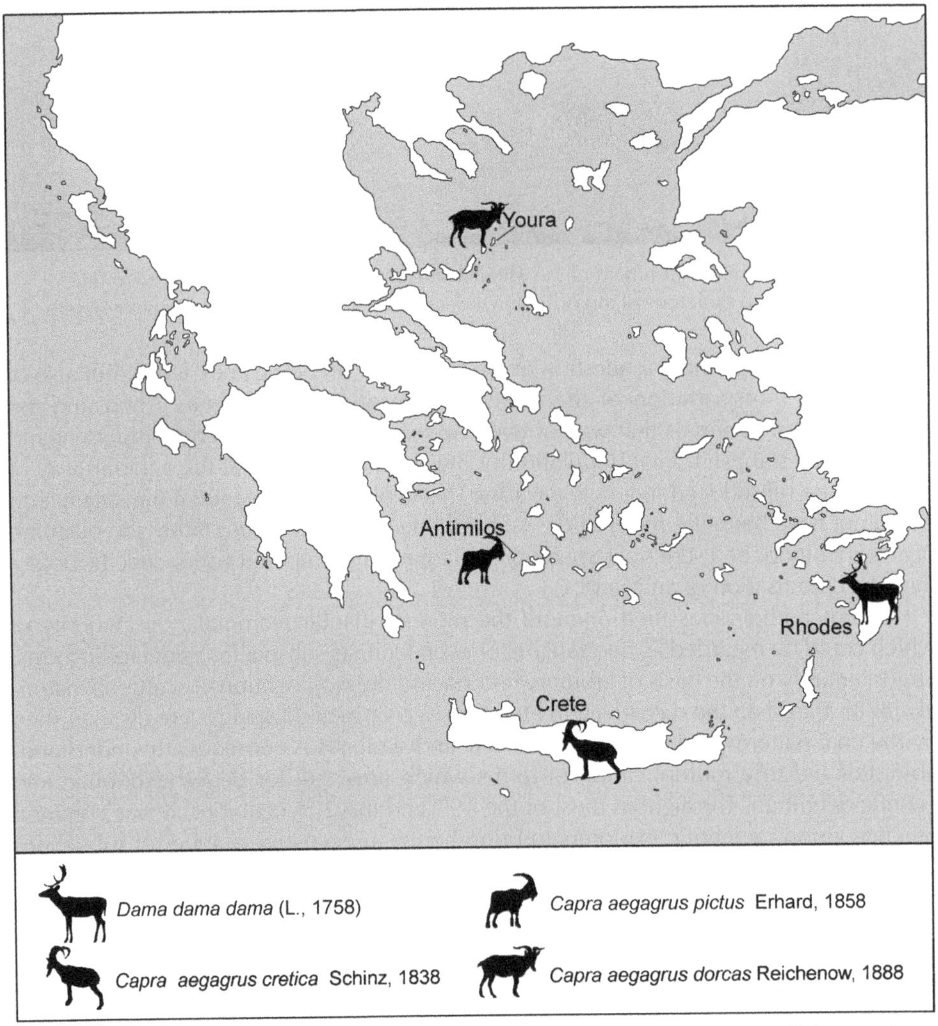

Fig. 143: The most ancient populations of ungulates of anthropochorous origin still dispersed on the Aegean islands.

Unlike the case of artistic monuments, here we are dealing with living organisms which interact with environments originally extraneous to their natural distribution, but which they have inhabited since ancient times. Their protection and their study can provide an opportunity for testing a range of different evolutionary theories. While the importance of these anthropochorous populations is not comparable, from a biological point of view, to that of veritable wild populations in their natural habitat, their elevated validity as environmental entities lies in their representing singular historical and cultural documents [29]. The exclusion of known anthropochorous taxa from conservation lists, followed by implementation of ulterior protective legislation, has been proposed as a necessary measure to maintain the uniqueness and richness of Mediterranean biodiversity [1075–1076]. Nevertheless, an excessively philological and "purist" approach to the compilation of the conservation documents can even prove to be detrimental, since, as we have seen, the primeval ecosystems of the Mediterranean islands were irretrievably lost thousands of years ago. In the more specific case of the endemic mammals, they are practically entirely extinct. Nor should we overlook the fact that many of these anthropochorous populations constitute some of the few available points of reference for an understanding of the biological and ecological characteristics of their wild continental ancestors, by now long extinct. The latest techniques for investigating population genetics have shown, for example, that the Rhodian fallow deer is very special, being of ancient lineage distinct even from the relic Anatolian population, the probable source of the stock introduced on Rhodes in Neolithic times [32]. According to other genetic studies, *Capra aegarus cretica* shows genetic affinities with domestic goats, although the phenotype of the Cretan goat resembles that of the Asiatic pasang [950, 1077]. Consequently these populations deserve to be protected and considered in terms of a veritable "cultural heritage". Just as human artefacts and sites of archaeological and historical interest are subject to conservation strategies, so we must also protect the populations of animals of ancient anthropochorous origin, acknowledging their importance as a common heritage of humanity. The study and observation of these populations offers fundamental information (compare with [32, 1077]) and allows us to advance intriguing hypotheses which may prove particularly important in terms of the study, conservation and management of the extant Mediterranean insular fauna.

Appendix A. Native domestic breeds

Then you will reach the island Thrinakria, where are pastured the cattle and the fat sheep of the sun god, Helios, seven herds of oxen, and as many beautiful sheepflocks, and fifty to each herd. There is no giving birth among them, nor do they ever die away, and their shepherdesses are gods, nymphs with sweet hair ...

(Homer, *The Odyssey*, XII: 130)

A.1 The Aegean bay horse

The barren uplands and the desolate stony wastes in the southern part of the island of Skyros, situated in the northern Sporades in the centre of the Aegean Seas (Greece), are home to a race of small, swift-limbed horses which is considered to be of ancient origins [6, 1079–1081] (▶Fig. 144). The morphology has recently been definitively classified

Fig. 144: The phenotypes of the so-called Skyrian pony fall within those of the Aegean bay breed, which is regarded as a miniaturized true horse. Island of Skyros. Photo by Marco Masseti.

under the denomination *Skyrian pony*, envisaging exemplars with an average height at the withers of 90–110 centimetres. However, although it is included in the category of ponies in the classification registers, this is actually a miniature horse. Effectively, ponies are characterised by a bodily circumference greater than their height, while the length of the head is generally the same as the distance between the withers and the tip of the shoulder, and the length of the spine equivalent to that between the withers and the croup [1082–1083].

The proportions of the Skyros pony are instead similar to those of a true horse. The breed is strong, frugal and robust, good-natured with a willing disposition and, considering its structure, a good jumper. It is fairly straight-shouldered, the fetlocks are normally equine and the hind legs lean. The coat is frequently distinguished by the presence of the dorsal line, which extends from the tail and is frequently accompanied by striping of the distal sections of the lower limbs. In individuals of bay colour, the hoofs, tail and mane are dark, with a black nose tip surrounded by a broad band of a very pale shade. These characteristics are considered to be the inheritance of the coat colourings of the wild forms and are also to be found in the last wild horse in existence, the *Equus przewalskii* Poliakov, 1881. On Skyros, a high percentage of the coats feature a dark bay colouring (well over 50%, according to the observations made in the course of the present study). The feet of the *Skyrian pony* are always dark, this being one of the essential requisites for the classification of the race. The morphological structure and the frugal temperament of this horse enable it to live in the wild throughout the year, in herds normally led by a stallion. It feeds on the sparse shrubs of the *friganà*, the vegetation which corresponds to the garrigue of the eastern Mediterranean; amidst this fragrant scrub the race reproduces in the wild on the sun-drenched stony uplands of Skyros. It appears that in recent times, because of their size, these ponies have been rarely used for riding. However, up to the 1960s they were captured seasonally, at the time of threshing, to be employed in the wheat harvesting [6].

Several authors, including Marinos and Symeonidis [104], have hypothesised that the reduced dimensions of the race were the result of adaptation to the island habitat. Palaeontological studies have effectively shown how, on islands, the tendency to develop dwarf forms may arise among large mammals, even in the space of a few millennia. The same authors consequently tend to identify an endemic form in the Skyros pony. Also called in to support this hypothesis, is an improbable descent from the race of *Hipparion* [1079], an equid fossil, tridactyl and unguligrade which lived in Europe in various forms up to and throughout most of the Pliocene (compare with [1084]).

In spite of the fact that the Skyros horse has been raised to the rank of a true race, equids with analogous characteristics are also recorded for other Mediterranean islands. There are in fact records of the past presence of ponies phenotypically similar to those of Skyros on Cyprus [1085], Skopelos and Euboea [1079], Syra [747, 1086], Naxos [401], Lesbos, Rhodes [1086–1088] and Cephalonia (compare with [1089]). Small ponies which are typologically similar are also occasionally to be found in continental Greece, without however having been identified as belonging to a specific race [1090]. In the past Lesbos was particularly famous for a robust and frugal equine race of small dimensions. These ponies, which were commonly known by the name of *midilides*, deriving from the name of the island's main town, nevertheless became extinct in the course of the 20th century [1091]. On Rhodes too, a pony of small dimensions was bred; Vezzani [1087] describes this as "... *a true pony, which is commonly known locally as Mitillì (possibly because believed to have originated from the island of Mitilene), with an average stature of 1.10 metres, lively and vivacious, and it too fairly rustic, austere and strong,".* During the Italian occupation of the Dodecanese (1912–1945), in particular, these animals became the object of great interest on the part of the zootechnics specialists, who even attempted an ethnic improvement [1086–1088, 1092]. Today, however, it is estimated that only a few individuals survive on the island, possibly reduced to no more than a dozen, who have found refuge in the region of Archangelos [1093]. Vezzani [1087] and Colombi [1086] were among the first to observe the phenotypic similarity of the ponies of Lesbos and Rhodes. Several horses which can be ascribed to the same phenotypes are still present on Samos and Cephalonia, while in the central Mediterranean exemplars with analogous characteristics are still to be found on the small island of Marettimo, in the Egadi archipelago (Sicily) [1094], and in Sardinia (▶Fig. 145).

In the latter island, according to Della Maria [1095–1096], the presence of horses of small dimensions on the Giara di Gesturi, a basalt upland dominating the hills of Marmilla in the region of Cagliari, dates to the 13th century, while for other authors, including Camboni [1097], their origins can be traced back to a more ancient period. In 1976 several exemplars were introduced into a protected area of Capo Caccia, to the west of Alghero [1098]. The racial improvements attempted, especially after the Second World War, have however resulted in a distortion of the original characteristics of the race in favour of the development of a form with a snub-nosed muzzle and featuring predominantly near-black coats. The dimensions, too, appear to have been affected, in favour of forms which are generally taller. Nevertheless, according to the testimony of the older residents, the Giara was in the past populated by ponies known as *mosca pia,* of a height of no more than 90 cm at the shoulder [1099–1100]. At the Istituto di Incremento Ippico (Horse Breeding Institute) in Fontana Burgos, however, there are still some pure-bred exemplars destined for the maintenance of the original ethnic characteristics. As far back as 1774, the naturalist Francesco Cetti [1101] described the wild horses of Sardinia as *"... small, with a short, bristling mane, and varying in colour, but mostly bay",* while, almost a century later, Cornalia [1102] recorded for these same ponies a height

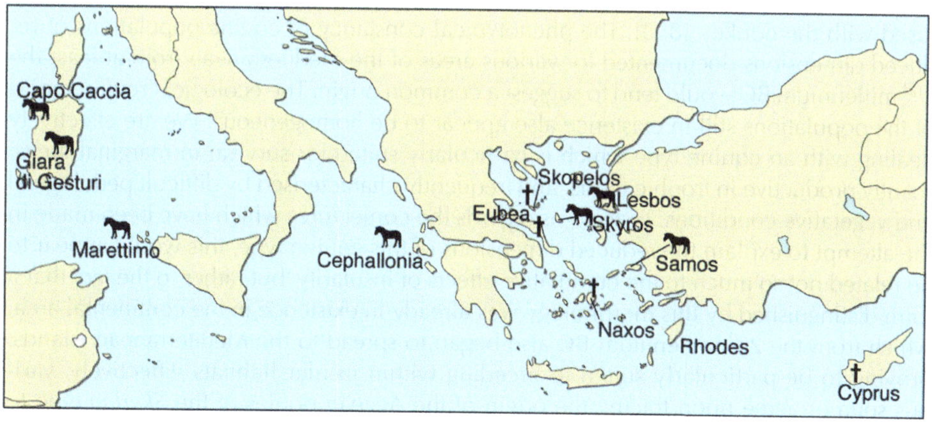

Fig. 145: In spite the fact that the Skyrian bay horse has been raised to the rank of a true – local – race, horses with analogous characteristics are also recorded from several other islands of the Mediterranean.

of no more than three feet (1 foot = 30.50 cm). There are no native horses in Sardinia, an island characterised exclusively by endemic Pleistocene mammals, among which no equid has ever figured. The generally accepted view is that the Giara ponies were introduced in the course of the Iron Age, or in a slightly earlier epoch [1103–1106], possibly by the Carthaginians. In support of this hypothesis, we could take into consideration the dimensional reconstruction calculated on certain finds originating from the *tofet* of Tharros, which are recorded among the most ancient equine remains (an entire metacarpus and vestigal tooth fragments) discovered in Sardinia [1107]. These would in fact appear to have belonged to exemplars measuring little over a metre high at the withers. According to Manconi [1108], the dimensions of the Sardinian domestic equids, discovered respectively in the Roman layers of a sacred district in Olbia (second half of the 2nd century BC) and in the necropolis of (Cuglieri, Oristano, 7th century AD), are comprised within the dimensional range of both the existing Giara ponies and of the donkey, *Equus africanus* Fitzinger, 1857. Other hypotheses have also been formulated about the origins of the ponies of the Giara di Gesturi. While some authors argue for an eventual origin from Numidia, at the time of the Carthaginian dominion of Sardinia, for others the bay horses instead came from the Aegean archipelago and were transported to Sardinia by Greek navigators [1100]. The ethnic homogeneity between the Sardinian horses and the ponies of Rhodes had already been observed by Taibel [1088]. Remains of equids of small dimensions have also been uncovered by the archaeological exploration of several Aegean sites of the late Bronze Age, such as Akrotiri on Santorini, and Phylakopi on the island of Milos [888]. The dimensional analysis of these has revealed measurements more congruous with those of a donkey than of a horse [858, 888]. Although considered indigenous to the southern Balkan peninsula in the course of the Pleistocene, the horse does not appear among the Greek fauna recorded for the Neolithic period and, from a strictly evolutionary point of view, the isolated finds of the Bronze Age would therefore appear to document a new arrival of the *Equus* genus, which had previously become extinct in Greece [315]. Even if the horse, in a pre-domesticated condition, would appear to have reached the Balkan peninsula sporadically during the early Bronze Age (Servia, Pentapolis, Kastanas), its presence is more widespread during the middle Bronze Age, when animals of small stature from an osteological point of view, are often confused with the donkey [858]. The phenotypical constancy of equine populations of reduced dimensions documented for various areas of the Mediterranean from at least the 2nd millennium BC would tend to suggest a common origin. The ecological requirements of the populations still in existence also appear to be homogeneous. We are effectively dealing with an equine type which is particularly suited for survival in marginal areas, scantly productive in trophic terms, and frequently characterised by difficult pedological and vegetative conditions. Instead, as regards the conjectures which have been made in the attempt to explain the reduced dimensions of this equine type, this would appear to be related not so much to the cumulative effects of insularity, but rather to the fact that a form distinguished by this morphology was already in existence in the continental area, which from the 2nd millennium BC also began to spread to the Mediterranean islands, proving to be particularly suited to breeding within insular habitats. Effectively, various sources agree upon tracing the origin of the Aegean ponies of the *Skyrian* type to continental Asia. Manetti [1092] had already observed how the so-called *ponies* of the Greek archipelago were "*undoubtedly of oriental origin*". Moreover, various considerably earlier literary documents had already recorded analogous information. For example, Francois Sieur de la Boullaye-le-Gouz, who visited Lesbos in 1653, wrote: «*Dans cette Isle se trouvent des chevaux sauvages forts et petits mais vigoreaux au possible, et estimes dans l'Asie.* » [1109].

In effect, we find ourselves dealing with an equine type which had already for some time been codified in the Near East and which, in view of its characteristics of robustness and frugality, was used in the colonisation of areas featuring particular conditions, such as many Mediterranean islands. In support of this we could recall that Vezzani [1087], speaking in general of the animals present on Rhodes in the 1920s, observed how their overall characteristics resembled those of southern Anatolia. The relatively recent discovery, in an area of the Near East not far from the shores of the eastern Mediterranean, of an equine race featuring characteristics similar to those of the *Skyrian pony* has cast new light on the origins of the bay horses of the Mediterranean islands. In 1965, on the southern shore of the Caspian Sea in Iran, a horse of small stature was discovered, which is currently considered as an exemplar of one of the most ancient races and even as the antecedent of the Arab horse [1110] (►Fig. 146). This is the so-called *Caspian pony*, no higher than 100–120 cm at the withers, and characterised by a small head, short ears and a long, arched neck. It is fine-limbed and gracefully proportioned [1111]. In this case too, the body is narrow and light, with proportions which are more congruous with those

Fig. 146: No higher than 100–120 cm at the withers, the so-called *Caspian pony* is characterised by a small head, short ears and a long, arched neck. It is fine-limbed and gracefully proportioned. From Alderson [1111].

of a miniature horse than a pony. It has the appearance, temperament and action of a thoroughbred horse [1111]. In Iran, the Caspian is a beast of burden. Some were exported to Britain in the 1970s and thereafter to several other countries. This rare native breed of northern Iran seems to date back to 3000 BC. Egyptian and Near Eastern artefacts from the pre-Christian era (between the middle of the 2nd millennium BC and 500 AD) show animals of small dimensions and refined shape which are very similar [1112–1113]. On the basis of ulterior archaeological evidence, we could therefore hypothesise that the diffusion of the *Caspian* type horse may have been launched in the proto-historic era in the Near East, spreading thence towards the West and the Mediterranean Basin.

Archaeological research recently carried out in the present-day settlement of Trianda (Ialyssos) on the island of Rhodes (Dodecanese, Greece), provided the complete skeleton of an adult horse buried together with several adult human beings in a cultural context referred to as the Late Bronze Age (Late Minoan IA). This finding corresponds chronologically to about the time of the disaster of the island of Santorini, circa 1630 BC (compare with [1114–1115]). According to the director of the excavation, Toula Marketou, 22nd Ephorate of Prehistoric and Classical Antiquities (Rhodes), since the Early Bronze Age the Dodecanese burial culture falls within the western Anatolian Bronze culture; but the finding is also important because it apparently represents the first known importation of the horse out of continental Asia Minor, in the Eastern Mediterranean islands and, at least, possibly in the Aegean.

To date, this event has been traditionally referred to an unspecified time during the Minoan period, corresponding to the first occurrence of some representatives of the genus *Equus* on the island of Crete (compare with [418]).

Measurements taken on the Late Bronze Age horse of Trianda reveal that the specimen does not exceed the height of one metre at the withers. Thus, it could be considered as the first certain horse bones, apparently also from a pony-sized breed, which come from Middle Helladic contexts of continental Greece, at Argissa, Nichoria and Lerna (circa 1800–1700 BC), and of which only the Nichoria context is definitely early [1116].

However, to return to the Trianda horse, it cannot be excluded that we are dealing with some of the earliest evidence of the artificial diffusion of the Aegean bay breed outside continental Asia Minor. The majority of the domestic breeds formerly occurring on the Mediterranean islands did not represent local – indigenous – populations, originating instead from small-sized livestock already available on the nearest mainland. According to Vezzani [1087], this was the case, for example, of the domestic breeds found on the Dodecanese islands at his time, which were the subject of regular importation from the opposite coast of Anatolia.

A.2 Other local and/or dwarf domestic breeds

Other breeds of domestic animals of reduced dimensions are known from some remote islets of the Aegean sea, such as the extinct race of Lilliputian cattle of Tilos, in the Dodecanese, that possibly survived up to the 1970s, and the donkeys and pigs of several other islands.

On the islet of Piperi, in the Northern Sporades, a type of longhorn grey cattle characterised by reduced body size was still bred in a free-ranging state up to few years ago (►Figs. 147 and 148). The longhorn grey cattle is an ethnic group regarded as a breed of very ancient origin. Its phenotypic patterns still characterise a domestic breed that

Fig. 147: A type of longhorn grey cattle, characterised by reduced body size, was still bred in a free-ranging state on the islet of Piperi, up to a few years ago. Photo courtesy of Anastasios Lekagis.

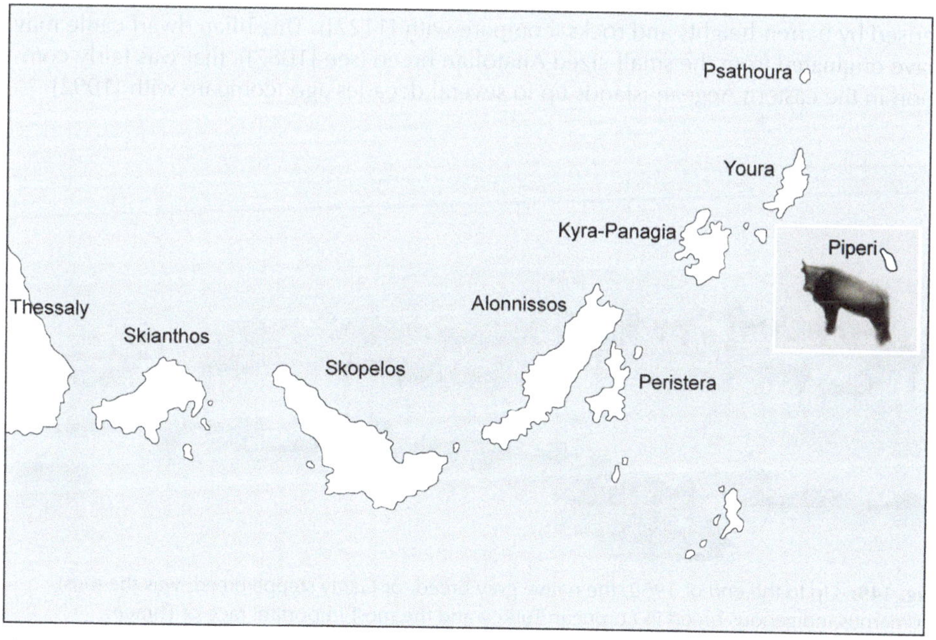

Fig. 148: Geographical location of the islet of Piperi, in the Northern Sporades.

has survived in central Italy, the so-called "Maremmana" cow, a strong and hardy race accustomed to living in the open and capable of surviving on poor grazing land. Maremmana bulls average 1 200 kg and measure up to 180 cm at the withers. To the same ethnic group belongs also the Northern Greece-Anatolian native grey breed, or Garay steppe breed (▶Fig. 149), and the so-called Katerini and Sykias races [1117]. Today, the Katerini, the larger type of Steppe cattle, can be considered to be saved, thanks to the engagement of one breeder (D. Dimos) and one municipality, whereas of the smaller Steppe cattle, the Sykia, remnants have now been found and currently a breeding programme is being established [1118]. According to Sojsal et al. [1119], the long horn grey cattle now represents only 4.2% of Turkish cattle, still occurring in Thrace, part of European Turkey, the region of Marmara, western Anatolia and partially also in central Anatolia. Up to the end of 1960, however, it was the most numerous indigenous breed in European Turkey and the most important race of Thrace [1120]. This is not to say, however, that cattle of the longhorn grey phenotypes are still to be found in Asia Minor [127], the majority of the Anatolian breeds being traditionally dominated by the phenotypes of shorthorn races [1087, 1092], such as the East Anatolian Red breed (compare with [1121]).

As reported by Marinos and Symeonidis [104], a small-sized domestic cow, endemic to Tilos was still bred on this island (▶Fig. 150). It was called *Lilliputian* ("λιλλιπουτεις" = lilliputeis) and according to the local people, it occurred up to around the first half of the 1970s. It was a dwarf breed of cattle, well adapted to the harsh and low trophic natural conditions of the island. In fact, the vegetation of Tilos is poor and scantly represented by Mediterranean phrygana, only in very few cases showing relics of a degraded maquis, made up of *Quercus coccifera* Linnaeus, and *Quercus macrolepis* Kotschy [63]. The eastern terebinth, *Pistacia terebinthus palaestina* (Boiss.), is also well represented in the cultivated areas. The *Lilliputians* were restricted to a few dozen of hectares, in the plain called "Misariàs", the rest of the island being characterised by barren heights and rocks (compare with [1122]). This Tilian dwarf cattle may have originated from the small-sized Anatolian breed (see [1087]), that was fairly common in the eastern Aegean islands up to several decades ago (compare with [1092]).

Fig. 149: Up to the end of 1960, the native grey breed, or Garay steppe breed, was the most numerous indigenous breed in European Turkey and the most important race of Thrace. From French [1120].

Fig. 150: A small-sized domestic cow, called Lilliputian, was still bred on Tilos up to the first half of the 1970s.

A very peculiar pig was also found on Tilos by Ghigi [394]. It was very small, with long, thick, dark hair and a sort of mane on the back. This pig appears to have been well adapted to the particular environment of the island, that is "... *opposite to the ordinary habitat of wild boars, since the entire surface of the island is possibly drier than any of the other islands of the Dodecanese*" [394]. According to Ghigi [394] and to Marinos and Symeonidis [104], small-sized domestic animals represent a good example of insular influence on size decrease and the effects of artificial selection, augmented by geographic isolation. However, in the case of the small pigs of Tilos, Taibel [816] demonstrated that they did not represent an endemic insular race, since they were subject to size increase when subjected to different more adequate environmental and breeding conditions. Thus Taibel [816] suggested referring to the Tilian dark pigs as a local population and/or to a peculiar phenotype of the domestic pig which was bred also on Rhodes and described by Vezzani [1087]. In the light of all this, it is possible to suggest that the "λιλλιπουτεις" cow formerly bred on Tilos may again have represented a local population.

Greece has always been considered the stronghold of donkeys. Nowadays, however, in the Dodecanese as in the rest of the Ionian and Aegean islands, donkeys are no longer as widespread as they were in the past. At the time of the Italian occupation of Rhodes, these beasts of burden demonstrated a substantial variation in size, ranging from normal dimensions to those typical of the Mediterranean dwarf varieties [1087] (▶Fig. 151), like the donkeys that are still reared on the island of Asinara, off the north-western coast of Sardinia [1123–1125] (▶Fig. 152). Sadly, Greek donkeys have never been divided into types, but only distinguished by two sizes [1118]. Nevertheless, even now in the majority of cases the coat colour of the Rhodian donkeys falls within the mixed phenotypes of the Nubian wild ass, *Equus africanus africanus* Fitzinger, 1857 (▶Fig. 153), and that of Somalia, *Equus africanus somaliensis* Noack, 1884, being characterised by a general greyish-blue colouration, with the marked dark shoulder stripe of the former ass combined with the leg stripes of the latter.

Several peculiar breeds of sheep still occur on the Ionian and Aegean islands. Zakynthos, for example, has a semi-coarse wooled breed which differs from other Greek sheep and, according to Georgiou [1126], is derived from the Bergamo sheep of Italy. Mason [1127] listed it with the Alpine breeds. The *Skopelos* breed or *Glossa*, a dairy breed of the Greek Island Zackel type (▶Fig. 154), was developed on the namesake island. Its ancestry is however obscure. According to Mason [1127], there is no record of sheep on Skopelos before 1800. Georgiou [1126] and Mason [1127] describe two varieties of the Greek Zackel on Crete: the *Sfakia* of the western lowlands, and the *Psiloritis* – also known as *Anogia*, from the namesake village on the slopes of Mount Ida – of central Crete. The *Sitia* breed of eastern Crete is regarded as a distinct variety [1128]. Also derived from the Greek Zackel is the *Levkimi* breed from southern Corfu [1127–1128]. The *Chios*, or *Sakiz*, is a semi fat-tailed breed from the island of the same name. Mason [1127] favoured its derivation from a cross between the Turkish *Kivircik* (finer wool) and *Dağliç* (fat-tailed).

Fig. 151: At the time of the Italian occupation of Rhodes, donkeys demonstrated a substantial variation in size, ranging from normal dimensions to those typical of the Mediterranean dwarf varieties. From Vezzani [1087].

Fig. 152: A small-sized race of donkey is still bred on the island of Asinara, off the northwestern coast of Sardinia (Italy). Photo by Marco Masseti.

Fig. 153: Nubian wild ass, *Equus africanus africanus* Fitzinger, 1857. Drawing by Silvia Cantagalli Masseti.

Fig. 154: Adult female of sheep of the Greek Island Zackel type, on the island of Skyros. Photo by Marco Masseti.

A.3 The water buffalo, *Bubalus bubalis* Linnaeus, 1758

It has been generally assumed that the water buffalo, *Bubalus bubalis* Linnaeus, 1758, was introduced into Europe in the course of the early Middle Ages [1129–1132] (▶Fig. 155). This assumption is however not supported by any reliable archaeozoological datum, since the bone finds of buffalo are extremely rare and late ([572]; compare with [1129]). According to Lombard [1133], the water buffalo was very likely introduced into Lower Mesopotamia in the course of the 8th century following the deportation of the Zott people, a nomadic human group originating from India: "*From here they moved on to northern Syria, where the buffalo became acclimatised in the marshes of the Orontes*". Epstein [1134] considered that it was the Arabs who imported the animal to Europe from Syria a little before the 10th century AD. Zeuner [723] believed that the first description of this Asiatic ungulate was given to the Western World by St. Willibald, an 8th century bishop of Eichstätt, in Bavaria. When visiting Palestine in 723 AD, he came across Indian buffaloes in the *ghor* of the Jordan Valley. He recounted his surprise at seeing these strange animals: "*Et ibi sunt armenta mirabilia longo dorso et brevis cruribus, magnis cornibus creati; omnes sunt unius coloris, osteae ... et quando estuale tempore magna solis caumatio de caelo terram urire solet, illa armenta tollentes se vadunt ad paludem et demergant se toto corpore nisi caput solum*" (= "They bred remarkable cattle here, long in the back, short in the leg and with huge horns; they are all the same greyish-purple colour ... so when summer comes, and the sun is very hot and scorches the land, these cattle move to the swamp and submerge their whole body, with only their head sticking out") [1135]. As Keller [1136] points out, St. Willibald had travelled through Italy and Sicily, where buffaloes are to be seen today, so they cannot have been known there then. His date is about a century after the spread of the Muslim religion to Mesopotamia [723]. It thus looks as if Indian buffaloes were at that

Fig. 155: Water buffaloes, *Bubalus bubalis* Linnaeus, 1758. Photo by Marco Masseti.

time used in Mesopotamia from where they had spread to the Jordan Valley. From there they may have reached Egypt in the Middle Ages, but they did not spread farther west in North Africa, nor did the Muslims take them to Spain. Another route, however, established itself along the Balkan peninsula, and this in advance of the Muslim conquest. It appears that the southern Slavs were familiar with the buffalo, which may have reached southern Russia by way of Persia. By around 1200 AD Bulgaria and Macedonia had buffaloes in plenty.

Up to a few decades ago, water buffaloes were raised in many wetlands of the southern Balkan peninsula, especially those of northern Greece, but today they are maintained only in the marshes of Lake Kerkini, a manmade lake created on the north-western edge of the Serres prefecture [1137–1139]. Water buffaloes were not only bred in the mainland areas of the Eastern Mediterranean basin but also on the islands, as is for example documented by an engraving by J. Cartwright, dated 1817 ([1139]; compare with [1140]) (▶Fig. 156). In this work, conserved in the Gennadios Library of the American School at Athens, we can see several cows and a buffalo swimming or standing in the sea water or grazing on land on a stretch of the islet of Vido, opposite the harbour of Corfu (compare with [1139, 1141–1142]). In the northern part of Rhodes, near the town there was a fresh water lake (*i limni tou Neochoriou* = the lake of Neo-Choris), south of the *Punta della Sabbia* dominated by the extant building of the Hydrobiological Station and Aquarium, and north of the original settlement of Neo-Choris, called *Borgo Nuovo* on the old Italian maps [1143]. This small lake was drained in the first half of the 1930s. Another swamp was also present south of the town, in the area now known as *Zephiros.* It may have been from one of these swamps that the buffaloes emerged which caused havoc in the old town of Rhodes. The event was even recorded in a short article in the local newspaper, *Il Messaggero di Rodi* on 12 December 1926: "*Three buffaloes strolling round Rhodes. Yesterday at around 8.30 three large buffalo were to be seen*

Fig. 156: Detail of an engraving by J. Cartwright, dated 1817, showing a stretch of the islet of Vido, opposite the harbour of Corfu. Courtesy of Gennadios Library of the American School at Athens.

strolling happily around the old streets of the walled city, generating extreme panic among the passers-by. Terrified by the thought that the animals might pay them an unwanted visit, all the shopkeepers hastened to shut up their shops, causing a great deal of noise that frightened the buffalo, which took to their heels and stampeded towards the Gate of the Arsenal. When the Royal Carabineers were informed that widescale panic had broken out in the town, a squad of soldiers armed with muskets, climbed up onto the bastions, where the beasts had taken shelter in the interim, to hunt them out. However the owners of the animals arrived and they were successfully captured without any painful or unpleasant" [1144]. Furthermore, considering what had previously been referred to by Bürchner [411], even at the end of the 19th century, these ungulates were also bred in the northernmost island of Leros, again in the Dodecanese, where it would appear, moreover, that the water buffalo was the only bovine present. Consequently, it is not surprising that buffaloes were bred in certain wetlands of the Dodecanese, between the end of the 19th and the first half of the 20th century. As late as 1921, Manetti [1092] noted that in southern Anatolia, in addition to the oxen, also deserving a special mention were the buffaloes that shared the most important work on the land, or rather it was often the buffalo that were assigned to the heavy labour of breaking up the harsh, clay soil which the small Anatolian oxen would have been unable to do on their own. At the time, the buffaloes were to be found throughout the coastal region, and in particular in the marshy, malaria-ridden areas, almost always kept in a free-ranging state. They are largely immune to piroplasmosis, and in these territories perform better than the oxen, in terms not only of work, but also of meat and milk. In Turkey, these animals are still bred in several wetland areas of the country, especially in south-western Anatolia.

Appendix B. Chiroptera reported from the Ionian and Aegean islands

Certain [animals], *such as the little owl and the bat come out at night, others live in the light of day.*

(Aristotle, *Historia animalium*, I, 488a)

A review of the available literature [59, 70, 88, 129, 297, 1013, 1145–1151] indicates that at least 27 species of bat occur in the area: the Blasius' horse bat, *Rhinolophus blasii* Peters, 1866, the Mediterranean horseshoe bat, *Rhinolophus euryale* Blanius, 1853, the greater horseshoe bat, *Rhinolophus ferrumequinum* Schreber, 1774, the lesser horseshoe bat, *Rhinolophus hipposideros* Bechstein, 1800, the Mehely's horse shoe bat, *Rhinolophus mehely* Matschie, 1901, the lesser mouse-eared bat, *Myotis blythii* Tomes, 1857, the long-fingered bat, *Myotis capaccinii* Bonaparte, 1837, the Geoffroy's bat, *Myotis emarginatus* E. Geoffroy, 1806, the greater mouse-eared bat, *Myotis myotis* Borkhausen, 1797, the whiskered bat, *Myotis mystacinus* Kuhl, 1817, the steppe whiskered bat, *Myotis aurascens* Kusjakin, 1935, the Natterer's bat, *Myotis nattereri* Kuhl, 1817, the Hanaki's dwarf bat, *Pipistrellus hanaki* Hulva and Benda, 2004, the Kuhl's pipistrelle, *Pipistrellus kuhlii* Kuhl, 1817, the Nathusius' pipistrelle, *Pipistrellus nathusii* Keyserling and Blasius, 1839, the common pipistrelle, *Pipistrellus pipistrellus* Schreber, 1774, the Leisler's bat, *Nyctalus leisleri* Kuhl, 1817, the Botta's serotine, *Eptesicus bottae* Peters, 1869, the serotine, *Eptesicus serotinus* Schreber, 1774, the barbastelle, *Barbastella barbastellus* Schreber, 1774, the Savi's pipistrelle, *Hypsugo savii* Bonaparte, 1837, the brown long-eared bat, *Plecotus auritus* Linnaeus, 1758, the grey long-eared bat, *Plecotus austriacus* J.B. Fischer, 1829, the Mediterranean long-eared bat, *Plecotus kolombatovici* Dulic, 1980, the Alpine long-eared bat, *Plecotus macrobullaris* Kuzjakin, 1965, the Schreibers' bat, *Miniopterus schreibersi* Kuhl, 1817, and the European free-tailed bat, *Tadarida teniotis* Rafinesque, 1814 (▶Fig. 157). All of them have been reported from the Aegean archipelago, whereas only 13 species of bat are known so far from the Ionian islands (▶Tab. 5).

No less than 17 species have been reported from Crete alone [129], the first published observation of local bats dating to the 18th century, when de Tournefort [677] found them in the Labyrinth, a famous ancient mine at Gortys in the southern part of Crete. The occurrence of the Hanaki's dwarf bat, *Pipistrellus hanaki*, has also been recently reported from the latter island (compare with [1180]). Described as a new species of bat from Cyrenaica (Djebel al Akhdar), in Libya, by Benda et al. [1181], it differs in terms of its chromosome number, large skull and teeth as well as a number of other characters from its nearest relatives *Pipistrellus pipistrellus* Schreber, 1774 and *Pipistrellus pygmaeus* Leach, 1825. On Crete, *Pipistrellus hanaki* is considered as vulnerable because the island holds a large percentage of the global population and it is a bat which depends on threatened habitats such as oak and chestnut forests or thickets

Fig. 157: Portrait of an adult of European free-tailed bat, *Tadarida teniotis* Rafinesque, 1814. Photo by Bruno Zava.

Tab. 5: Checklist of the bat species reported on the Ionian and Aegean islands.

Species	Ionian	Aegean	References
1. *Rhinolophus blasii*	1•	1•	[50, 129, 301, 1146, 1148, 1152–1153]
2. *Rhinolophus euryale*	2•	2•	[284, 1148, 1154]
3. *Rhinolophus ferrumequinum*		3•	[129, 1148, 1155–1157]
4. *Rhinolophus hipposideros*	3•	4•	[129, 284, 1148, 1157–1158]
5. *Rhinolophus mehely*		5•	[1148, 1159–1160]
6. *Myotis blythii*	4•	6•	[50, 58, 129, 355, 1148, 1161–1162]
7. *Myotis capaccinii*	5•	7•	[129, 1146, 1148, 1152, 1157, 1163]
8. *Myotis emarginatus*	6•	8•	[129, 284, 1148, 1164]
9. *Myotis myotis*	7•	9•	[284, 1148, 1165]
10. *Myotis mystacinus*	8•	10•	[70, 129, 1155, 1166]
11. *Myotis aurascens*		11•	[129]
12. *Myotis nattereri*	9•	12•	[1155, 1157]
13. *Pipistrellus hanaki*		13•	[129]
14. *Pipistrellus kuhlii*	10•	14•	[50, 129, 301, 1167]

(*Continued*)

Tab. 5: (*Continued*)

Species	Ionian	Aegean	References
15. *Pipistrellus nathusii*		15•	[129]
16. *Pipistrellus pipistrellus*		16•	[1157, 1168]
17. *Nyctalus leisleri*		17•	[129]
18. *Eptesicus bottae*		18•	[1149, 1169–1170]
19. *Eptesicus serotinus*		19•	[129, 1157, 1171]
20. *Barbastella barbastellus*		20•	[1172]
21. *Hypsugo savii*		21•	[129, 1157, 1173]
22. *Plecotus auritus*		22•	[1147, 1174–1175]
23. *Plecotus austriacus*	11•	23•	[129, 1176–1177]
24. *Plecotus kolombatovici*		24•	[129]
25. *Plecotus macrobullaris*		25•	[129]
26. *Miniopterus schreibersi*	12•	26•	[50, 129, 284, 1148, 1178]
27. *Tadarida teniotis*	13•	27•	[50, 59, 129, 1179]

with old trees and wetlands [1182]. Its Cretan population differs morphologically and genetically from the common *Pipistrellus hanaki* in Libya, which suggests that it should be placed in a separate endemic taxon. The occurrence on Crete of another newly described species, the Alpine long-eared Bat, *Plecotus macrobullaris* Kuzjakin, 1965, has also been recently recorded. The presence of this species in Greece has been confirmed in only a few other areas of Epirus, and Fthiotida [1183]. Recently, a new bat species has been reported from the Eastern Aegean islands (Dodecanese). In fact, the Botta's serotine, *Eptesicus bottae* Peters, 1869, a species which in recent years has been shown to be widespread over the coastal areas of Anatolia [1184], was recorded for the first time on Rhodes by von Helversen [1150]. In Europe, this species is currently found only on this island, where it appears to be fairly common [1169]. In the Eastern Aegean territories of its range, the Botta's serotine appears to be confined to coastal plains below about 200 m above sea level. Its presence in other islands of the Aegean is also possible [1170].

In general, however – and apart from the Hanaki's dwarf bat which appears to be endemic to the lands bordering the Libyan Sea, that is to say Crete and the Cyrenaica, but may also occur elsewhere – we can observe that there are no other endemic bat species in the Ionian and Aegean islands, neither at present nor in the Upper Pleistocene. In fact, the Hanaki's dwarf bat appears to be the only endemic bat of the area. Unlike other insular complexes, such as the Macaronesian islands, in the Eastern Atlantic Ocean [1185], the excessive closeness of the islands to the mainland and to each other plays a role, that in the case of bats, is opposite to that of the non-flying terrestrial mammals.

Appendix C. Cetaceans reported from the Ionian sea and the Aegean Seas

Reports of cetaceans from the Ionian and Aegean Seas are not scarce. A review of the literature [1148, 1150, 1186–1198] indicates that, at least, 8 species (7 and 1 respectively of the suborders Odontoceti and Mysticeti) are constantly present in the area, with another 4 cetaceans less regularly recorded, such as the common mink whale, *Balaenoptera acutorostrata* Lacépède, 1804, the killer whale, *Orcinus orca* Linnaeus, 1758, the false killer whale, *Pseudorca crassidens* Owen, 1846 [1188] (▶Fig. 158), and the Gervais' beaked whale, *Mesoplodon europaeus* Gervais, 1855, suggesting an unexpectedly high species diversity of these organisms [1191–1192, 1198] (▶Tab. 6).

The bottle-nosed dolphin, *Tursiops truncates* Montagu, 1821 (▶Fig. 159), occurs in all the coastal areas, straits and gulfs of Greece and eastern Turkey, as well as around and between islands in the Ionian Sea and from the Thracian to the Libyan Sea [1194, 1202]. An isolated population exists in the Amvrakikos Gulf.

The common dolphin, *Delphinus delphis* Linnaeus, 1758, is instead considered an endangered species and occurs in the inner Ionian Sea, the Gulf of Corinth, the Thracian Sea, the Northern Sporades, the Northeast Aegean (between the Greek islands and the Turkish coasts), the Thermaikos Gulf, the Gulf of Euboea, the Pagasitikos Gulf, the Saronic Gulf, the Dodecanese and the Cyclades [1194, 1201].

Fig. 158: Among the cetacean species living in the Ionian and Aegean Seas, the occurrence of the false killer whale, *Pseudorca crassidens* Owen, 1846, is occasional. Photo by Marco Masseti.

Tab. 6: Literature sources for reports of cetaceans from the Ionian and Aegean Seas.

Species	References
Mysticeti	
Balaenoptera acutorostrata	[1198]
Balaenoptera physalus	[1186–1187, 1189–1190, 1195]
Odontoceti	
Pseudorca crassidens	[1188, 1192, 1194, 1195]
Stenella ceruleoalba	[1186, 1189–1190, 1194–1195, 1199]
Delphinus delphis	[1186–1187, 1189–1190, 1194–1195, 1200–1201]
Tursiops truncatus	[1148, 1186, 1189–1190, 1194–1195, 1202]
Grampus griseus	[1187, 1189–1190, 1194–1195, 1198, 1203]
Phocoena phocoena	[1148, 1187, 1194, 1204]
Physeter macrocephalus	[1186–1187, 1189, 1194–1195, 1205]
Mesoplodon europaeus	[1198]
Ziphius cavirostris	[1186, 1189–1190, 1194–1195, 1198]

Fig. 159: The bottle-nosed dolphin, *Tursiops truncates* Montagu, 1821, is widespread all around the Ionian and Aegean Seas. Photo by Marco Masseti.

The striped dolphin, *Stenella ceruleoalba* Meyen, 1833, is the most common cetacean of the Mediterranean Greek and Turkish Seas, where it occurs in all pelagic waters and also in coastal areas with deep waters or a steep slope [1194, 1199]. This cetacean appears to be the species of dolphin most frequently represented in Aegean Bronze Age art [1070] (▶Fig. 160).

The Cuvier's beaked whale, *Ziphius cavirostris* G. Cuvier, 1823 (▶Fig. 161), is probably less rare than it was considered in the past [1192, 1198].

The sperm whale, *Physeter macrocephalus* Linnaeus, 1758, and the fin whale, *Balaenoptera physalus* Linnaeus, 1758, have been observed in the southern Aegean, off the Peloponnese, Crete and the Turkish coasts, but there are reports of the occasional occurrence of sperm whales also in the northern waters. The Hellenic Trench (from the western Ionian islands and the Peloponnese to the south of Crete and south-east of Rhodes) is possibly one of the most important areas for *Physeter macrocephalus* in the Mediterranean. The species has been reported in the Myrtoon Sea and in areas in the Aegean with steep slopes and considerable depths – especially in the area between the Northern Sporades islands and the peninsula of Chalkidiki [1205].

In the Mediterranean Sea the common mink whale, *Balaenoptera acutorostrata* Lacépède, 1804, is regarded as an occasional "visitor" [1206]. Strandings of the species have been reported on more than one occasion from the Eastern Mediterranean basin, from the Greek island of Skiathos [1207], the Israeli coast [1208], and the beach of Erdemli, near Mersin [1198], which is the first record from the Turkish coast.

Fig. 160: The image of a striped dolphin, *Stenella ceruleoalba* Mayen, 1833, in the decoration of a Late Helladic I blade, circa 1500 BC (Athens, National Museum). This cetacean appears to be the species of dolphin most frequently represented in Aegean Bronze Age art.

Fig. 161: Living Cuvier's beaked whale, *Ziphius cavirostris* Cuvier, 1823, stranded on the beach of Faliraki (north-eastern Rhodes), in 1996. Photo by Mariolina Corsini-Foka.

Individuals of the Risso's dolphin, *Grampus griseus* G. Cuvier, 1812, were reported for the first time by Carpentieri et al. [1189], in the marine channel between Rhodes and Karpathos, but the species occurs in all the Greek seas (at depths of 200–1 700 m and at a distance of 1–32 km from the coast). It does not appear to be common anywhere, with the exception of the area between the Myrtoon Sea and Antikythera island [1203]. There is a recent record of a stranded specimen from the Gulf of Fethiye [1198].

Although the harbour porpoise, *Phocoena phocoena* Linnaeus, 1758, is known to occur in the Black Sea and the Sea of Marmara off the coast of Turkey [1195], according to Fraguedakis-Tsolis and Chondropoulos [1148], its presence in the Greek seas is very limited and occasional, mainly confined to the Thracian and the northern Aegean Seas. Nevertheless, stranded individuals have been recorded from Euboea, the Saronic Gulf, and Gökçeada (Imbros). The Black Sea harbour porpoises, *Phocoena phocoena relicta* Abel, 1905, of the Aegean Sea, the Sea of Marmaras and the Black Sea are genetically differentiated from those of the Atlantic Ocean. Based on inferred calculations, the Aegean subpopulation could theoretically number from a minimum of 30 to a possible maximum of 450 individuals based on Black Sea abundance data [1204]. Harbour porpoises in the Thracian and northern Aegean seas are important from a conservation perspective since this species does not inhabit any other part of the Mediterranean Sea [1193]. The size and geographical isolation of this subpopulation makes it extremely vulnerable.

At the beginning of 2009, a beaked whale, *Mesoplodon* sp., was stranded live in Fethiye and was rescued and returned to the sea through the efforts of local people

[1209]. Later, based on the photographs taken, it was identified as possibly a Gervais' beaked whale, *Mesoplodon europaeus* Gervais, 1855, which would make this 2009 stranding in Fethiye the second case in the entire Mediterranean after that in Livorno (Italy), 2001 [1206]. Gervais' beaked whales occur between the temperate and tropical waters in the Atlantic Ocean [1210], but are rare in the Mediterranean Sea [1198].

Five other species – the white whale, *Delphinapterus leucas* Pallas, 1776, the Blainville's beaked whale, *Mesoplodon densirostris* Blainville, 1817, the long-finned pilot whale, *Globicephala melas* Traill, 1809, the killer whale, *Orcinus orca* Linnaeus, 1758, and the blue whale, *Balaenoptera musculus* Linnaeus, 1758 – have been erroneously included in the Ionian and Aegean cetacean fauna in the past due to mistaken assumptions, false identifications or lack of supporting evidence [1198]. The occasional occurrence of the killer whale, (compare with [1187]), or pilot whale in the Ionian and Aegean Seas should be still regarded as unconfirmed.

Appendix D. The monk seal, *Monachus monachus* Hermann, 1779

"... then the seals came thronging out of the sea and lay side by side near the breaking billows"

(Homer, *The Odyssey*. IV: 447–449)

Apart from a multitude of aquatic organisms including the aforementioned rare cetaceans, the Ionian and Aegean Seas still principally sustain the largest Mediterranean population of the universally protected Mediterranean monk seal, *Monachus monachus* Hermann, 1779 [1211] (▶Fig. 162). The distribution of this species is today restricted to the Mediterranean and north-western Africa with its largest colony at Cabo Blanco, in Mauritania: it ranges from Turkey and the western coast of the Balkan peninsula to Madeira and the Canary archipelagos [1212–1213]. Up to a few decades ago, it was

Fig. 162: During the last decades, several of the reports of Mediterranean monk seals, *Monachus monachus* Hermann, 1779, in the Ionian and Aegean Seas were about newborn pups. Photo © Panaghiotis Dendrinos.

still present in the waters of the Black Sea which represented the northernmost limit of the species' distribution (compare with [1214]), and where it is thought to have recently become extinct. The core range of the Mediterranean monk seal is now, however, the Ionian and the Aegean Seas [1215] (▶Fig. 163). This pinniped is still widely distributed throughout coastal and insular Greece and Turkey [1200, 1213, 1216–1217] and important sub-populations can be found in the Northern Sporades, Kimolos, Gyaros, Karpathos and the Turkish peninsula of Dilek, off Samos. The distribution of the species is currently scattered all along the 15 000 km coastline of the Greek mainland and islands, and is found mainly on rocky, isolated and inaccessible areas. The minimum population size estimate for Greece is 179 adult individuals, which represents 39.77% to 51.14% of the global total estimate [1217]. This seal breeds mainly in caves, often those that have only underwater entrances, and has become mainly nocturnal, although it is believed to be essentially a daylight-loving creature better suited to pupping on open sandy beaches; disturbance and hunting have forced it to alter its habitats, although these could change again within better protected areas [1218]. According to Dendrinos and Adamantopoulou [1219], the rarity of the species and the great difficulty of its habitat accessibility render the study of its biology and ecology extremely complex. The presence of the Mediterranean monk seal in the Aegean waters was already recorded in the 1940s: the Italian zoologist Enrico Tortonese who visited the Dodecanese archipelago during the last World War, observed that the animals sometimes entered the harbour of Rhodes [562, 801], an event which continues occasionally to occur today (▶Fig. 164).

While it may be now regarded as a relatively obscure species, the Mediterranean monk seal played an important role in many ancient myths and legends, indicating that it was once well-known to the human inhabitants of the Mediterranean basin (compare with [1220]). Today, however, we know very little of the biology and ecology of the species, since studies carried out on the animal in the wild are practically inexistent.

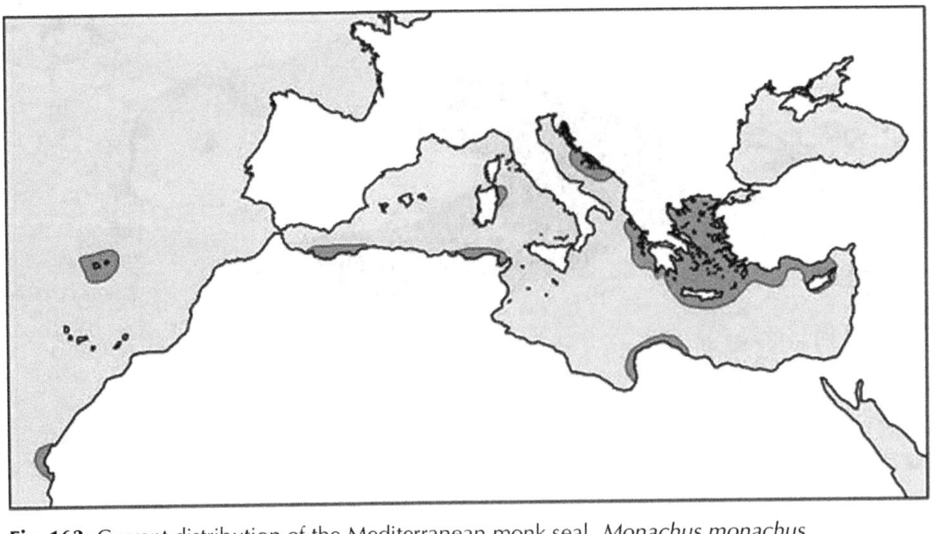

Fig. 163: Current distribution of the Mediterranean monk seal, *Monachus monachus* Hermann, 1779.

Fig. 164: As in the past, even now the Mediterranean monk seals occasionally enter the harbour of Rhodes. This three year-old female, originating from Turkey and named *Badem* ("almond" in Turkish), remained there for several days in January 2010, sleeping on the anchored fishing boats. She was later also sighted at Simi. Photo by Mariolina Corsini-Foka.

Experience deriving from observation in captivity is equally rare, since very few monk seals have been reared in zoological enclosures for long periods (compare with [1221]). In this context, we should mention the experience of breeding adult and juvenile monk seals at the Hellenic Center for Marine Research and Hydrobiological Station of Rhodes since the 1960s [1221–1224].

In the second half of the 1970s, Marchessaux and Duguy [1225] estimated that the entire population of *Monachus monachus* currently occurring in Greek waters did not exceed 310 individuals. About twenty years later a study carried out by Adamantopoulou et al. [1226] reported the observation of 540 live seals and 70 dead specimens, together with the sightings of newborn pups in a number of coastal areas. The high frequency of seal sightings in conjunction with evidence of breeding in specific areas, such as Samos, Fourni, Kythera, Kasos, Karpathos, Pelion, northern Euboea, and Milos, suggested that there may be important (sub)populations which ought to be taken into consideration in the Turkish-Hellenic strategy for the protection of the species.

According to Dendrinos et al. [1217], major threats to the species are habitat deterioration and loss caused by coastal development, deliberate killing and accidental entanglement in fishing gear, decreased food availability due to overfishing, marine pollution and stochastic events, such as disease outbreaks. Conservation measures for the species focus on the establishment of protected marine areas, rescue and the rehabilitation of orphaned or injured seals, environmental education and public awareness-raising [1213]. The species is strictly protected under Greek law, European directives and international conventions.

D.1 Honoured with its ancient owner: a monk seal buried in Rhodes over 2000 years ago

A few years ago, archaeological excavations on Rhodes unearthed an ancient burial containing both monk seal and human remains. The seal skeleton measured 1.70 metres (▶Fig. 165). It was buried near an ancient wall in the vicinity of the island's commercial harbour and the archaeologists investigating the site believe that the burial is more than 2000 years old and have therefore referred it to the 2nd century BC [1227–1219] (▶Fig. 166).

The archaeologist in charge of the excavations at the time of the discovery, Lefteris Platon, said that *"the ancient Greeks were in the habit of burying their beloved animals, mainly dogs, but this is the very first case of a buried seal"* [1227]. He suggested that the pinniped may have frequented the area, tame or semi-wild, evoking the love of people who conferred formal burial rites upon it after death. It is interesting to note that, in the

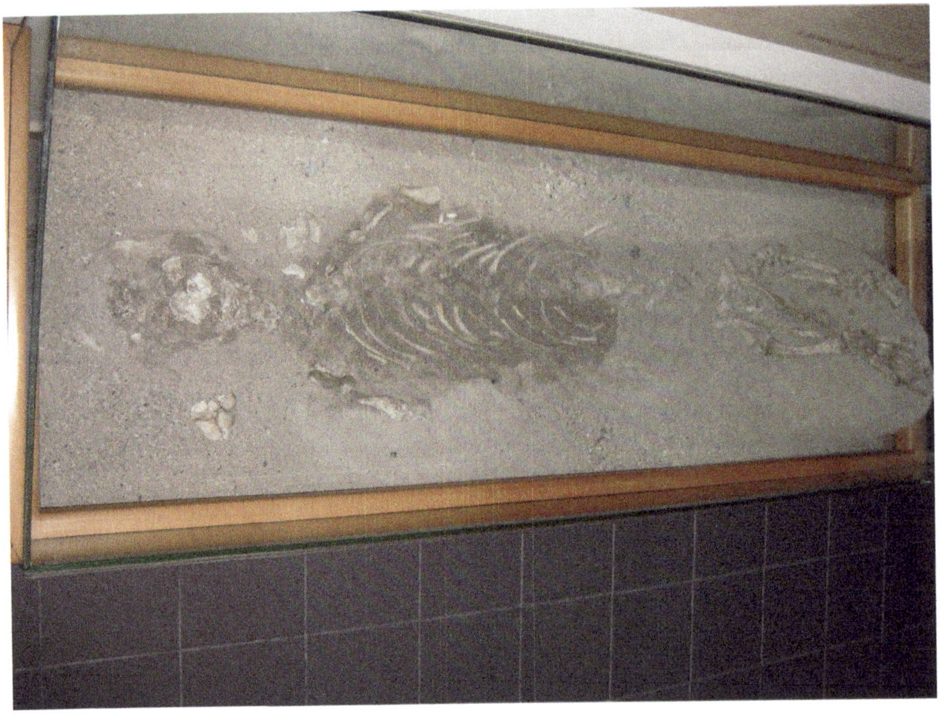

Fig. 165: The monk seal skeleton buried with human remains in the Hellenistic necropolis, aside the harbour of Rhodes. Courtesy of the 22nd Ephorate of Prehistoric and Classical Antiquity, Rhodes and of the Hellenic Centre for Marine Research, Hydrobiological Station of Rhodes.

Fig. 166: An aspect of the Hellenistic necropolis of Rhodes in the vicinity of the extant island's commercial harbour. Photo by the 22nd Ephorate of Prehistoric and Classical Antiquity, Rhodes.

majority of the most ancient cases known to date of inhumation of human beings along with other mammals, the latter almost always prove to be subadults [1229], as in the case of this Hellenistic monk seal of the Rhodian necropolis. As far as is presently known, the fact that the seal was honoured with ritual burial represents a unique discovery in the history of archaeology. Various different cases of the burial of animals with their supposed owners are known since prehistoric times. The possibility of hierarchically imposing ourselves, with ease, on the subadults of the wild species is a cognition which must have been acquired very early on by the prehistoric human populations. However, it is only starting from certain chronologies that this approach appears to have been used rationally to the point of evolving into the process of domestication. In the light of current archaeological knowledge, this process, as we have already had occasion to observe, appears to have taken place only when the human communities began to manifest the need for a permanent sedentary life. An example of this is the discovery in Ein Mallaha (Israel) of a dog pup found together with an inhumated human being, dated to the 8th millennium BC (compare with [163, 1230]). We cannot, however, exclude that in more remote prehistoric chronologies, an imposition of a permanent behavioural character may have already been experimented with, through the process of imprinting on species such as the wolf and other taxa, at times when the economic conditions of the human populations of hunters and gatherers were not yet such as to favour the development of an ongoing and permanent relationship with the animal. Effectively, the development of even the least rational form of breeding demands conditions of both cultural and geographic stability. In analogous cultural conditions it has been even possible to document the most ancient association known to date between man and the cat, *Felis silvestris* Schreber, 1775, in the case of the felid specimen buried along with a human from the site of Shillourokambos on the island of Cyprus, demonstrating that a close relationship had developed by about 9500 years ago [774].

Dog bones (▶Fig. 167), and the remains of 5–6 human adults, were also found in the same burial ground. The skeleton of the monk seal has been conserved by the Archaeological Service and is now on display at the museum of the Hydrobiological Station of Rhodes.

Fig. 167: Dog bones were also found in the same burial ground as the monk seal, in the Hellenistic necropolis of Rhodes. Photo by the 22nd Ephorate of Prehistoric and Classical Antiquity, Rhodes.

Bibliography

[0001] Alcover JA. Note on the origin of the present mammalian fauna from the Balearic and Pityusic Islands. Misc Zool 1980; 6: 141–9.

[0002] Vigne J-D. Les Mammifères post-glaciaires de Corse. Etude archéozoologique. XXVIe supplément à «Gallia Préhistoire». Paris, France: Éditions du Centre national de la recherche scientifique [CNRS], 1988.

[0003] Sanders EAC, Reumer JWF. The influence of prehistoric and Roman migrations on the vertebrate fauna of Menorca (Spain). British Archaeological Reports International Series 1984; 229: 119–44.

[0004] Blondel J, Vigne J-D. Space, time, and man as determinants of diversity of birds and mammals in the Mediterranean region. In: Ricklefs RE, Schluter D, eds. Species diversity in ecological communities. Chicago, IL, USA: Chicago University Press, 1993: 135–46.

[0005] Masseti M. Post-Pleistocene variations of the non-flying terrestrial mammals on some Italian islands. Supplemento alle Ricerche di Biologia della Selvaggina 1993; 21: 209–17.

[0006] Masseti M. Holocene and anthropochorous wild mammals of the Mediterranean islands. Anthropozoologica 1998; 28: 3–20.

[0007] Masseti M. The non-flying terrestrial mammals of the Mediterranean islands: an example of the role of the biological invasion of alien species in the homogenisation of biodiversity. Proceedings of the Workshop on Invasive Alien Species on European Islands and Evolutionary Isolated Ecosystems and Group of Experts on Invasive Alien Species (Horta/Azores, 10–12 October 2002). Strasbourg, France: Directorate of Culture and of Cultural and Natural Heritage, Council of Europe, 2002: T-PVS/IAS, 2: 4–6.

[0008] Masseti M. Holocene endemic and non-endemic mammals of the Aegean islands. In: Kotjabopoulou E, Hamilakis Y, Halstead P, Gamble C, Elefanti P, eds. Zooarchaeology in Greece. Recent advances. British School at Athens Studies 2003; 9: 47–57.

[0009] Masseti M. Mammals of the Mediterranean islands: homogenisation and the loss of biodiversity. Mammalia 2009; 73: 169–202.

[0010] Cheylan G. Endemism and speciation in Mediterranean mammals. Vie Milieu 1990; 40(2–3): 137–43.

[0011] Vigne J-D. L'emprise de l'homme depuis la préhistoire. In: Vigne J-D, ed. Îles. Vivre entre ciel et mer. Paris, France: Éditions Nathan; Muséum national d'Histoire naturelle, 1997: 89–103.

[0012] Vogel P, Hutterer R, Sarà M. The correct name, species diagnosis and distribution of the Sicilian shrew. Bonn Zool Beitr 1989; 40: 243–8.

[0013] Sarà M. The Sicilian (*Crocidura sicula*) and the Canary (*C. canariensis*) shrew (Mammalia, Soricidae): peripheral isolate formation and geographic variation. Boll Zool 1995; 62: 173–82.

[0014] Sarà M, Vitturi R. *Crocidura* populations (Mammalia, Soricidae) from the Sicilian-Maltese insular area. Hystrix It J Mamm NS 1996; 8(1–2): 121–32.

[0015] Reumer JWF. Notes on the Soricidae (Insectivora, Mammalia) from Crete. I. The Pleistocene species *Crocidura zimmermanni*. Bonn Zool Beitr 1986; 37(3): 161–71.

[0016] Reumer JWF, Payne S. Notes on the Soricidae (Insectivora, Mammalia) from Crete. II. The shrew remains from Minoan and Classical Kommos. Bonn Zool Beitr 1986; 37(3): 173–82.

[0017] Vogel P, Maddalena T, Schembri PJ. Cytotaxonomy of shrews of the genus *Crocidura* from Mediterranean islands. Vie Milieu 1990; 40(2–3): 124–9.

[0018] Cucchi T, Orth A, Auffray J-C, Renaud S, Fabre L, Catalan J, Hadjisterkotis E, Bonhomme F, Vigne J-D. A new endemic species of the subgenus *Mus* (Rodentia, Mammalia) on the Island of Cyprus. Zootaxa 2006; 1241: 1–36.

[0019] Bonhomme F, Orth A, Cucchi T, Hadjisterkotis E, Vigne J-D, Auffray J-C. Découverte d'une nouvelle espèce de souris sur l'île de Chypre. Comptes Rendus Biologies 2004; 327(5): 501–7.

[0020] Abiadh A, Colangelo P, Capanna E, Lamine-Cheniti T. Morphometric analysis of six Gerbillus species (Rodentia, Gerbillinae) from Tunisia. Comptes Rendus Biologies 2010; 333: 680–687.

[0021] Abiadh A, Chetoui M, Lamine-Cheniti T, Capanna E, Colangelo P. Molecular phylogenetics of the genus Gerbillus (Rodentia, Gerbillinae): Implications for systematics, taxonomy and chromosomal evolution. Molecular Phylogenetics and Evolution 2010; 56: 513–528.

[0022] Cockrum EL, Vaughan TC, Vaughan PJ. A review of North African short-tailed Gerbils (*Dipodillus*). With description of new taxon from Tunisia. Mammalia 1976; 40: 313–26.

[0023] Masseti M, Vianello F. Importazioni preistoriche di mammiferi alloctoni nelle isole del Mar Tirreno centro-settentrionale. Rivista di Scienze Preistoriche 1991; 43(1–2): 275–92.

[0024] Giokas S, Legakis A, Sfenthourakis S. Biogeography in a country of islands. In: Abstracts of the 9th International Congress on the Zoogeography and Ecology of Greece and Adjacent Regions [ICZEGAR] (Thessaloniki, 22–25 May 2002). Hellenic Zoological Society. Thessaloniki, Greece: University Studio Press, 2002: 40.

[0025] Legakis A. Zoological research in the Dodecanese. In: Masseti M, ed. Island of deer. Natural history of the fallow deer of Rhodes and of the vertebrates of the Dodecanese (Greece). City of Rhodes: Environment Organization, 2002: 31–34.

[0026] Flux JEC, Fullagar PJ. World distribution of the Rabbit *Oryctolagus cuniculus* on islands. Mammal Rev 1992; 22(3–4): 151–205.

[0027] Bousbouras D, Ioannidis Y, Matsakis J. Quelques informations récentes sur le daim (*Dama dama*) de l'île de Rhodos. Biol Gallo-Hell 1991; 18(1): 7–12.

[0028] Masseti M. The fallow deer, *Dama dama* L, 1758, in the Aegean region. Contributions to the Zoogeography and Ecology of the Eastern Mediterranean Region 1999; 1 (Supplement): 17–30.

[0029] Masseti M, ed. Island of deer. Natural history of the fallow deer of Rhodes and of the vertebrates of the Dodecanese (Greece). City of Rhodes: Environment Organization, 2002.

[0030] Masseti M. Concerning the mammalian species currently regarded as endemic to the Mediterranean islands. In: Prigioni C, Sforzi A, eds. Abstracts of the 5th European Congress of Mammalogy [ECM] (Siena, 21–26 September 2007). Hystrix It J Mamm NS 2007; I-2 Suppl: 427.

[0031] Masseti M, Cavallaro A, Pecchioli E, Vernesi C. Artificial occurrence of the fallow deer, *Dama dama dama* (L, 1758), on the island of Rhodes (Greece): insight from mtDNA analysis. Human Evolution 2006; 21(2): 167–76 [Special Issue on the Anthropochorous Fauna (Ed. by M Masseti)].

[0032] Masseti M, Pecchioli E, Vernesi C. Phylogeography of the last surviving populations of Rhodian and Anatolian fallow deer (*Dama dama dama* L, 1758). Biol J Linn Soc 2008; 93(4): 835–44.

[0033] Papageorgiou N. Food preferences and survival of the agrimi (*Capra aegagrus cretica*) on Crete. MS thesis. East Lansing, MI, USA: Michigan State University, 1972.

[0034] Papageorgiou N. Population energy relationships of the agrimi (*Capra aegagrus cretica*) on Theodorou Island, Greece. Unpublished PhD dissertation. East Lansing, MI, USA: Michigan State University, 1974.

[0035] Papageorgiou N. Population energy relationships of the Agrimi of Theodorou Island, Greece. Mammalia depicta (Beihefte zur Zeitschrift für Säugetierkunde) 1979; 11: 1–56.

[0036] Husband P. The ecology and behavior of the Cretan agrimi following population reduction. Unpublished PhD dissertation. East Lansing, MI, USA: Michigan State University, 1977.

[0037] Husband TP, Davis PB. Population dynamics of the Cretan agrimi. Boston, MA, USA: School for Field Studies, Northeastern University, 1983.
[0038] Husband TP, Davis PB. Ecology and behaviour of the Cretan agrimi. Canadian Journal of Zoology 1984; 62(3): 411–20.
[0039] Husband TP, Brown Jr JH. The biology and management of the Cretan agrimi. Biol Gallo-Hell 1994; 22: 255–6.
[0040] Husband TP, Davis PB, Brown Jr JH. Population measurement of the Cretan agrimi. J Mammal 1986; 67(4): 757–9.
[0041] Logan GT. A study of the conservation biology of the Cretan wild goat. MSc thesis. Rhode Island (RI), USA: University of Rhode Island, 1989.
[0042] Sfougaris AI. Wild goats (*Capra aegagrus*) in Greece: distribution and status. Biol Gallo-Hell 1994; 22: 233–40.
[0043] Sfougaris AI. The distribution, ecology and management of goats *Capra aegagrus* in Greece. An outline. In: Lovari S, Shackleton DM, eds. Caprinae news (Caprinae: Specialist Group of the Species Survival Commission [SSC] of IUCN [International Union for Conservation of Nature]), 1995: No. 8/9, 5–9.
[0044] Logan GT, Brown JH, Husband TP, Nicholson MC. Conservation biology of the Cretan agrimi. Biol Gallo-Hell 1994; 22: 241–6.
[0045] Sfougaris AI, Lymberakis P. *Capra aegagrus* (Erxleben, 1777). In: Legakis A, Maragou P, eds. The Red Data Book of Endangered Animals of Greece (in Greek with English summary). Athens, Greece: Ministry of the Environment; Hellenic Zoological Society, 2009: 373–5.
[0046] Barome P-O, Lymberakis P, Monnerot M, Gautun J-C. Cytochrome *b* sequences reveal *Acomys minous* (Rodentia, Muridae) paraphyly and answer the question about the ancestral karyotype of *Acomys dimidiatus*. Mol Phylogenet Evol 2001; 18: 37–46.
[0047] Giagia-Athanasopoulou EB, Rovatsos MTH, Mitsainas GP, Martimianakis S, Lymberakis P, Angelou L-XD, Marchal JA, Sánchez A. New data on the evolution of the Cretan spiny mouse, *Acomys minous* (Rodentia, Murinae), shed light on the phylogenetic relationships in the *cahirinus* group. Biol J Linn Soc 2011; 10(3): 499–509.
[0048] Niethammer G, Niethammer J. Zur Variabilität der Kehlzeichnung beim Steinmarder, *Martes foina* (Erxleben, 1777). Z Saugetierkunde 1967; 32(3): 185–7.
[0049] Niethammer J. Die Säugetiere von Korfu. Bonn Zool Beitr 1962; 13: 1–49.
[0050] Ioannidis Y. Πανίδα της Ζακύνθου. Συμβολή στη Ε.Π.Μ. την δυτική ακτή της νήσου Ζακύνθου. Unpublished manuscript. Athens, Greece: WWF Greece, 2000.
[0051] Legakis A, Sfenthourakis S, Adamopoulou C. A contribution to the knowledge of the vertebrate fauna of Milos island (Kyklades, Greece). Newsletter of the Hellenic Zoological Society 1997; 30: 5–7.
[0052] Niethammer J. Kleinsäuger von Kithira, Giechenland. Säugetierkundliche Mitteilungen 1971; 4: 363–5.
[0053] Kominos T. Κυθηρα: Στο δρόμο των πουλιών Athens, Greece: Hellenic Ornithological Society, 1995.
[0054] Zimmermann K. Die Rodentia Kretas. Z Saugetierkd 1953; 17: 21–51.
[0055] de Beaux O. Ricerche faunistiche nelle isole italiane dell'Egeo. Mammiferi. Archivio Zoologico Italiano 1929; 12–13[1928–1929] (7): 135–54.
[0056] Paragamian K. Barn owl (*Tyto alba*) pellet analysis from Karpathos island, Greece. In: Abstracts of the 8th International Congress on the Zoogeography and Ecology of Greece and Adjacent Regions [ICZEGAR] (Kavala/Greece, 17–21 May 1999). Athens, Greece: Hellenic Zoological Society, 1999: 112.
[0057] Festa E. Escursioni Zoologiche del Dr. Enrico Festa nell'Isola di Rodi. XI. Mammiferi. Bollettino dei Musei di Zoologia ed Anatomia comparata della R Università di Torino 1914; 686: 1–21.
[0058] von Wettstein O. Die Säugetiere der Ägäis, nebst einer Revision des Rassenkreises von *Erinaceus europaeus*. Ann Nathist Mus Wien 1942; 12: 245–78.

[0059] Pieper H. Über einige bemerkenswerte Kleinsäuger-Funde auf den Inseln Rhodos und Kos. Acta Biologica Hellenica 1966; 1: 21–8.

[0060] Masseti M. The non-flying terrestrial mammals of the Dodecanese islands. In: Masseti M, ed. Island of deer. Natural history of the fallow deer of Rhodes and of the vertebrates of the Dodecanese (Greece). City of Rhodes: Environment Organization, 2002: 89–98.

[0061] Angelici FM, Pinchera F, Riga F. First record of *Crocidura* sp, *Mus domesticus* and notes on the Mammals of Astipalaia Island (Dodecanese, Greece). Mammalia 1992; 56(1): 159–61.

[0062] Angelici FM, Riga F. Primi dati sull'alimentazione del Barbagianni *Tyto alba* nell'isola di Stampalia [Astypalaia] (Dodecaneso, Grecia). In: Atti del 6° Convegno Italiano di Ornitologia (Torino, 8–11 October 1991). Torino, Italy: Museo regionale di Scienze naturali Torino, 1994: 419–20.

[0063] Masseti M, Sarà M. Non-volant Terrestrial mammals on Mediterranean Islands: Tilos (Dodecanese, Greece), a Case Study. Bonn Zool Beitr 2002; 51(4): 261–8.

[0064] Niethammer J. Gewöllinhalte der Schleiereule (*Tyto alba*) von Kos und aus Südwestanatolien. Bonn Zool Beitr 1989; 40: 1–9.

[0065] van Laar V, Daan S. The Etruscan Shrew, *Suncus etruscus* (Savi, 1822), found on Samos, Greece. Z Saugetierkd 1967; 32: 174–5.

[0066] Dimitropoulos A, Dimaki M, Ioannidis I. ζώα της Σάμου. Αναφορά στην πανίδα του νησιού. Πνευματικό Ίδρυμα Σάμου «Νικόλαος Δημητρίου». Athens, Greece: Βιβλιοθήκη Επιστημονικών Εκδόσεων 14, 1998. English Version: Dimitropoulos A, Dimaki M, Ioannidis I. The animals of Samos. Contribution to the island fauna. Cultural Society of Samos, 1998.

[0067] Kock D. Zur Säugetierfauna der Insel Chios, Ägäis (Mammalia). Senckenberg Biol 1974; 55: 1–19.

[0068] Giagia EB, Ondrias JC. Karyological analysis of eastern European hedgehog *Erinaceus concolor* (Mammalia, Insectivora) in Greece. Mammalia 1980; 44(1): 59–71.

[0069] Spinthakis V, Houlis D, Choremi J. The birds of the island of Chios, Greece. Homeropoulis/Chios, Greece: Municipality of Homeropoulis, 1993.

[0070] Ondrias JC. Die Säugetiere Griechenlands. Säugetierkundliche Mitteilungen 1965; 13: 109–27.

[0071] Ondrias JC. The taxonomy and geographical distribution of the rodents of Greece. Säugetierkundliche Mitteilungen 1966; 14: 1–136.

[0072] Pieper H. Zur Säugetiernahrung des Uhus (*Bubo bubo*) auf der griechischen Insel Lesbos. Vogelwelt 1981; 102(2): 55–6.

[0073] Vogel P, Sofianidou TS. The shrews of the genus *Crocidura* on Lesbos, an eastern Mediterranean island. Bonn Zool Beitr 1996; 46(1–4): 339–47.

[0074] Özkan B. Gökçeada ve Bozcaada Adalarının Kemiricileri [Rodents on the islands of Gökçeada and Bozcaada]. Unpublished PhD thesis. Edirne, Turkey: Fen Bilimleri Enstitüsü, Trakya Üniversitesi, 1995.

[0075] Özkan B. Gökçeada ve Bozcaada Kemirici Faunası (Mammalia; Rodentia) [Rodent fauna of Gökçeada and Bozcaada (Mammalia, Rodentia)]. Turk J Zool 1999; 23(1): 133–47.

[0076] Özkan B, Kryštufek B. Wood mice, *Apodemus* of two Turkish islands: Gökçeada and Bozcaada. Folia Zool 1999; 48(1): 17–24.

[0077] Vohralík V, Sofianidou T. Small mammals (Insectivora, Rodentia) of Thrace, Greece. Acta Universitatis Carolinae Biologica 1992; 36: 341–69.

[0078] Prigioni C. Distribution of mammals in Albania. Hystrix It J Mamm NS 1996; 8(1–2): 67–73.

[0079] Antěra, M. First records of Microtus (Pitymys) felteni in Albania. Z Saugetierkd 1991; 56: 380.

[0080] Rovatsos MT, Mitsainas GP, Paspali G, Oruci S, Giagia-Athanasopoulou EB. Geographical distribution and chromosomal study of the underground vole *Microtus thomasi* in Albania and Montenegro. Mamm Biol 2011; 76: 22–7.

[0081] Prigioni C, Bogliani G, Barbieri F. The otter *Lutra lutra* in Albania. Biol Conserv 1986; 36: 375–83.

[0082] Grémillet X. Field survey of *Lutra lutra* on Corfu island (Greece). IUCN Otter Spec Group Bull 1993; 8: 39–42.

[0083] Fistani A. First record of a fossil *Cervus elaphus* L, 1758, from Albania. Z Saugetierkd 1991; 56: 248–51.

[0084] Bego F, Kryštufek B, Paspali G, Rogozi E. Small terrestrial mammals of Albania: annotated list and distribution. Hystrix It J Mamm NS 2008; 19(2): 83–101.

[0085] Bego F. Kontribut ne njohjen e gjitareve te vegjel (Mammalia: Insectivora, Chiroptera dhe Rodentia) te Shqiperise: te dhena taksonomike, biogjegrafike and bioekologjike [Contribution to the knowledge of small mammals (Mammalia: Insectivora, Chiroptera and Rodentia) of Albania: taxonomic data, biogeography and bioecology]. Unpublished PhD thesis. Tirana, Albania: University of Tirana [Universiteti i Tiranes], 1997.

[0086] Bego F. Te dhena te reja mbi Gjitaret (Mammalia) e vendit tone: statusi i njohje, i perhapjes gjeografike dhe statusi i ruajtjes dhe rrezikimit te tyre [New data on mammals (Mammalia) of our country: status of knowledge, geographic distribution, conservation status and risk]. Studime Biologjike 2001; numer special 5–6: 274–9.

[0087] Bego F. Te dhena mbi komunitetet e gjitareve te vegjel (Mammalia: Rodentia dhe Insectivora) te fushes se Vurgut (Messopotam) dhe Myzeqese (Apolloni) [Data on communities of small mammals (Mammalia: Rodentia and Insectivora) in the plains of Vurgut (Messopotam) and Myzeqe (Apolloni)]. Akademia e Shkencave: Studime Biologjike 2003; 7: 123–31.

[0088] Mitchell-Jones AJ, Amori G, Bogdanowicz W, Kryštufek B, Reijnders PJH, Spitzenberger F, Stubbe M, Thissen JBM, Vohralík V, Zima J (eds.). The Atlas of European Mammals. London, UK: Academic Press, 1999.

[0089] Giannatos G, Albayrak T, Erdoğan A. Status of the Caracal in Protected Areas in Southwestern Turkey. Cat News 2006; 45: 23–4.

[0090] Masseti M. Carnivores of Syria. In: Neubert E, Amr Z, Taiti S, Gümüs BA, eds. Animal Biodiversity in the Middle East. Proceedings of the First Middle Eastern Biodiversity Congress (Aqaba, Jordan: 20–23 October 2008). ZooKeys 2009; 31: 229–52 (DOI: 10.3897/zookeys.31.170).

[0091] de Marinis AM, Masseti M. Mammalian fauna of the Termessos National Park, Turkey. In: Neubert E, Amr Z, Taiti S, Gümüs BA, eds. Animal Biodiversity in the Middle East. Proceedings of the First Middle Eastern Biodiversity Congress (Aqaba, Jordan: 20–23 October 2008). ZooKeys 2009; 31: 221–8 (DOI: 10.3897/zookeys.31.171).

[0092] Kryštufek B, Tvrtković N, Paunović M, Özkan B. Size variation in the Northern white-breasted hedgehog, *Erinaceus roumanicus*: latitudinal cline and the island rule. Mammalia 2009; 73(4): 299–306.

[0093] Kryštufek B, Vohralík V. Mammals of Turkey and Cyprus. Introduction, Checklist, Insectivora. Koper, Slovenia: Knjižnica Annales Majora, 2001.

[0094] Kryštufek B, Vohralík V. Mammals of Turkey and Cyprus. Rodentia I. Sciuridae, Dipodidae, Gliridae, Arvicolidae. Koper, Slovenia: Založba Annales (Knjižnica Annales Majora), 2005.

[0095] Kryštufek B, Vohralík V. Mammals of Turkey and Cyprus. Rodentia II: Cricetinae, Muridae, Spalacidae, Calomyscidae, Capromydae, Hystricisae, Casoridae. Koper, Slovenia: Založba Annales (Knjižnica Annales Majora), 2009.

[0096] Pieroni M. Il fuoco imperfetto. Florence [Firenze], Italy: Phasar Edizioni, 2006.

[0097] Handrinos G, Akriotis T. The birds of Greece. London, UK: Christopher Helm/A and C Black, 1997.

[0098] Higgins MD, Higgins R. A geological companion to Greece and the Aegean. London, UK: Duckworth, 1996.

[0099] Lincoln RJ, Boxshall GA. The Cambridge illustrated dictionary of Natural History. Cambridge, UK; New York, NY, USA: Cambridge University Press, 1987.

[0100] Psarianos P. Über das Vorkommen von *Hippopotamus* auf Kephallinia (Griechenland). Praktiká Akademías Athinón 1953; 28: 408–12.
[0101] Caloi L, Kotsakis T, Palombo MR. La fauna a vertebrati terrestri del Pleistocene delle isole del Mediterraneo. Geologica Romana 1986; 25(6): 235–56.
[0102] Sordinas A. Investigations of the prehistory of Corfu during 1964–1966. Balkan Studies 1969; 10: 393–424.
[0103] Marinos G. Επί της παρακτίου θαλασσάς γεωλογίας του Ιονίου και του Αιγαίου πελάγους. ΙΩΚΑΕ (Ινστιτούτο Ωκεανογραφικών και Αλιευτικών Ερευνών [Ι.ΩΚ.Α.Ε]), περίοδος Γ, 1971: 10α, 9–16.
[0104] Marinos G, Symeonidis NK. Island populations of dwarf mammals of the Aegean Archipelago during Quaternary. Ann Geol Pays Hell 1977; 28: 352–67 (in Greek with English summary).
[0105] Dermitzakis MD, Sondaar PY. The importance of fossil mammals in reconstructing paleogeography with special reference to the Pleistocene Aegean Archipelago. Ann Geol Pays Hell 1978; 29: 808–40.
[0106] Kotsakis T. Insular and non-insular vertebrate fossil faunas in the Eastern Mediterranean islands. Atti dei Convegni Lincei 1990; 85: 289–334.
[0107] van Andel TH, Shackleton J. Late Paleolithic and Mesolithic coastlines of Greece and the Aegean. Journal of Field Archaeology 1982; 9: 445–54.
[0108] Mavridis F. Early island archaeology and the extinction of endemic fauna in the eastern Mediterranean: problems of interpretation and methodology. In: Kotjabopoulou E, Hamilakis Y, Halstead P, Gamble C, Elefanti P, eds. Zooarchaeology in Greece. Recent Advances. British School at Athens Studies 2003; 9: 65–74.
[0109] Lorenzini R, Lovari S, Masseti M. The rediscovery of the Italian roe deer: genetic differentiation and management implications. Ital J Zool 2002; 69(4): 367–79.
[0110] Peters J, Helmer D, von den Driesch A, Saña Segui M. Early animal husbandry in the northern Levant. Paléorient 1999; 25(2): 27–47.
[0111] Schmidt, K. Frühe Tier- und Menschenbilder vom Göbekli Tepe; Kampagnen 1995–98. Ein kommentierter Katalog der Grossplastik und der Reliefs. Istanbuler Mitteilungen 1999; 49: 5–21.
[0112] Uerpmann H-P. Probleme der Neolithisierung des Mittelmeersraumes. In: Blume H, Frey W, eds. Tübinger Atlas des Vorderen Orients [TAVO]. Beihefte. Reihe B: Geisteswissenschaften. B28. Wiesbaden, Germany: Dr Ludwig Reichert, 1979.
[0113] Lewthwaite J. Three steps to leaven: applicazione del modello di disponibilità al Neolitico italiano. In: Atti della XXVI Riunione Scientifica dell'Istituto Italiano di Preistoria e Protostoria [IIPP], "Il Neolitico in Italia" (Firenze, 7–10 November 1985). Florence [Firenze], Italy: Istituto Italiano di Preistoria e Protostoria, 1987: 89–102.
[0114] Binder D. Aspects de la néolithisation dans les aires padané, provençale et ligure. In: Aurenche O, Cauvin J, eds. Néolithisation. British Archaeological Reports International Series 1989; 516: 199–226.
[0115] Guilaine J. La Mer Partagée. La Méditerranée Avant l'Écriture (7000–2000 avant Jesus-Christ). Paris, France: Hachette, 1994.
[0116] Orliac M. L'homme marin. In: Vigne J-D, ed. Îles. Vivre entre ciel et mer. Paris, France: Éditions Nathan; Muséum national d'Histoire naturelle, 1997: 39–53.
[0117] Masseti M, Darlas A. Pre-Neolithic man and other mammals on the Eastern Mediterranean islands. Arkeos 1999; 5: 189–204.
[0118] Guilaine J, Briois F, Coularou J, Vigne J-D, Carrère I. Shillourokambos et les debuts du Néolithique à Chypre. Espacio, Tiempo y Forma, Série I, Prehistoria y Arquelogía 1996; 9: 159–71.
[0119] Guilaine J, Briois F, Vigne J-D, Carrère I. Découverte d'un Néolithique précéramique ancien chypriote (fin 9e, début 8e millénaires cal BC), apparenté au PPNB ancien/moyen du levant nord. CR Acad Sci II A 2000; 330: 75–82.

[0120] Peltenburg E, Colledge S, Croft P, Jackson A, McCartney C, Murray MA. Agro-pastoralist colonization of Cyprus in the 10th millennium BP: initial assessments. Antiquity 2000; 74: 844–53.

[0121] Guilaine J, Le Brun A, eds. Le Néolithique de Chypre. Actes du Colloque International (Nicosia [Nicosie], 17–19 May 2001). Bulletin de Correspondance Hellénique, vol suppl 2003; 43: 17–9.

[0122] Blondel J, Aronson J. Biology and wildlife of the Mediterranean Region. Oxford, UK; New York, NY, USA: Oxford University Press, 1999.

[0123] Hammer CU, Clausen HB, Friedrich WL, Taubert H. The Minoan eruption of Santorini in Greece dated back to 1645 BC? Nature 1997; 328: 517–19.

[0124] Warren P. Absolute dating of the Bronze Age eruption of Thera (Santorini). Nature 1984; 308: 492–3.

[0125] Sparks RSJ, Sigurdsson H, Watkins ND. The Thera eruption and Late Minoan-IB destruction on Crete. Nature 1978; 271: 91.

[0126] Doumas C. The prehistoric eruption of Thera and its effects. The evidence from Rhodes. In: Diez S, Papachristodoulou I, eds. Archaeology in the Dodecanese. Copenhagen, Denmark: Department of Near Eastern and Classical Antiquities, The National Museum of Denmark, 1988: 34–8.

[0127] Masseti M. Ancient historical faunae of continental and insular Asia Minor, and their relations with the western Mediterranean, with particular reference to the Italian peninsula. International Journal of Anthropology, 2007; 22(3–4): 177–95.

[0128] Rackham O, Moody J. The making of the Cretan landscape. Manchester, UK; New York, NY, USA: Manchester University Press, 1996.

[0129] Benda P, Georgiakakis P, Dietz C, Hanák V, Galanaki K, Markantonatou V, Chudárková A, Hulva P, Horáček I. Bats (Mammalia: Chiroptera) of the Eastern Mediterranean and Middle East. Part 7. The bat fauna of Crete, Greece. Acta Soc Zool Bohem 2008; 72: 105–90.

[0130] Legakis A, Kypriotakis Z. A biogeographical analysis of the island of Crete, Greece. J Biogeogr 1994; 21: 441–5.

[0131] Tsachalidis EP, Hadjisterkotis E. Current distribution and population status of wild boar (*Sus scrofa* L.) in Greece. Acta Silvatica et Lignaria Hungarica 2009; 5: 153–7.

[0132] Dapper O. Description Exacte Des Isles De L'Archipel. De l'Île de Crete ou Candie. Amsterdam, Netherlands: George Gallet, 1703.

[0133] Azzaroli A. Il significato delle faune insulari quaternarie. Le Scienze 1971; 30: 84–93.

[0134] Azzaroli A. Considerazioni sui mammiferi fossili delle isole mediterranee. Boll Zool 1977; 44: 201–11.

[0135] Sondaar PY. Palaeozoogeography of the Pleistocene mammals from the Aegean. In: Strid A, ed. Evolution in the Aegean. Opera botanica 1971; 30: 60–70.

[0136] Sondaar PY. Insularity and its effect on mammal evolution. In: Hecht MK, Goody PC, Hecht BM, eds. Major Patterns in Vertebrate Evolution. New York, NY, USA: Plenum Press 1977, 671–707.

[0137] Masseti M, Mazza P. Is there any palaeontological "treatment" for the "insular syndrome"? Vie Milieu 1996; 46(3–4): 355–63.

[0138] Sondaar PY, Boekschoten GJ. Quaternary mammals in the South Aegean island arc; with notes on other fossil mammals from the coastal regions of the Mediterranean. I–II. K Ned Akad van Wet-B 1967; 70: 556–76.

[0139] Sondaar PY, de Vos J, Dermitzakis MD. Late Cenozoic faunal evolution and paleogeography of the South Aegean island arc. Mod Geol 1986; 10: 249–59.

[0140] Wallace AR. The geographical distribution of animals. New York, NY, USA: Harper, 1876.

[0141] Ghigi A. La vita degli animali. II. Mammiferi e uccelli sedentari delle terre continentali. Torino, Italy: Unione Tipografico-Editrice Torinese, 1950.

[0142] Dermitzakis MD, Theodorou GE. Glossary of Geological Terms. Athens, Greece: Despina Mavrommaty, 1994.

[0143] Alcover JA, Campillo X, Macias M, Sans A. Mammal species of the world: additional data on insular mammals. Am Mus Novit 1998; 3248: 1–29.

[0144] Vigne J-D. The large "true" Mediterranean islands as a model for the Holocene human impact on the European vertebrate fauna? Recent data and new reflections. In: Archäologie in Eurasien. Volume 6. Benecke N, ed. The Holocene History of the European vertebrate Fauna. Modern Aspects of Research. Papers presented at an international workshop (Berlin, 6–9 April 1998). Rahden/Westfalen, Germany: Marie Leidorf, 1999: 295–322.

[0145] Gorman M. Island ecology. London, UK: Chapman and Hall, 1979.

[0146] Felten H, Storch G. Kleinsäuger von den Italienischen Mittelmeer-Inseln Pantelleria und Lampedusa (Mammalia). Senckenberg Biol 1970; 51: 227–90.

[0147] Krapp F. Gli *Apodemus* d'Italia in rapporto alla situazione europea del genere. Supplemento alle Ricerche di Biologia della Selvaggina 1984; 9: 51–63.

[0148] Kurtén B. Pleistocene mammals of Europe. London, UK: Weidenfeld & Nicolson, 1968.

[0149] Symeonidis NK, Bachmayer F, Zapfe H. Entdeckung von Zwergelefanten auf der Insel Rhodos. Ann Nathist Mus Wien 1974; 78: 193–202.

[0150] Bachmayer F, Symeonidis N, Seemann R, Zapfe H. Die Ausgrabungen in der Zwergelefantenhöhle "Charkadio" auf der Insel Tilos (Dodekanes, Griechenland) in den Jahren 1974 und 1975. Ann Nathist Mus Wien 1976; 80: 113–44.

[0151] Bachmayer F, Symeonidis N, Zapfe H. Die Ausgrabungen in der Zwergelefantenhöhle der Insel Tilos (Dodekanes, Griechenland) im Jahr 1983. Sitzungsber Österr Akad Wiss Mathem Naturwiss Kl Abt 1, 1984; 193(6–10): 321–8.

[0152] Symeonidis N, Bachmayer F, Zapfe H. Grabungen in der Zwergelefanten-Höhle Charkadio auf der Insel Tilos (Dodekanes, Griechenland). Ann Nathist Mus Wien 1973; 77: 133–9.

[0153] Theodorou GE. Die fossilen Zwergelefanten der Höhle "Charkadio" auf der Insel Tilos (Dodekanes, Griechenland). Anz Österr Akad Wiss Mathem Naturwiss Kl 1983; 120: 83–5.

[0154] Theodorou GE. Environmental factors affecting the evolution of island endemics: the Tilos example from Greece. Mod Geol 1988; 13: 183–8.

[0155] Theodorou GE, Symeonidis NK. The excavations of the last ten years at Charkadio cave on Tilos Island, Dodekanese, Greece. In: Caravetta G, Gioia P, Mussi M, Palombo MR, eds. La terra degli elefanti/The world of elephants. Proceedings of the 1st International Congress (Rome, 16–20 October 2001). Rome [Roma], Italy: Comune di Roma; Consiglio Nazionale delle Ricerche; Ministero per i Beni e le Attività Culturali; Università degli Studi di Roma "La Sapienza", 2001: 514–8.

[0156] Poulakakis N, Theodorou GE, Zouros E, Mylonas M. Molecular phylogeny of the extinct Pleistocene dwarf elephant *Palaeoloxodon antiquus falconeri* from Tilos island, Dodekanisa, Greece. J Mol Evol 2002; 55: 364–74.

[0157] Theodorou G, Symeonidis N, Stathopoulou E. *Elephas tiliensis* n. sp. from Tilos island (Dodecanese, Greece). Hellenic Journal of Geosciences 2007; 42: 19–32.

[0158] Vaufrey R. Les éléphants nains des îles méditerranéennes et la question des isthmes pléistocènes. Mémoire des Archives de l'Institut de Paléontologie Humaine 1929; 6: 1–220.

[0159] Papp A. Die paläogeologische Entstehung der Ägäis nach dem derzeitigen Stand unserer Erkentnisse. In: von Wettstein O, ed. Herpetologia Aegaea. Sitzungsber Österr Akad Wiss Mathem Naturwiss Kl Abt 1, 1953; 162: 815–18.

[0160] Mitzopoulos MK. Über einen pleistozänen Zwergelefanten von der Insel Naxos (Kykladen). Praktiká Akademías Athinón 1961; 36: 332–40.

[0161] Boekschoten GJ, Sondaar PY. The Pleistocene of Katharo Basin (Crete) and its *Hippopotamus*. Bijdr Dierkd 1966; 36(8): 17–44.

[0162] Johnson DL. Problems in the land vertebrate zoogeography of certain islands and the swimming powers of elephants. J Biogeogr 1980; 7: 383–98.

[0163] Davis SJM. The archaeology of animals. London, UK: B T Batsford, 1987.

[0164] Honea K. Prehistoric remains on the island of Kythnos. Am J Archaeol 1975; 79(3): 277–9.

[0165] Anastasakis GC, Dermitzakis M. Post-Middle-Miocene paleogeography evolution of the Central Aegean Sea and detailed Quaternary reconstruction of the region. Its possible influence on the distribution of the Quaternary mammals of the Cyclades islands. Neues Jahrb Geol P-M 1990; 22: 1–16.

[0166] Cherry JF. The first colonization of the Mediterranean islands: a review of recent research. Journal of Mediterranean Archaeology 1990; 3: 145–221.

[0167] Reese DS, Belluomini G, Ikeya M. Absolute dates for the Pleistocene fauna of Crete. In: Reese DS, ed. Pleistocene and Holocene fauna of Crete and its first settlers. Madison, WI, USA: Prehistory Press, 1996: 47–51.

[0168] Kuss SE. Die pleistozänen Säugetierfaunen der ostmediterranen Inseln. Ihr Alter und ihre Herkunft. Berichte der Naturforschenden Gesellschaft zu Freiburg i. Br. [im Breisgau] 1973; 36: 49–71.

[0169] Kuss SE. Abfolge und Alter der pleistozänen Säugetierfaunen der Insel Kreta. Berichte der Naturforschenden Gesellschaft zu Freiburg i. Br. [im Breisgau] 1970; 60: 35–83.

[0170] Schlager N. Archäologische Geländeprospektion Südostkretas. Erste Ergebnisse. Österreichisches Archäologisches Institut Berichte und Materialien 1991: 23–6.

[0171] Mol D, de Vos J, van den Berg GD, Sondaar PY. The taxonomy and ancestry of the fossil elephants of Crete: faunal turnover and a comparison with proboscidean faunas of Indonesian islands. In: Reese DS, ed. Pleistocene and Holocene fauna of Crete and its first settlers. Madison, WI, USA: Prehistory Press, 1996: 81–98.

[0172] Caloi L, Palombo MR. Functional aspects and ecological implications in hippopotami and cervids of Crete. In: Reese DS, ed. Pleistocene and Holocene fauna of Crete and its first settlers. Madison, WI, USA: Prehistory Press, 1996: 125–51.

[0173] Masseti M. Deer and dwarf elephants of the Dodecanese. The Holocene redefinition of the faunal ecosystems of Rhodes and other nearby islands. In: Karageorghis V, Giannikouri A, eds. Conservation and preservation of the cultural and natural heritage of the large islands of the Mediterranean. Athens, Greece: Ministry of Culture; Archaeological Institute of Aegean Studies; Foundation Anastasios G Leventis, 2006: 113–22.

[0174] Boekschoten GJ, Sondaar PY. On the fossil mammalia of Cyprus. K Ned Akad van Wet-B 1972; 75: 306–38.

[0175] Symeonidis N, Theodorou GE. New findings of fossil elephants on Crete island. Ann Geol Pays Hell 1982; 31: 113–29.

[0176] Caloi L, Kotsakis T, Palombo MR, Petronio C. The Pleistocene dwarf elephants of Mediterranean islands. In: Shoshani J, Tassy P, eds. The Proboscidea. Evolution and Palaeoecology of Elephants and Their Relatives. Oxford, UK; New York, NY, USA; Tokyo, Japan: Oxford University Press, 1996: 234–9.

[0177] Herridge VL, Lister AM. The evolution of dwarf elephants in the Mediterranean. In: Prigioni C, Sforzi A, eds. Abstracts of the 5th European Congress of Mammalogy [ECM] (Siena, 21–26 September 2007). Hystrix It J Mamm NS 2007; I-2 Suppl: 424.

[0178] Kuss SE. Die pleistozänen Hirsche der ostmediterranen Inseln Kreta, Kasos, Karpathos und Rhodos (Griechenland). Berichte der Naturforschenden Gesellschaft zu Freiburg i. Br. [im Breisgau] 1975; 65: 25–79.

[0179] Kotsakis T, Petronio C, Sirna G. The Quaternary vertebrates of the Aegean islands: palaeogeographical implications. Ann Geol Pays Hell 1979; 30: 31–64.

[0180] Halstead P, Jones G. Bioarchaeological remains from Kalythies Cave, Rhodes. In: Sampson A, ed. The Neolithic period in the Dodecanese. Athens, Greece: Ministry of Culture, 1987: 135–52.

[0181] Kuss SE. *Hippopotamus creutzburgi* parvus n. ssp, ein pleistozänes Zwergflusspferd von der Insel Kreta. Berichte der Naturforschenden Gesellschaft zu Freiburg i. Br. [im Breisgau] 1975; 65: 5–23.

[0182] Capasso Barbato L, Kotsakis T, Petronio C. Nuovi resti e osservazioni filogenetiche su *Hippopotamus creutzburgi* Boekschoten and Sondaar, 1966 (Hippopotamidae, Mammalia) del Pleistocene di Creta (Grecia). Geologica Romana 1982; 21: 67–77.

[0183] Spaan A. *Hippopotamus creutzburgi*: the case of the Cretan hippopotamus. In Reese DS, ed. Pleistocene and Holocene fauna of Crete and its first settlers. Madison, WI, USA: Prehistory Press, 1996: 99–110.

[0184] van der Geer A, Lyras G, de Vos J, Dermitzakis M. Evolution of island mammals. Adaptation and extinction of placental mammals on islands. Oxford, UK: Wiley-Blackwell, 2010.

[0185] Theodorou GE, Dermitzakis MD. Απολιθωμένα Σπονδυλωτά σε Σπήλαια της Κρήτης. Αποτελέσματα των Ερευνών και Προοπτικές. (Fossil vertebrates in Cretan caves. Results and perspectives.) Πεπραγμένα του ΣΤ′ Διεθνούς Κρητολογικού Συνέδριου (Chania/ Crete, 24–30 August 1986). Volume B. Chania, 1991: 377–92.

[0186] Kotsakis T. Gli *Apodemus (Muridae, Rodentia)* fossili delle isole mediterranee. Supplemento alle Ricerche di Biologia della Selvaggina 1984; 9: 25–49.

[0187] Malatesta A. Dwarf deer and other late Pleistocene fauna of the Simonelli Cave in Crete. Quaderni dell'Accademia Nazionale dei Lincei 1980; 249: 1–97.

[0188] Schüle W. Mammals, vegetation and the initial human settlement of the Mediterranean islands: a palaeoecological approach. J Biogeogr 1993; 20: 399–412.

[0189] Masseti M, Zava B. The red deer of Lampedusa (Pelagian islands, Italy): literary references and osteological evidence. Archives of Natural History 2002; 29: 51–66.

[0190] Held S. Early prehistoric island archaeology in Cyprus: configurations of formative culture growth from the Pleistocene/Holocene boundary to the mid-3rd millennium BC. Unpublished PhD thesis. London, UK: Institute of Archaeology, University College, University of London, 1989.

[0191] Fish FE. Aquatic locomotion. In: Tomasi TE, Horton TH, eds. Mammalian energetics: interdisciplinary views of metabolism and reproduction. Ithaca, NY, USA: Cornell University Press, 2001: 34–63.

[0192] Coughlin BL, Fish FE. Hippopotamus underwater locomotion: reduced gravity movements for a massive mammal. J Mammal 2009; 90(3): 675–9.

[0193] Estes RD. The Behavior Guide to African Mammals. Berkeley, CA, USA: University of California Press, 1992.

[0194] Eltringham SK. The hippos: Natural History and Conservation. London, UK: T & AD Poyser, 1999.

[0195] Nowak RM. Walker's mammals of the world. Volumes 1–2. Baltimore, Maryland, USA: Johns Hopkins University Press, 1999.

[0196] Wall WP. The correlation between high limb-bone density and aquatic habits in recent mammals. J Paleontol 1983; 57(2): 197–207.

[0197] Simpson GG. Mammals and land bridges. Journal of the Washington Academy of Sciences 1940; 30: 137–63.

[0198] Sondaar PY. The island sweepstakes. Why did pygmy elephants, dwarf deer and large mice populate the Mediterranean? Nat Hist 1976; 95(9): 50–7.

[0199] Theodorou GE, Roussiakis SI, Athanassiou A, Giaourtsakis I, Panayides I. A Late Pleistocene endemic genet (Carnivora, Viverridae) from Aghia Napa, Cyprus. Bulletin of the Geological Society of Greece 2007; 40: 201–8.

[0200] Cox CB, Moore PD. Biogeography. An ecological and evolutionary approach. 5th Edition. Oxford/London/Edinburgh, UK: Blackwell Scientific Publications, 1993.

[0201] Lomolino MV. Body size evolution in insular vertebrates: generality of the island rule. J Biogeogr 2005; 32(10): 1683–99.

[0202] Sanchiz FB. Algunas Batracofaunas Pleistocénicas de Islas del Mediterraneo Oriental. In: Actas II Reunión Iberoamericana sobre Conservación y Zoología de Vertebrados (Cáceres/ Spain, 15–20 June 1980), 59–69.

[0203] Koukouli-Chrysanthaki C, Weisgerber G. Παλαιολιθικό ορυχείο ώχρας στη Θάσο. Αρχαιολογία και Τέχνες 1996; 60: 82–9.

[0204] Storch G. Eine mittelpleistozäne Nager-Fauna von der Insel Chios, Ägäis. Senckenberg Biol 1975; 56: 165–89.

[0205] Kuss SE, Storch G. Eine Säugetierfauna (Mammalia, Artiodactyla, Rodentia) des älteren Pleistozäns von der Insel Kalymnos (Dodekanes, Griechenland). Neues Jahrb Geol P-M 1978; 206–27.

[0206] van der Meulen AJ, van Kolfschoten T. Review of the Late Turolian to Early Biharian mammal faunas from Greece and Turkey. Memorie della Società Geologica Italiana 1986; 31: 201–11.

[0207] von Koenigswald W, Fejfar O, Tchernov E. Revision einiger alt- und mittelpleistozäner Arvicoliden (Rodentia, Mammalia) aus dem östlichen Mittelmeergebiet (Ubeidiya, Jerusalem und Kalymnos-Xi). Neues Jahrb Geol P-A 1992; 184: 1–23.

[0208] Kotsakis T, Barisone G. Paleobiogeografia dei micromammiferi cenozoici dell'Anatolia. Biogeographia 2000; 21: 579–92.

[0209] Gliozzi E, Abbazzi L, Argenti P, Azzaroli A, Caloi L, Capasso Barbato L, Di Stefano G, Esu D, Ficcarelli G, Girotti O, Kotsakis T, Masini F, Mazza P, Mezzabotta C, Palombo MR, Petronio C, Rook L, Sala B, Sardella R, Zanalda E, Torre D. Biochronology of selected mammals, molluscs and ostracods from the Middle Pliocene to the Late Pleistocene in Italy. The state of the art. Riv Ital Paleont Strat 1997; 103(3): 369–88.

[0210] Mazza P. Interrelations between Pleistocene hippopotami of Europe and Africa. Bollettino della Società Paleontologica Italiana 1991; 30(2): 153–86.

[0211] Maiuri A. Esplorazione di grotte con avanzi preistorici nell'isola di Calimno. In: Clara Rhodos. Studi e materiali pubblicati a cura dell'Istituto storico-archeologico di Rodi. 12 Volumes, 1928–1942. Volume 1, 1928: 104–17.

[0212] Howald G, Donlan CJ, Galván JP, Russell JC, Parkes J, Samaniego A, Wang Y, Veitch D, Genovesi P, Pascal M, Saunders A, Tershy B. Invasive rodent eradication on islands. Conservation Biology 2007; 21(5): 1258–68.

[0213] Jacobsen TW. 17.000 anni di preistoria greca. Le Scienze 1976; 98: 68–81.

[0214] Perlès C. Des navigateurs méditerranéens il y a 10,000 ans. Recherche 1979; 10: 82–3.

[0215] Cherry JF. Pattern and process in the earliest colonization of the Mediterranean islands. P Prehist Soc 1981; 47: 41–68.

[0216] Cherry JF. Palaeolithic Sardinians? Some questions of evidence and method. In: Tykot RH, Andrews TK, eds. Sardinia in the Mediterranean: a footprint in the sea. Sheffield, UK: Sheffield Academic Press, 1992: 43–56.

[0217] Simmons AH. Humans, island colonisation and Pleistocene extinctions in the Mediterranean: the view from Akrotiri Aetokremnos, Cyprus. Antiquity 1991; 65: 857–69.

[0218] Renfrew C, Aspinall A. Aegean obsidian and Franchthi Cave. In: Perlès C, ed. Les Industries lithiques taillées de Franchthi (Argolide, Grèce) [The Chipped Stone Industries of Franchthi (Argolide, Greece)]. Volume 2: Les Industries du Mésolithique et du Néolithique Initial, Fascicle 5 [The Mesolithic and Early Neolithic Industries]. Bloomington, IN, USA: Indiana University Press, 1990: 257–70.

[0219] Cherry JF. Four problems in Cycladic prehistory. In: Davis JL, Cherry JF, eds. Papers in Cycladic Prehistory. Los Angeles, CA, USA: UCLA Institute of Archaeology, Monograph 14, 1979: 22–47.

[0220] Simmons AH. Extinct pygmy hippopotamus and early man in Cyprus. Nature 1988; 333: 554–7.

[0221] Simmons AH. Faunal Extinction in an Island Society: Pygmy Hippopotamus Hunters of Cyprus. New York, NY, USA: Kluwer Academic, 1999.

[0222] Payne S. Faunal change at Franchthi Cave from 20,000 BC to 3,000 BC. In: Clason AT, ed. Archaeozoological Studies. Amsterdam, Netherlands: North Holland, 1975: 120–31.

[0223] Shackleton JC, van Andel TH, Runnels CN. Coastal palaeogeography of the central and western Mediterranean during the last 125 000 years, and its archaeological implications. Journal of Field Archaeology 1984; 11: 307–14.

[0224] Fedele F. Malta: origini e sviluppo del popolamento preistorico. In: Fradkin Anati A, Anati E, eds. Missione a Malta. Milan [Milano], Italy: Jaca Book, 1988: 51–90.

[0225] Pennacchioni M. Correnti marine di superficie e navigazione durante il Neolitico. In: Atti del XIII Congresso Internazionale dell'UISPP [Union Internationale des Sciences Préhistoriques et Protohistoriques] (Forlì/Italy, 8–14 September 1996). Volume 3. Forlì, Italy: Abaco Edizioni, 1996: 379–88.

[0226] van der Geer A, Lyras G. Field trip guidebook. In: Abstracts of the 9th Annual Meeting of the European Association of Vertebrate Palaeontologists [EAVP] (Herakleion/Crete, 14–19 June 2011). Natural History Museum of Crete; Nederlands Centrum voor Biodiversiteit Naturalis, 2011.

[0227] MacArthur RH, Wilson EO. The Theory of Island Biogeography. Princeton, NJ, USA: Princeton University Press, 1967.

[0228] Patton M. Islands in time. London, UK: Routledge, 1996.

[0229] Kopaka K, Matzanas C. Palaeolithic industries from the island of Gavdos, near neighbor to Crete in Greece. Antiquity 2009; 83: 321.

[0230] Strasser TF, Thompson N, Panagopoulo E, Karkanas P, Runnels CN, McCoy FW, Murray PM, Wegmann KW. Stone Age Seafaring in the Mediterranean. Evidence from the Plakias Region for Lower Palaeolithic and Mesolithic Habitation of Crete. Hesperia 2010; 79: 145–90.

[0231] Vigne J-D. L'extinction holocène du fond de peuplement mammalien indigène des îles de Méditerranée occidentale. Mémoires de la Société géologique de France 1987; 150: 167–77.

[0232] Vigne J-D, Valladas H. Evolution of the man-made environment in Northern Corsica (Western Mediterranean) during the last 2,500 years, from the small mammal fossils of Monte di Tuda. Journal of Archaeological Science 1996; 23: 199–215.

[0233] Pieper H. Eine neue Mesocricetus-Art. Stuttgarter Beiträge zur Naturkunde Serie B 1984; 107: 1–9.

[0234] Reumer JWF. Shrews (Soricidae) on islands, with special reference to *Crocidura zimmermanni* from Crete. In: Reese DS, ed. Pleistocene and Holocene fauna of Crete and its first settlers. Madison, WI, USA: Prehistory Press, 1996: 173–9.

[0235] Ambrosetti P, Azzaroli A, Kotsakis T. Mammiferi del Plio-Pleistocene delle isole. In: Parisi G, ed. I vertebrati fossili italiani. Verona, Italy: Comune di Verona 1980, 243–8.

[0236] Theodorou GE. Pleistocene elephants from Crete (Greece). Mod Geol 1985; 10: 235–42.

[0237] Böhme W, Eisentraut M. Zur weiteren Dokumentation des Zwergelefanten (*Loxodonta pumilio* Noack, 1906). Zeitschrift des Kölner Zoos 1990; 33(4): 153–8.

[0238] Bachmayer F, Symeonidis N. Eigenartige Abspaltungen von Stosszähnen der Zwergelefanten aus der Höhle "Charkadio" auf der Insel Tilos – Artefakte? Ann Geol Pays Hell 1975; 26: 320–3.

[0239] Masseti M. Did endemic dwarf elephants survive on Mediterranean islands up to protohistorical times? In: Caravetta G, Gioia P, Mussi M, Palombo MR, eds. La terra degli elefanti/The world of elephants. Proceedings of the 1st International Congress (Rome, 16–20 October 2001). Rome [Roma], Italy: Comune di Roma; Consiglio Nazionale delle Ricerche; Ministero per i Beni e le Attività Culturali; Università degli Studi di Roma "La Sapienza", 2001: 402–6.

[0240] de Garis Davies N. Paintings from the tomb of Rekh-mi-re at Thebes. New York, NY, USA: Platin Press, 1935.

[0241] Rosen B. Mammoths in ancient Egypt? Nature 1994; 369: 364.

[0242] White W. Painted out. Nature 1994; 370: 604.

[0243] Lister AM. Mammoths in miniature. Nature 1993; 362: 288–9.

[0244] Scullard HH. The elephant in the Greek and Roman world. Cambridge, UK: Thames and Hudson, 1974.

[0245] Osborn DJ, Osbornová J. The mammals of ancient Egypt. Warminster, UK: Aris and Phillips, 1998.

[0246] Hatt RT. The mammals of Iraq. Miscellaneous publications (University of Michigan. Museum of Zoology), 1959; 106: 1–113.

[0247] Buitenhuis H. Archaeozoological aspects of Late Holocene economy and environment in the Near East. In: Bottema S, Brentjes SG, van Zeist W, eds. Man's role in the shaping of the Eastern Mediterranean landscape. Rotterdam, Netherlands: Balkema, 1990: 195–205.

[0248] Gabolde M. Les elephants de Niyi d'après les sources égyptiennes. In: Béal J-C, Goyon J-C, eds. Des ivoires et des cornes dans les mondes anciens (Orient-Occident). Lyon, France: Université Lunière 2000; 2: 129–39.

[0249] Pirenne J. Histoire de la Civilisation de l'Égypte Ancienne. Deuxième cycle, de la fin de l'ancien empire à la fin du nouvel empire 2200–1085 av J-C. Neuchâtel, Switzerland: Editions de la Baconnière, 1962.

[0250] Aldred, C. Egyptian art. London, UK: Thames and Hudson, 1984.

[0251] Evans A. The Palace of Minos: a comparative account of the successive stages of the early Cretan civilization as illustrated by the discoveries at Knossos. London, UK: Macmillan, 1921–1936. Volume II, part II, 1928.

[0252] Stubbings FH. Mycenean Pottery from the Levant. Cambridge, UK: Cambridge University Press, 1951.

[0253] Benzi M. Problems of the Mycenean expansion in the South-Eastern Aegean. In: de Miro E, Godart L, Sacconi A, eds. Atti e memorie del Secondo Congresso Internazionale di Micenologia (Roma/Napoli, 14–20 October 1991). Rome [Roma], Italy: Gruppo editoriale internazionale 1996, 947–78.

[0254] Niemeier W-D. The Minoans in the Eastern Aegean and in Cyprus. In: Karageorghis V, Stampolidis N, eds. Eastern Mediterranean: Cyprus – Dodecanese – Crete 16th–6th century BC. Proceedings of the International Conference (Rethymnon/Crete, May 1997). Rethymnon, Greece: The University of Crete; Nicosia, Cyprus: AG Levendis Foundation, 1998: 29–45.

[0255] Sampson A. Topographical survey of prehistoric sites in the Dodecanese. In: Keller DR, Rupp DW, eds. Archaeological survey in the Mediterranean area. British Archaeological Reports International Series 1983; 155: 283–5.

[0256] Pearce J. Albinism in the cinereous shrew. J Mammal 1934; 15: 67.

[0257] Balli A. Osservazioni biologiche su *Talpa europea* L. Riv Biol 1940; 29: 35–55.

[0258] von Lehmann E. Über die Kleinsäuger der La Sila (Kalabrien). Zool Anz 1961; 167: 213–29.

[0259] Fons R, Catalan J, Poitevin F. Càs d'albinisme chez deux Insectivores Soricidae: *Suncus etruscus* (Savi, 1822) et *Neomys fodiens* (Pennant, 1771). Z Saugetierkd 1983; 48: 117–22.

[0260] Michalak I. Colour pattern in *Neomys anomalus*. Acta Theriologica 1983; 28(2): 25–32.

[0261] Chętnicki W, Fedyk S, Bajkowska U. Cases of Coat Colour Anomalies in the Common Shrew, *Sorex araneus* L. Folia biologica (Kraków) 2007; 55(1–2): 73–6.

[0262] Pucek M. Cases of white spotting in shrews. Acta Theriologica 1964; 9: 367–8.

[0263] Crowcroft P. The life of the shrew. London, UK: Max Reinhardt, 1957.

[0264] Corbet GB. The frequency of albinism of the tail tip in British mammals. P Zool Soc Lond 1963; 140: 327–30.

[0265] Corbet GB, Harris S, eds. Handbook of the British Mammals. 3rd Edition. Oxford, UK: Blackwell Scientific Publications, 1991.

[0266] Gelling M. Partial albinism in the common shrew *Sorex araneus*. Mammal Rev 2003; 33(2): 189–90.

[0267] Masseti M. The economic role of *Sus* in early human fishing communities. In: Albarella U, Dobney K, Ervynck A, Rowley-Conwey P, eds. Pigs and humans. 10,000 years of interaction. Oxford, UK; New York, NY, USA: Oxford University Press, 2007: 156–70.

[0268] Hutterer R. Order Erinaceomorpha. In: Wilson DE, Reeder D-AM, eds. Mammal species of the world. A taxonomic and geographic reference. 3rd Edition, volume 1. Baltimore, MD, USA: The Johns Hopkins University Press, 2005: 212–9.

[0269] Corbet GB. The mammals of the Palaearctic Region. A taxonomic review. London, UK: British Museum (Natural History); Ithaca, NY, USA: Cornell University Press, 1978.

[0270] Harrison DL, Bates PJJ. The Mammals of Arabia. Sevenoaks, Kent, UK: Harrison Zoological Museum, 1991.

[0271] Reeve N. Hedgehogs. London, UK: T & AD Poyser, 1994.

[0272] Giagia-Athanasopoulou EB, Ondrias C. The distribution of the Eastern Hedgehog *Erinaceus concolor bolkay* in continental Greece. Säugetierkundliche Mitteilungen 1995; 36(4): 154–73.

[0273] Giagia-Athanasopoulou EB, Markakis G. Multivariate analysis of morphometric characters in the Eastern hedgehog *Erinaceus concolor* from Greece and adjacent areas. Z Saugetierkd 1996; 61: 129–39.

[0274] Lapini L. *Erinaceus concolor* (Martin, 1838). In: Mitchell-Jones AJ, Amori G, Bogdanowicz W et al., eds. The Atlas of European Mammals. London, UK: Academic Press, 1999: 36–7.

[0275] Douglas G. Zur Fauna Santorins. Zool Anz Leipzig 1892; 15: 453–5.

[0276] Bate DMA. On the mammals of Crete. P Zool Soc Lond 1905; 2: 315–23.

[0277] Bate DMA. The mammals of Crete. In: Trevor-Battye A, ed. Camping in Crete. London, UK: Witherby and Co, 1913: 254–6.

[0278] Schaschl H, Lymberakis P, Suchentrunk F. On allozyme and cyt-b gene characteristics of Cretan hedgehogs, *Erinaceus concolor nesiotes* Bate, 1906. Mamm Biol 2002; 67: 257–67.

[0279] Miller GS. Catalogue of the mammals of Western Europe (Europe exclusive of Russia). London, UK: British Museum (Natural History), 1912.

[0280] Giagia EB. Κυτταροταξινομική μελέτη του ακανθόχοιρου *Erinaceus concolor M.* εις την Ελλάδα. Thesis (Διδακτορική Διατριβή). Patras, Greece: University of Patras, 1977.

[0281] Holz H, Niethammer J. *Erinaceus concolor* (Martin, 1838) – Weissbrustigel, Ostigel. In: Niethammer J, Krapp F, eds. Handbuch der Säugetiere Europas. Band 3/I. Insectivora/ Primates [Insektenfresser/Herrentiere]. Wiesbaden, Germany: AULA-Verlag, 1990: 50–64.

[0282] Catsadorakis G. Προκαταρκτική καταγραφή της πανίδας και διάφορη διαχειριστική θεμάτων του Εθνικού Δρυμού Αίνου (Κεφαλλονιά). Unpublished manuscript for Διεύθυνση Δασών Κεφαλληνίας (Cephalonia, Greece: Forest Department), 1985.

[0283] Giagia-Athanasopoulou E. Mammals. In: Efthymiatou-Katsouni N, ed. Dedication to the National Park of Ainos. Cephalonia and Ithaca, Greece: Museum of Natural History, 1998: 159–61.

[0284] Mitsainas GP, Tryfonopoulos GA, Katsouni N, Chondropoulos BP, Giagia-Athanasopoulou EB. A study on the vertebrate fauna of the Ionian islands Kefalonia and Ithaki. In: Abstracts of the 10th International Congress on the Zoogeography and Ecology of Greece and Adjacent Regions [ICZEGAR] (Patras, 26–30 June 2006). Hellenic Zoological Society; Dept of Biology, University of Patras; Hellenic Ornithological Society; Hellenic Ministry of Rural Development and Food, 2006: 83.

[0285] Bonetti A. Ταξιδεύοντας στην Ελλάδα. 40 φυσικοί παράδεισοι. Athens, Greece: Road Εκδόσεις, 2008.

[0286] Pantis I, Zoi M, Katselidis K, Koleyris A, Basilakis K, Dumoulin D. Το Εθνικό Θαλάσσιο Πάρκο Ζακύνθου. Athens, Greece: Εκδοτικός Οργανισμός Λιβάνη, 2004.

[0287] Niethammer J. Die Igel Neuseelands. Zool Anz 1969; 183: 151–5.

[0288] von Wettstein O. Bemerkungen über einige Säuger des griechisch-kleinasiatischen Raumes. Sitzungsber Österr Akad Wiss Mathem Naturwiss Kl Abt 1, 1966; 175: 357–62.

[0289] Matsakis J, ed. Programme "Sporades du Nord". Unpublished manuscript. Athens, Greece: University of Athens, 1985.

[0290] Werner F. Beiträge zur Kenntnis der Fauna Griechenlands, namentlich der ägäischen Inseln. Sitzungsber Österr Akad Wiss Mathem Naturwiss Kl Abt 1, 1928; 137: 284–95.

[0291] Akriotis T. Food ecology of five sympatric owls. Unpublished manuscript. First Degree dissertation. Oxford, UK: Wolfson College, University of Oxford, 1981.
[0292] Lindermayer A. Euboea. Eine naturhistorische Skizze. Moscow [Moscau/Moskau], Russia: Buchdruckerei der Kaiserlichen Universität, 1855: 3–35. [Electronic version: http://bvbm1.bib-bvb.de/view/custom/DisplayPersistentURL.jsp?pid=961212 (accessed May 5, 2012).] [Also cited as: Bulletin de la Société Impériale de Naturalistes de Moscou, XXVIII. Moscow, Russia: Imprimerie de l'Université Impériale.]
[0293] Sfougaris AI, ed. Περιβαλλον προστασία περιοχή Μεγάλο και Μικρό Λιβάρι – Δέλτα Ξηριά – Υδρόχαρες Δάσος Αγίου Νικόλαου. Prefecture of Euboea, Department of the Environment; Volos, Greece: University of Thessaly, 2000.
[0294] Bolkay SJ. Additions to the Mammalian Fauna of the Balkan Peninsula. Glasnik Zemaljskog muzeja Bosne i Hercegovine [Journal of the National Museum of Bosnia and Herzegovina] 1926; 38: 159–79.
[0295] Schmalfuss H. Santorin. Leben auf Schutt und Asche. Ein naturkundlicher Reiseführer. Weikersheim, Germany: Josef Margraf Scientific Publishers, 1991.
[0296] von Wettstein O. Die Insectivora von Kreta. Z Saugetierkd 1953; 17: 4–13.
[0297] Zimmermann K. Das Gesamtbild der Säuger-Fauna Kretas. Z Saugetierkd 1953; 17: 67–9.
[0298] Catsadorakis G. The vertebrate animals of Samaria National Park (Crete, Greece). Biol Gallo-Hell 1994; 22: 9–22.
[0299] Cattaneo A. Osservazioni sui rettili delle isole egee di Karpathos e Kasos (Dodecaneso meridionale) (*Reptilia*). Il Naturalista siciliano S 4 2010; 34(1–2): 29–47.
[0300] Zimmermann K. Bemerkungen über die Wildziege Kretas. Z Saugetierkd 1953, 17: 66.
[0301] Demetropoulos A. Specially protected areas in Rhodes. Tunis, Tunisia: Report Regional Activity Centre for Specially Protected Areas (RAC/SPA), 1993.
[0302] Arapis T, Basilakis K, Zogaris S, Mastroyiannis A, Misiakos K, Panista M, Pappa X. Special environmental study of the island of Tilos. Athens, Greece: Municipality of Megalo Chorio of Tilos, 2000.
[0303] Cattaneo A. Osservazioni sull'ofidiofauna delle isole egee di Leros e Patmos (Dodecaneso) (*Reptilia Serpentes*). Il Naturalista siciliano S 4 2008; 32(1–2): 201–19.
[0304] Cheylan G. Compte-rendu de la table ronde: répartition géographique et statut des mammifères menacés dans les îles méditerranéennes. Bull Ecol 1988; 19(2–3): 481–4.
[0305] Jounis G. The ecosystems of Samos. In: Ecotouristic guide of the Aegean Samos. Natural History Museum of the Aegean; Paleontological Museum of Samos; Zimalis Foundation, 1995: 12–9.
[0306] Ioannidis Y, Dimaki M, Dimitropoulos A. Θηλαστικά στη Σάμο. Μια πρώτη αναφορά. Samiakes Meletes 1996; 2: 465–8.
[0307] Axiotis M. Η πανίδα στη Λέσβο. Mitilini/Lesbos, Greece: Ιδιωτική Έκδοση, 1995.
[0308] Adelsen C. Hedgehogs sighted on Aegean isle of Gokceada. British Hedgehog Preservation Society Newsletter 1993; 21: 6.
[0309] Reyniers A. Consommation et interdits: l'animal dans l'alimentation tsigane. Anthropozoologica, Second Numéro Spécial 1988: 199–205.
[0310] Vigne J-D. Biogéographie insulaire et anthropozoologie des sociétés néolithiques méditerranéennes: hérisson, renard et micromammifères. Anthropozoologica 1988; 8: 31–52.
[0311] Bellakhdar J. La pharmacopée marocaine traditionelle. Médicine arabe ancienne et savoirs populaires. Paris, France: Ibis Press, 1997.
[0312] Masseti M. Erinaceomorfi, Soricomorfi, Lagomorfi e Roditori endemici, ed antropocori dell'Italia continentale, ed insulare. In: Amori G, Contoli L, Nappi A, eds. Fauna d'Italia. Mammalia II. Erinaceomorpha, Soricomorpha, Lagomorpha, Rodentia (Volume 44). Milan [Milano], Italy: Edizioni Calderini de il Sole 24 Ore, 2008: 33–65.
[0313] Rackham H, ed. Pliny. Natural History [Naturalis historia]. Volume 3. Libri VIII–IX. Second edition. Loeb Classical Library. Cambridge, MA, USA: Harvard University Press; London, UK: William Heinemann, 1983.

[0314] Borghini A, Giannarelli E, Marcone A, Ranucci G, eds. Storia Naturale. II. Antropologia e Zoologia, di Gaio Plinio Secondo, Libri 7–11. Torino, Italy: Giulio Einaudi Editore, 1983.
[0315] Trantalidou K. Agriculture, animal husbandry, hunting, fishing. In: Papathanassopoulos GA, ed. Neolithic culture in Greece. Athens, Greece: Nicholas P Goulandris Foundation; Museum of Cycladic Art, 1996: 95–102.
[0316] Yannouli E. Aspects of animal use in prehistoric Macedonia, Northern Greece. Examples from the Neolithic and the Early Bronze Age. Unpublished PhD thesis. Cambridge, UK: University of Cambridge, 1994.
[0317] Gejvall N-G. Lerna. A preclassical site in the Argolid. Volume I. The Fauna. Princeton, NJ, USA: American School of Classical Studies at Athens (ASCSA), 1969.
[0318] Hutterer R. *Sorex minutus* (Linnaeus, 1758). In: Mitchell-Jones AJ, Amori G, Bogdanowicz W et al., eds. The Atlas of European Mammals. London, UK: Academic Press, 1999: 54–5.
[0319] Hutterer R. Order Soricomorpha. In: Wilson DE, Reeder D-AM, eds. Mammal species of the world. A taxonomic and geographic reference. 3rd Edition, volume 1. Baltimore, MD, USA: The Johns Hopkins University Press, 2005: 220–311.
[0320] Zima J, Slivková L, Andreas M, Benda P, Reiter A. Karyotypic status of shrews (*Sorex*) from Thrace, European Turkey. Z Saugetierkd 1997; 62: 315–7.
[0321] Zima J, Lukácová L, Macholán M. Chromosomal evolution in shrews. In: Wójcik JM, Wolsan M, eds. Evolution of shrews. Białowieża, Poland: Mammal Research Institute, Polish Academy of Sciences, 1998: 175–218.
[0322] Spitzenberger F. *Neomys anomalus* (Cabrera, 1907) – Sumpfspitzmaus. In: Niethammer J, Krapp F, eds. Handbuch der Säugetiere Europas. Band 3/I. Insectivora/Primates [Insektenfresser/Herrentiere]. Wiesbaden, Germany: AULA-Verlag, 1990: 317–33.
[0323] Spitzenberger F. *Neomys anomalus* (Cabrera, 1907). In: Mitchell-Jones AJ, Amori G, Bogdanowicz W et al., eds. The Atlas of European Mammals. London, UK: Academic Press, 1999: 58–9.
[0324] Krapp F. *Crocidura leucodon* (Hermann, 1780) – Feldspitzmaus. In: Niethammer J, Krapp F, eds. Handbuch der Säugetiere Europas. Band 3/I. Insectivora/Primates [Insektenfresser/Herrentiere]. Wiesbaden, Germany: AULA-Verlag, 1990: 465–84.
[0325] Krapp F. *Crocidura leucodon* (Hermann, 1780). In: Mitchell-Jones AJ, Amori G, Bogdanowicz W et al., eds. The Atlas of European Mammals. London, UK: Academic Press, 1999: 64–5.
[0326] Papapavlou K, Sfenthourakis S. Small mammals from the northern Dodecanese archipelago, Greece. Newsletter of the Hellenic Zoological Society 2000; 33: 7–8.
[0327] Ondrias JC. Die Ussuri-Großspitzmaus, *Crocidura lasiura* Dobson, 1890, der Agäischen Insel Lesbos. Z Saugetierkd 1969; 34: 353–8.
[0328] Felten H, Spitzenberger F, Storch G. Zur Kleinsäugerfauna West-Anatoliens. Teil II. Senckenberg Biol 1973; 54(4–6): 227–90.
[0329] Akriotis T, Papamichael G, Symeonidis M. Temporal and spatial variation in the diversity of small mammals in Mediterranean habitats on the island of Lesbos (Greece). In: Abstracts of the 9th International Congress on the Zoogeography and Ecology of Greece and Adjacent Regions [ICZEGAR] (Thessaloniki, 22–25 May 2002). Hellenic Zoological Society. Thessaloniki, Greece: University Studio Press, 2002: 2.
[0330] Wolsan M, Hutterer R. A List of the Living Species of Shrews. In: Wójcik JM, Wolsan M, eds. Evolution of Shrews. Białowieża, Poland: Mammal Research Institute, Polish Academy of Sciences 1998: 425–48.
[0331] Vogel P, Maddalena T, Catzeflis F. A contribution to the taxonomy and ecology of shrews (*Crocidura zimmermanni* and *C. suaveolens*) from Crete and Turkey. Acta Theriologica 1986; 31: 537–45.
[0332] LeBlanc L. The accumulation and significance of micromammals in an Albanian cave site. In: Buitenhuis H, Bartosiewicz L, Choyke AM, eds. Archaeozoology of the Near East III. Proceedings of the 3rd International Symposium on the Archaeozoology of

southwestern Asia and adjacent areas [ASWA] (Budapest, 1–5 September 1996). Groningen, Netherlands: ARC [Archaeological Research & Consultancy] Publicaties 18, 1998: 160–165.
[0333] Poitevin F, Catalan FJ, Fons R, Croset H. Biologie évolutive des populations ouest-européennes des crocidures (Mammalia, Insectivora). I. Critères d'identification et répartition biogéographique de *Crocidura russula* (Hermann, 1780) et *Crocidura suaveolens* (Pallas, 1811). Rev Ecol-Terre Vie 1986; 41: 299–314.
[0334] Poitevin F, Catalan FJ, Fons R, Croset H. Biologie évolutive des populations ouest-européennes des crocidures (Mammalia, Insectivora). II. Ecologie comparée de *Crocidura russula* (Hermann, 1780) et *Crocidura suaveolens* (Pallas, 1811) dans le midi de la France et en Corse: rôle probable de la compétition dans le partage des milieux. Rev Ecol-Terre Vie 1987; 42: 39–58.
[0335] Catalan G, Poitevin F, Fons R, Guerasimof S, Croset H. Biologie évolutive des populations ouest-européennes de crocidures (Mammalia, Insectivora). III. Structures génétiques des populations continentals et insulaires de *Crocidura russula* (Hermann, 1780), et de *Crocidura suaveolens* (Pallas, 1811). Mammalia 1988; 52: 387–400.
[0336] Besenecker H, Spitzenberger F, Storch G. Eine holozäne Kleinsäuger-Fauna von der Insel Chios, Ägäis. Senckenberg Biol 1972; 53: 145–77.
[0337] Tvrtković N, Dulić B, Mrakovčić M. Distribution of Insectivora and Rodentia on the northeast Adriatic coast (Yugoslavia). Acta Zool Fennica 1985; 170: 201–3.
[0338] Dubey S, Cosson J-F, Magnanou E, Vohralík V, Benda P, Frynta D, Hutterer R, Vogel V, Vogel P. Mediterranean populations of the lesser white-toothed shrew (*Crocidura suaveolens* group): an unexpected puzzle of Pleistocene survivors and prehistoric introductions. Mol Ecol 2007; 16: 3438–52.
[0339] Hoffmann RS. Noteworthy shrews and voles from the Xizang-Qinghai Plateau. In: Genoways HH, Baker RJ, eds. Contributions in Mammalogy: A Memorial Volume Honoring Dr J Knox Jones Jr. Lubbock, TX, USA: Museum of Texas Tech University, 1996: 155–68.
[0340] Ondrias JC. Η πανίδα των θηλαστικών της Ελλάδας (The Mammal Fauna of Greece). Athens, Greece: University of Athens, 1967.
[0341] Vlasák P, Niethammer J. *Crocidura suaveolens* (Pallas, 1811) – Gartenspitzmaus. In: Niethammer J, Krapp F, eds. Handbuch der Säugetiere Europas. Band 3/I. Insectivora/Primates [Insektenfresser/Herrentiere]. Wiesbaden, Germany: AULA-Verlag, 1990: 397–428.
[0342] Peus F. Flöhe aus dem Mittelmeergebiet (Insecta, Siphonaptera). II. Griechenland. Mitteilungen aus dem Museum für Naturkunde in Berlin. Zoologisches Museum und Institut für Spezielle Zoologie (Berlin) 1958; 34(1): 135–71.
[0343] Ondrias JC. Contribution to the knowledge of *Crocidura suaveolens* (Mammalia, Insectivora) from Greece, with a description of a new subspecies. Z Saugetierkd 1970; 35: 371–81.
[0344] Niethammer J. Zur Verbreitung und Taxonomie griechischer Säugetiere. Bonn Zool Beitr 1974; 25(1–3): 28–55.
[0345] Alivizatos H, Goutner V. Winter diet of the barn owl (*Tyto alba*) and long-eared owl (*Asio otus*) in northeastern Greece: a comparison. J Raptor Res 1999; 33(2): 160–3.
[0346] Hutterer R. Der Status von *Crocidura ariadne* Pieper, 1979 (Mammalia: Soricidae). Bonn Zool Beitr 1981; 32: 3–12.
[0347] Payne S. The small mammals [Appendix 5.1]. In: Shaw JW, Shaw MC, eds. Kommos I (1). The Kommos Region, Ecology, and Minoan Industries. Princeton, NJ, USA: Princeton University Press, 1995: 278–90.
[0348] Vardakis M, Lymberakis P, Mylonas M. The terrestrial small mammals of Crete (Poster). In: Abstracts of the 10th International Congress on the Zoogeography and Ecology of Greece and Adjacent Regions [ICZEGAR] (Patras, 26–30 June 2006). Hellenic Zoological Society; Dept of Biology, University of Patras; Hellenic Ornithological Society; Hellenic Ministry of Rural Development and Food, 2006: 121.

[0349] Dubey S, Zaitsev M, Cosson J-F, Abdukadier A, Vogel P. Pliocene and Pleistocene diversification and multiple refugia in a Eurasian shrew (*Crocidura suaveolens* group). Mol Phylogenet Evol 2006; 38: 635–47.

[0350] Alivizatos H, Goutner V, Zogaris S. Contribution to the study of the diet of four owl species (Aves, Strigiformes) from mainland and island areas of Greece. In: Abstracts of the 9th International Congress on the Zoogeography and Ecology of Greece and Adjacent Regions [ICZEGAR] (Thessaloniki, 22–25 May 2002). Hellenic Zoological Society. Thessaloniki, Greece: University Studio Press, 2002: 4.

[0351] Masseti M. Uomini e (non solo) topi. Gli animali domestici e la fauna antropocora. Florence [Firenze], Italy: Università degli Studi di Firenze; Firenze University Press, 2002.

[0352] Alivizatos H, Goutner V, Zogaris S. Contribution to the study of the diet of four owl species (Aves, Strigiformes) from mainland and island areas of Greece. Belg J Zool 2005; 135(2): 109–18.

[0353] Richter H. Eine Serie *Crocidura gueldenstaedti* (Pallas, 1811) (Mammalia, Insectivora) von der griechischen Insel Samos. Beaufortia (Series of Miscellaneous Publications. Zoological Museum of the University of Amsterdam) 1966; 13(157): 109–15.

[0354] Catzeflis F, Maddalena T, Hellwing S, Vogel P. Unexpected findings on the taxonomic status of east Mediterranean *Crocidura russula* auct. (Mammalia, Insectivora). Z Saugetierkd 1985; 50: 185–201.

[0355] Crucitti P. Micromammiferi raccolti nelle missioni zoologiche della S.R.S.N. nella penisola ellenica (Mammalia). In: Crucitti P, ed. Atti del Convegno sul tema Zoologia Ellenica (Roma, 17 May 1986). Rome [Roma], Italy: Società Romana di Scienze Naturali [SRSN]. Mentana, Italy: Tip Marini Gabriella, 1987: 48–50.

[0356] Vohralík V, Sofianidou T. Small mammals (Insectivora, Rodentia) of Macedonia, Greece. Acta Universitatis Carolinae Biologica 1985; 29: 319–54.

[0357] Vogel P. *Crocidura zimmermanni* (Wettstein, 1953). In: Mitchell-Jones AJ, Amori G, Bogdanowicz W et al., eds. The Atlas of European Mammals. London, UK: Academic Press, 1999: 74–5.

[0358] Pieper H. *Crocidura zimmermanni* (Wettstein, 1953). In: Niethammer J, Krapp F, eds. Handbuch der Säugetiere Europas. Band 3/I. Insectivora/Primates [Insektenfresser/Herrentiere]. Wiesbaden, Germany: AULA-Verlag, 1990: 453–60.

[0359] Vesmanis I, Kahmann H. Morphometrische Untersuchungen an Wimperspitzmäusen (*Crocidura*) 4. Bemerkungen über die Typusreihe der kretaischen *Crocidura russula zimmermanni* Wettstein, 1953 im Vergleich mit *Crocidura gueldenstaedti canae* (Miller, 1909). Säugetierkundliche Mitteilungen 1978; 26(3): 214–22.

[0360] Vogel P. Der Karyotyp der Kretaspitzmaus *Crocidura zimmermanni* Wettstein, 1953 (Mammalia, Insectivora). Bonn Zool Beitr 1986; 37: 35–8.

[0361] Dubey S, Kazuhiro K, Parapanov R, Ribi M, Hutterer R, Vogel P. Molecular phylogenetics reveal Messinian, Pliocene, and Pleistocene colonizations of islands by North African shrews. Mol Phylogenet Evol 2008; 47: 877–82.

[0362] Zimmermann K, von Wettstein O, Siewert H, Pohle H. Die Wildsäuger von Kreta. Z Saugetierkd 1953; 17: 1–72.

[0363] Libois R, Fons R. *Suncus etruscus* (Savi, 1822). In: Mitchell-Jones AJ, Amori G, Bogdanowicz W et al., eds. The Atlas of European Mammals. London, UK: Academic Press, 1999: 76–7.

[0364] Kock D, Malec F, Storch G. Rezente und subfossile Kleinsäuger aus dem Vilayet Elaziğ, Ostanatolien. Z Saugetierkd 1972; 37: 204–29.

[0365] Obuch J. On the food of the eagle-owl (*Bubo bubo*) and tawny owl (*Strix aluco*) in the eastern part of Turkey. Tichodroma 1994; 7: 7–16.

[0366] Amori G, Masseti M. Does the occurence of predators on Central Mediterranean islands affect the body size of micromammals? Vie Milieu 1996; 46(3–4): 205–11.

[0367] Spitzenberger F. Erstnachweise der Wimperspitzmaus (*Suncus etruscus*) für Kreta und Kleinasien und die Verbreitung der Art im südwestasiatischen Raum. Z Saugetierkd 1970; 35(2): 107–13.

[0368] Spitzenberger F. Die Säugetierfauna Zyperns. I: Insectivora und Rodentia. Ann Nathist Mus Wien 1978; 81: 401–41.

[0369] Pieper H. Zur Verbreitung einiger Kleinsäuger auf Kreta. Z Saugetierkunde 1976; 41(5): 274–7.

[0370] Springhorn R, Kachel G. Erster Lebendnachweis der Etruskerspitzmaus (Insectivora, Soricidae) auf Kreta. Z Saugetierkunde 1981; 46(1): 55–6.

[0371] Cattaneo A. Osservazioni sull'erpetofauna dell'isola greca di Kos (Sporadi Meridionali) con un inedito caso di simpatria microinsulare fra due specie affini di colubridi: *Hierophis caspius* (Gmelin) e *Hierophis jugularis* (L.) [Observations on the herpetofauna of the Greek island of Kos (Southern Sporades) with an unpublished case of microinsular sympatry of two closely related colubridae species: *Hierophis caspius* (Gmelin) and *Hierophis jugularis* (L.). Atti del Museo di Storia Naturale della Maremma 2005; 21: 89–103.

[0372] Kryštufek B. *Talpa stankovici* (V. Martino and E. Martino, 1931). In: Mitchell-Jones AJ, Amori G, Bogdanowicz W et al., eds. The Atlas of European Mammals. London, UK: Academic Press, 1999: 88–9.

[0373] Ellerman JR, Morrison-Scott TCS. Checklist of Palaearctic and Indian mammals 1758 to 1946. London, UK: British Museum (Natural History), 1951.

[0374] Petrov B. Taxonomy and distribution of Moles (Genus *Talpa*, Mammalia) in Macedonia. Acta Musei Macedonici Scientiarum Naturalium 1971; 12(6): 117–38.

[0375] Todorović M, Soldatović B, Dundjerski Z. Kariotipske odlike populacije roda *Talpa* iz Makedonije i Crne Gore [Karyotype characteristics of the population of the genus *Talpa* from Macedonia and Montenegro]. Arhiv bioloških nauka (Beograd) [Archives of biological sciences] 1972; 24: 131–9.

[0376] Todorović M, Dundjerski Z, Soldatović B. Distribution of the mole species (*Talpa*) in the south-west region of the Balkans. Biol Gallo-Hell 1987; 13: 119–26.

[0377] Kryštufek B. The taxonomy of blind moles (*Talpa caeca* and *T. stankovici*, Insectivora, Mammalia) from south-western Europe. Bonn Zool Beitr 1994; 45(1): 1–16.

[0378] Colangelo P, Loy A, Annesi F, Kryštufek B, Capanna E. Molecular phylogeny of the European moles of the genus *Talpa* Linnaeus, 1758: systematic and biogeographic implications. In: Prigioni C, Sforzi A, eds. Abstracts of the 5th European Congress of Mammalogy [ECM] (Siena, 21–26 September 2007). Hystrix It J Mamm NS 2007; I-2 Suppl: 391.

[0379] Niethammer J. *Talpa stankovici* (V. Martino and E. Martino, 1931) – Balkan-Maulwurf. In: Niethammer J, Krapp F, eds. Handbuch der Säugetiere Europas. Band 3/I. Insectivora/Primates [Insektenfresser/Herrentiere]. Wiesbaden, Germany: AULA-Verlag, 1990: 141–4.

[0380] Tsounis G, Sfikas G. Ecotouristic Guide to Greece (in Greek). Athens, Greece: General Secretariat of Youth; Hellenic Society for the Protection of Nature, 1993.

[0381] Tryfonopoulos GA, Thanou EG, Fraguedakis-Tsolis SE, Chonfropoulos BP. New data on the distribution and genetic structure of Greek moles of the genus *Talpa* (Mammalia, Talpidae). J Zool Systemat Evol Res 2009; 48(2): 188–93.

[0382] Homolka M, Zima J. *Lepus europaeus* (Pallas, 1778). In: Mitchell-Jones AJ, Amori G, Bogdanowicz W et al., eds. The Atlas of European Mammals. London, UK: Academic Press, 1999: 166–7.

[0383] Hoffmann RS, Smith AT. Order Lagomorpha. In: Wilson DE, Reeder D-AM, eds. Mammal species of the world. A taxonomic and geographic reference. 3rd Edition, volume 1. Baltimore, MD, USA: The Johns Hopkins University Press, 2005: 185–211.

[0384] de Beaufort F. La faune des mammifères de Grèce: caractéristiques, endémisme, particularismes. Biol Gallo-Hell 1991; 18: 99–106.

[0385] Osborn DJ. The hare, porcupine, beaver, squirrels, jerboas, and dormice of Turkey. Mammalia 1964; 28(4): 573–92.

[0386] Ristow D. Πανίδα. In: Karetsos G, Brophas G, Dipolous P, eds. Dionysades Islands (in Greek). Athens, Greece: Ministry of the Environment 1995: 24–39.

[0387] Bodson L. Données antiques de zoogéographie. L'expansion des Léporidés dans la Méditerranée classique. Naturalistes Belges 1978; 59: 66–81.

[0388] Masseti M, de Marinis AM. Prehistoric and Historic Artificial Dispersal of Lagomorphs on the Mediterranean Islands. In: Alves PC, Ferrand N, Hackländer K, eds. Lagomorph Biology: Evolution, Ecology, and Conservation. Berlin/Heidelberg, Germany: Springer, 2008: 13–24.

[0389] Niethammer J. Über einige Hasenschädel (*Lepus europaeus*) von der griechischen Insel Paximada vor Kreta. Säugetierkundliche Mitteilungen 1992; 34: 43–9.

[0390] de Beaux O. *Lepus europaeus ghigii*, subsp. nova. (Nota preliminare.) Bollettino dei Musei di Zoologia ed Anatomia comparata della R Università di Genova 1927; 7(17): 2.

[0391] Sert H, Suchentrunk F, Erdoğan A. Genetic diversity within Anatolian brown hares (*Lepus europaeus* Pallas, 1778) and differentiation among Anatolian and European populations. Mamm Biol 2005; 70(3): 171–86.

[0392] Mamuris Z, Moutou KA, Stanatis C, Sarafudou T, Suchentrunk F. Y DNA and mitochondrial lineages in European and Asian populations of the brown hare (*Lepus europaeus*). Mamm Biol 2009; 75: 233–42.

[0393] Sert H, Ben Slimen H, Erdoğan A, Suchentrunk F. Mitochondrial HVI sequence variation in Anatolian hares (*Lepus europaeus* Pallas, 1778). Mamm Biol 2009; 74: 286–97.

[0394] Ghigi A. Risultati generali e conclusioni. In: Ghigi A, Issel R, Brian A, Santucci R, Citterio V, Alzani F. Ricerche faunistiche nelle isole italiane dell'Egeo. Archivio Zoologico Italiano 1929 (1928–1929, VII); 12–13: 423–84.

[0395] Niethammer G. Zur Vogelwelt Kretas nach Winterbeobachtungen. Anzeiger der Ornithologischen Gesellschaft in Bayern 1966; 7(Sonderheft): 726–32.

[0396] Austria, Archduke Ludwig Salvator of [Erzherzog Ludwig Salvator von Österreich-Toskana]. Anmerkungen über Levkas. Prague [Prag/Praha], Austria-Hungary: Heinrich Mercy Sohn, 1908. [Internet resource and electronic version: http://www.ludwig-salvator.com/biblio.htm and http://www.ludwigsalvator.com/digi/levkas/levkas.htm (accessed May 5, 2012).]

[0397] Antoniou A, Magoulas A, Platis P, Papamatheakis J, Kotoulas G. Divergence of *Lepus europaeus* populations in Greece: Is it only a matter of isolation by distance and geographic barriers? In: Alves PC, Ferrand N, eds. Abstracts of the 2nd World Lagomorph Conference (Vairão, 26–31 July 2004). Vairão (Porto), Portugal: Research Center in Biodiversity and Genetic Resources (CIBIO), University of Porto, 2004: 22.

[0398] Jameson R. Notes on the natural History and Statistics of the Island of Cerigo and its dependencies. Edinburgh New Philosophical Journal 1837; 22: 62–9.

[0399] Jameson R. Naturgeschichte der Insel Cerigo. Isis oder Encyklopädische Zeitung/Isis von Oken 1838; 31(2): 127–130. [Electronic version: Biodiversity Heritage Library (BHL): http://www.biodiversitylibrary.org/page/27396479 (accessed May 5, 2012).]

[0400] Kasapidis P, Suchentrunk F, Magoulas A, Kotoulas G. The shaping of mitochondrial DNA phylogeographic patterns of the brown hare (*Lepus europaeus*) under the combined influence of Late Pleistocene climatic fluctuations and anthropogenic translocations. Mol Phylogenet Evol 2005; 34: 55–66.

[0401] Erhard D. Fauna der Cykladen. Die Wirbelthiere der Cykladen, nebst einem Anhange über deren Pflanzendecke. Leipzig, Germany: Voigt und Günther, 1858.

[0402] Anastassiou T. Kea: History, sightseeing. Hermoupolis/Syros, Greece, 1994.

[0403] Sfyroera SN. Aegina. The first capital of modern Greece. Athens, Greece: Ellinika Grammata, 2002.

[0404] de Heldreich T. La faune de Grèce. Exposition Universelle de Paris en 1878. Athens [Athènes], Greece: Imprimerie de la Philocalie, 1878.

[0405] Dafis S, Papastergiadou E, Georghiou K, Babalonas D, Georgiadis T, Papageorgiou M, Lazaridou T, Tsiaoussi V. Directive 92/43/EEC. The Greek Habitat Project Natura 2000: an overview. Thessaloniki, Greece: The Goulandris Natural History Museum; Athens: University of Athens; Thessaloniki: University of Thessaloniki; Patras: University of Patras, 1996.

[0406] Barrett-Hamilton GEH. The hares of Crete and of Cyprus. J Nat Hist Ser 7 1903; 11(61): 126–7.

[0407] Xirouchakis S. The golden eagle *Aquila chrysaetos* in Crete. Distribution, population status and conservation problems. Avocetta 2001; 25(2): 275–81.

[0408] Suchentrunk F, Mamuris Z, Sfougaris AI, Stamatis C. Biochemical Genetic Variability in Brown Hares (*Lepus europaeus*) From Greece. Biochemical Genetics 2003; 41(5–6): 127–40.

[0409] Karetsos G, Brofas G, Dimopoulos P, Ristow D, Wink M. Αναγνώριση και Αξιολόγηση Βιοτόπων Ορνιθοπανίδας. Unpublished report. Athens, Greece: Ministry of the Environment, 1995.

[0410] Melas EM. The islands of Karpathos, Saros and Kasos in the Neolithic and Bronze Age. Studies in Mediterranean Archaeology. Gothenburg [Göteborg], Sweden: Paul Äströms Förlag, 1985: 68.

[0411] Bürchner L. Die Insel Leros. Jahresbericht des Königlichen Theresien-Gymnasiums in München 1898: 1–48.

[0412] Zwilling EA. Vom Urahn zum Gorilla – Eines Jägers Wanderjahre. Giessen [Gießen], Germany: Brühlscher Verlag, 1949.

[0413] Maltezou A. The Springs and Streams of Samothrace. In: Gerakis PA, Koutrakis ET, eds. Greek Wetlands. Athens, Greece: Commercial Bank of Greece, 1996: 93–8.

[0414] Suchentrunk F, Ben Slimen H, Stamatis C, Sert H, Scandura M, Apollonio M, Mamuris Z. Molecular approaches revealing prehistoric, historic, or recent translocations and introductions of hares (genus *Lepus*) by humans. Human Evolution 2006; 21(2): 151–66 [Special Issue on the Anthropochorous Fauna (Ed. by M Masseti)].

[0415] Trantalidou K. From Mesolithic Fishermen and Fish Hunters to Neolithic Goat herders: The transformation of an island economy in the Aegean. In: Sampson A, ed. The Cyclops cave on Youra, Alonnessos. Volume 2. Philadelphia, PA, USA: INSTAP Academic Press, 2011: 53–150.

[0416] Trantalidou K. Animal Bones and Animal Representations at Late Bronze Age Akrotiri. In: Sherratt S, ed. The Wall Paintings of Thera: Proceedings of the 1st International Symposium, Volume 2 (Thera, 30 August–4 September 1997). Piraeus, Greece: The Thera Foundation – Petros M Nomikos and The Thera Foundation, 2000: 709–35.

[0417] Reese DS, Rose MJ, Payne S. The Minoan Fauna [Chapter 5]. In: Shaw JW, Shaw MC, eds. Kommos I (1). The Kommos Region, Ecology and Minoan Industries. Princeton, NJ, USA: Princeton University Press, 1995: 163–291.

[0418] Jarman MR. Human influence in the development of the Cretan mammalian fauna. In: Reese DS, ed. Pleistocene and Holocene fauna of Crete and its first settlers. Madison, WI, USA: Prehistory Press, 1996: 211–29.

[0419] Trantalidou K. The organic materials. In: Marangou L, Renfrew C, Doumas C, Gavalas G, eds. Markiani, Amorgos. An early Bronze Age fortified settlement. Overview of the 1985–1991 investigations. British School at Athens, Supplementary Volume 2006; 40: 223–83.

[0420] Trantalidou K. Faunal remains from the earliest strata of the Cave of Cyclope, Youra. In: Galanidou N, Perlès C, eds. The Greek Mesolithic. Problems and perspectives. British School at Athens Studies 2003; 10: 143–72.

[0421] Trantalidou K. Dietary adaptations of coastal people in the Aegean archipelago during the Mesolithic period: the macrofaunal assemblages of Maroulas on Kythnos. In: Sampson A, Kaczanowska M, Kozlowski JK, eds. The Prehistory of the Island of Kythnos (Cyclades, Greece) and the Mesolithic settlement at Maroulas. Cracow [Kraków], Poland: Polish Academy of the Arts [Polska Akademia Umiejętności], 2010: 163–77.

[0422] van Andel TH. Landscape and people of the Franchthi Region. Part 1. The Landscape. In: Jacobsen TW, ed. Excavations at Franchthi Cave, Greece. Bloomington, IN, USA: Indiana University Press, 1987: 1–62.

[0423] Trantalidou K. Animal remains related to sacred areas on the Cycladic islands Amorgos and Tenos, during the Geometric Period: two distinct examples bearing evidence of sacrificial rites and bone working activities. In: Mazarakis Ainian A, ed. The "Dark Ages" revisited. Papers of an International Conference (Volos/Greece, 14–17 June 2007). Volume 2. Volos, Greece: University of Thessaly Press, 2011: 1015–61.

[0424] López Martínez N. Revisión sistemática y bioestratigráfica de los Lagomorpha (Mammalia) del Terciario y Cuaternario inferior de España. Tesis Doctoral. Facultad de Ciencias Geológicas, Universidad Complutense de Madrid, 1977. [Also published as: Revisión sistemática y bioestratigráfica de los Lagomorpha (Mammalia) del Terciario y Cuaternario inferior de España. Zaragoza, Spain: Memorias del Museo Paleontológico de la Universidad de Zaragoza 1989; 3: 1–342.]

[0425] Rogers PM, Arthur CP, Soriguer RC. The rabbit in continental Europe. In: Thompson HV, King CM, eds. The European rabbit. The history and biology of a successful colonizer. Oxford, UK; New York, NY, USA; Tokyo, Japan: Oxford University Press, 1994: 22–63.

[0426] Homolka M, Zima J. *Oryctolagus cuniculus* (Linnaeus, 1758). In: Mitchell-Jones AJ, Amori G, Bogdanowicz W et al., eds. The Atlas of European Mammals. London, UK: Academic Press, 1999: 172–3.

[0427] Callou C. De la garenne au clapier. Étude archéozoologique du Lapin en Europe occidentale. Paris, France: Publications Scientifiques du Muséum national d'Histoire naturelle, 2003.

[0428] Kaetze P, Niedermeier J, Masseti M. *Oryctolagus cuniculus* (Linné, 1758) – Europäisches Wildkaninchen. In: Krapp F, ed. Handbuch der Säugetiere Europas. Band 3/II. Lagomorpha. Wiebelsheim, Germany: AULA-Verlag, 2003: 187–289.

[0429] Masseti M. Note paleontologiche, genetiche e archeozoologiche sul coniglio selvatico, *Oryctolagus cuniculus*. In: Trocchi V, Riga F, eds. I Lagomorfi in Italia. Linee guida per la conservazione e gestione. Rome [Roma], Italy: Ministero per le Politiche Agricole e Forestali; Ozzano dell'Emilia (Bologna), Italy: Istituto Nazionale per la Fauna Selvatica "Alessandro Ghigi", 2005: 11–2.

[0430] Callou C. Reconstructing the history of the European rabbit *Oryctolagus cuniculus*. Contribution of the Zooarchaeology. In: Alves PC, Ferrand N, eds. Abstracts of the 2nd World Lagomorph Conference (Vairão, 26–31 July 2004). Vairão (Porto), Portugal: Research Center in Biodiversity and Genetic Resources (CIBIO), University of Porto, 2004: 99.

[0431] Flux JEC. World distribution. In: Thompson HV, King CM, eds. The European Rabbit. Oxford, UK; New York, NY, USA; Tokyo, Japan: Oxford University Press, 1994: 8–21.

[0432] Kowalski K, Rzebik-Kowalska B. Mammals of Algeria. Wrocław/Warsaw [Warszawa]/ Cracow [Kraków], Poland: Institute of Systematics and Evolution of Animals, Polish Academy of Sciences; Ossolineum, 1991.

[0433] Posner SD. Biological diversity and tropical forests in Tunisia. Washington, DC, USA; Tunis, Tunisia: The Washington DC and Tunis Offices of the Agency for International Development, 1988.

[0434] Vigne J-D. Données préliminaires sur l'histoire du peuplement mammalien de l'îlot de Zembra (Tunisie). Mammalia 1988; 52(4): 567–74.

[0435] Rezig M. Animaux de Tunisie. Tunis, Tunisia: Centre National Universitaire de Documentation Scientifique et Technique, 1999.

[0436] Vigne J-D, Thibault JC, Cheylan G. L'archipel des Lavezzi: presentation, histoire des peuplements, problématiques archéozoologiques. In: Vigne J-D, ed. L'île Lavezzi. Hommes, animaux, archéologie et marginalité (XIIIe–XXe siècles, Bonifacio, Corse). Paris, France: Éditions du Centre national de la recherche scientifique [CNRS], 1994: 25–41.

[0437] Masseti M, de Marinis AM. La diffusione antropocora del coniglio selvatico sulle isole mediterranee. In: Trocchi V, Riga F, eds. I Lagomorfi in Italia. Linee guida per la conservazione e gestione. Rome [Roma], Italy: Ministero per le Politiche Agricole e Forestali; Ozzano dell'Emilia (Bologna), Italy: Istituto Nazionale per la Fauna Selvatica "Alessandro Ghigi", 2005: 67.

[0438] Bell BD. The effects of goats and rabbits on breeding seabirds: methods of eradication and control. Boletim do Museu Municipal do Funchal Suppl 1995; 4: 83–9.

[0439] Martín A. Rabbit Eradication on Montaña Clara (Canary Islands, Spain). Proceedings of the Workshop on Invasive Alien Species on European Islands and Evolutionary Isolated Ecosystems and Group of Experts on Invasive Alien Species (Horta/Azores, 10–12 October 2002). Strasbourg, France: Directorate of Culture and of Cultural and Natural Heritage, Council of Europe, 2002: T-PVS/IAS, 2: 14–6.

[0440] Courchamp F, Chapuis J-L, Pascal M. Mammal invaders on islands: impact, control and control impact. Biol Rev 2003; 78: 347–83.

[0441] Chapuis J-L, Frenot Y, Lebouvier M. Recovery of native plant communities after eradication of rabbits from the subantarctic Kerguelen Islands, and influence of climate change. Biol Conserv 2004; 117: 167–79.

[0442] Genovesi P. Eradications of invasive alien species in Europe: a review. Biol Invasions 2005; 7(1): 127–33.

[0443] Stephanides T. Island Trails. London, UK: Macdonald, 1973.

[0444] Pauli FW. Die Insel Chios. In: Mittheilungen der Geographischen Gesellschaft in Hamburg. Hamburg, Germany: L Friederichsen und Co, 1881: 99–114.

[0445] Tsounis G. Ένα μνημεία της φύσης. Επτά Ημέρες, 23 October 1994: 19.

[0446] Sfougaris AI, Nastis AS, Papageorgiou NK. Food resources and quality for the introduced Cretan wild goat or agrimi *Capra aegagrus cretica* on Atalandi island, Greece, and implications for ecosystem management. Biol Conserv 1996; 78: 239–45.

[0447] Barbour S. Small Mammals. In: Sporades Islands Expedition [19]82. Edinburgh, UK: University of Edinburgh, 1982: 73.

[0448] Dimaki M. Antikythira island complex. Newsletter Societas Hellenica Herpetologica (SHH) [ca. 2005]; 4: 1–4.

[0449] Cattaneo A. I serpenti delle isole greche di Kythnos e Kea (Cicladi occidentali). Atti della Società Italiana di Scienze Naturali e del Museo Civico di Storia Naturale di Milano 1990; 131(11): 209–19.

[0450] Niethammer G. Die Einbürgerung von Säugetieren und Vögeln in Europa. Hamburg/Berlin, Germany: Paul Parey, 1963.

[0451] Boback AW. Das Wildkaninchen. Die Neue Brehm-Bücherei. Wittenberg Lutherstadt, Germany: A Ziemsen, 1970.

[0452] Koster SH, Grether H. Population and habitat characteristics of the Cretan wild goat on Dia and Pantes islands and the Gorge of Samaria. Mimeographed report. Boston: School for Field Studies, 1985.

[0453] Foufopoulos J. The reptile fauna of the Northern Dodecanese (Aegean Islands, Greece). Herpetozoa 1997; 10(1–2): 3–12.

[0454] Akriotis T, Kardakari N, Meliadou A. The avifauna of Alyki and Hortarolimni lagoons, Limnos Island. Newsletter of the Hellenic Zoological Society 1996; 27: 3–4.

[0455] Cattaneo A. The herpetofauna of the Aegean islands of Thassos, Samothraki and Lemnos. Bollettino del Museo Civico di Storia Naturale di Venezia 2001; 52: 155–81.

[0456] Tiné S, Traverso A. Poliochni, the earliest town in Europe. Athens, Greece: Archaeological Society of Athens, 2001.

[0457] Sorrentino C. Poliochni: il materiale faunistico. In: Doumas CG, La Rosa V, eds. Poliochni e l'antica età del bronzo nell'Egeo Settentrionale. Abstracts of the International Congress of Scuola Archeologica Italiana di Atene (Athens, 22–25 April 1996), 1997: 157–67.

[0458] Nobis G. Karthago: eine antike Weltstadt im Blickfeld der klassischen Archäozoologie. Tier und Museum 1992; 3(1): 1–11.

[0459] Gauci-Maistre J. "Tax-Xiber" – the indigenous rabbit of Malta. Cahiers Options Méditerranéennes 1999; 41: 183–7.

[0460] García y Bellido A. Veinticinco estampas de la España antigua. Madrid, Spain: Editorial Espasa-Calpe, 1967.

[0461] Toynbee JMC. Animals in Roman life and art. Ithaca, NY, USA: Cornell University Press, 1973.

[0462] Barrett-Hamilton GEH. A history of British mammals. Volume 2. London, UK: Gurney and Jackson, 1912.

[0463] Albarella U. La fauna. In: Arthur P, ed. L'Isola e il Santo. La chiesa di San Costanzo alla Marina Grande di Capri, scavi 1990. Naples [Napoli], Italy: Editoriale Scientifica, 1992: 53–8.

[0464] Gavish L. Preliminary observations on the behavior and ecology of the free living populations of the subspecies *Sciurus anomalus syriacus* (golden squirrel) on Mount Hermon, Israel. Isr J Zool 1993; 39: 275–80.

[0465] Hecht-Markou P. Beschreibung, geographische Verbreitung, Biotope und Ortswechsel des *Sciurus anomalus* Güldenstädt, 1785 auf der Insel Lesbos (Griechenland). Annales Musei Goulandris 1994; 9: 429–44.

[0466] Gavish L, Gurnell J. *Sciurus anomalus* (Güldenstädt, 1785). In: Mitchell-Jones AJ, Amori G, Bogdanowicz W et al., eds. The Atlas of European Mammals. London, UK: Academic Press, 1999: 176–7.

[0467] Mendelssohn H, Yom-Tov Y. Fauna palaestina. Mammalia of Israel. Jerusalem, Israel: The Israel Academy of Sciences and Humanities, 1999.

[0468] Thorington Jr RW, Hoffmann RS. Family Sciuridae. In: Wilson DE, Reeder D-AM, eds. Mammal species of the world. A taxonomic and geographic reference. 3rd Edition, volume 1. Baltimore, MD, USA: The Johns Hopkins University Press, 2005: 754–818.

[0469] Hecht-Markou P. Das Markieren des Lebensraumes von *Sciurus anomalus* auf der Insel Lesbos. Annales Musei Goulandris 1999; 10: 201–21.

[0470] Scotti G. L'arcipelago del Quarnero. Natura, storia, arte, turismo. Milan [Milano], Italy: Mursia, 1980.

[0471] Masseti M. Natural and anthropochorous squirrels and dormice of the Mediterranean Region. Hystrix It J Mamm NS 2005; 16(1): 3–26.

[0472] Musser GG, Carleton MD. Superfamily Muroidea. In: Wilson DE, Reeder D-AM, eds. Mammal species of the world. A taxonomic and geographic reference. 3rd Edition, volume 2. Baltimore, MD, USA: The Johns Hopkins University Press, 2005: 894–1531.

[0473] Vohralík V. *Cricetulus migratorius* (Pallas, 1773). In: Mitchell-Jones AJ, Amori G, Bogdanowicz W et al., eds. The Atlas of European Mammals. London, UK: Academic Press, 1999: 204–5.

[0474] Harrison DL. The mammals of Arabia, Lagomorpha-Rodentia. Volume 3. London, UK: Ernest Benn, 1972.

[0475] Ondrias JC. Statut taxonomique actuel des rongeurs de Grèce. Mammalia 1961; 25(1): 22–8.

[0476] Zima J. *Microtus arvalis* (Pallas, 1778). In: Mitchell-Jones AJ, Amori G, Bogdanowicz W et al., eds. The Atlas of European Mammals. London, UK: Academic Press, 1999: 228–9.

[0477] Gorman ML. Orkney and Guernsey voles *Microtus arvalis*. In: Corbet GB, Harris S, eds. The Handbook of British Mammals. Oxford/London, UK; Boston, MA, USA: Blackwell Scientific Publications, 1991: 208–11.

[0478] Masseti M. Homeless mammals from the Ionian and Aegean islands. Bonn Zoological Bulletin 2010; 57(2): 367–73.

[0479] Kryštufek B. *Microtus guentheri* (Danford and Alston, 1880). In: Mitchell-Jones AJ, Amori G, Bogdanowicz W et al., eds. The Atlas of European Mammals. London, UK: Academic Press, 1999: 238–9.

[0480] Schlitter DA. African rodents of special concern: A preliminary assessment. IUCN/SSC Rodent Specialist Group. In: Lidicker Jr WZ, ed. Rodents. A World Survey of Species of Conservation Concern. Gland, Switzerland: IUCN [International Union for Conservation of Nature/Union Internationale pour la Conservation de la Nature et de les Ressources (UICN)], 1989: 33–9.
[0481] Stamatopoulos C, Ondrias J. First record of the Levant vole *Microtus guentheri* Danford and Alston, 1880 in Lesbos island, Greece. Säugetierkundliche Mitteilungen 1995; 36: 53–9.
[0482] Wolf H. Säugetierfauna Bulgariens. Izvestiya na Tsarskitia prirodonauchni instituti v Sofiya [Mitteilungen aus den Königlichen Naturwissenschaftlichen Instituten in Sofia, Bulgarien; Bulletin des Institutions royales d'histoire naturelle à Sofia, Bulgarie] 1940; 13: 153–68.
[0483] Fraguedakis-Tsolis SE, Chondropoulos BP, Stamatopoulos CV, Giokas S. Morphological variation of the five vole species of the genus *Microtus* (Mammalia, Rodentia, Arvicolinae) occurring in Greece. Acta Zool 2009; 90: 254–64.
[0484] Niethammer J. *Microtus felteni* (Malec and Storch, 1963). In: Niethammer J, Krapp F, eds. Handbuch der Säugetiere Europas. Band 2/I. Rodentia II. Wiesbaden, Germany: Akademische Verlagsgesellschaft, 1982: 438–41.
[0485] Kryštufek B. *Microtus subterraneus* (de Sélys-Longchamps, 1836). In: Mitchell-Jones AJ, Amori G, Bogdanowicz W et al., eds. The Atlas of European Mammals. London, UK: Academic Press, 1999: 250–1.
[0486] Giagia-Athanasopoulou EB, Stamatopoulos C. Geographical distribution and interpopulation variation in the karyotypes of *Microtus* (*Terricola*) *thomasi* (Rodentia, Arvicolidae) in Greece. Caryologia 1997; 50: 303–15.
[0487] Rovatsos MT, Mitsainas GP, Stamatopoulos C, Giagia-Athanasopoulou EB. First reports of XXY aneuploidy in natural populations of Thomas' pine vole *Microtus thomasi* (Rodentia: Arvicolidae) from Greece. Mamm Biol 2008; 73(5): 342–9.
[0488] Mitsainas GP, Rovatsos MT, Giagia-Athanasopoulou EB. Heterochromatin study and geographical distribution of *Microtus* species (Rodentia, Arvicolinae) from Greece. Mamm Biol 2010; 75: 261–9.
[0489] Giagia EB. Karyotypes of "44-Chromosomes" *Pytimis* Species (Rodentia, Mammalia) and their Distribution in Southern Greece. Säugetierkundliche Mitteilungen 1985; 32: 169–73.
[0490] Niethammer J. Über griechische Nager im Museum A. Koenig in Bonn. Ann Nathist Mus Wien 1986; 88–89: 245–56.
[0491] Kratochvíl J. Der Status der Populationen der Gattung *Pitymys* aus Attika (Rodentia, Mammalia). Zoologické Listy 1971; 20: 107–206.
[0492] Giagia EB. Karyological analysis of the vole of the genus *Pitymis* (Rodentia, Mammalia) from Southern Greece. Biol Gallo-Hell 1986; 12: 529–30.
[0493] Niethammer J. *Microtus thomasi* (Barrett-Hamilton, 1903) – Balkan-Kurzohrmaus. In: Niethammer J, Krapp F, eds. Handbuch der Säugetiere Europas. Band 1. Wiesbaden, Germany: Akademische Verlagsgesellschaft, 1978: 484–90.
[0494] Niethammer J. *Microtus thomasi* (Barrett-Hamilton, 1903) – Balkan-Kurzohrmaus. In: Niethammer J, Krapp F, eds. Handbuch der Säugetiere Europas. Band 2/I. Rodentia II. Wiesbaden, Germany: Akademische Verlagsgesellschaft, 1982: 485–90.
[0495] Kryštufek B. *Microtus thomasi* (Barrett-Hamilton, 1903). In: Mitchell-Jones AJ, Amori G, Bogdanowicz W et al., eds. The Atlas of European Mammals. London, UK: Academic Press, 1999: 254–5.
[0496] Thanou E, Tryfonopoulos G, Chondropoulos B, Fraguedakis-Tsolis S. MtDNA suggests the ongoing differentiation of Greek *Microtus (T.) thomasi* (Rodentia, Arvicolidae). In: Abstracts of the 10th International Congress on the Zoogeography and Ecology of Greece and Adjacent Regions [ICZEGAR] (Patras, 26–30 June 2006). Hellenic Zoological Society; Dept of Biology, University of Patras; Hellenic Ornithological Society; Hellenic Ministry of Rural Development and Food, 2006: 109.

[0497] Atallah SI. Mammals of the eastern Mediterranean region: their ecology, systematics and zoogeographical relationships. Part 2. Säugetierkundliche Mitteilungen 1978; 26: 1–50.

[0498] Amori G. *Chionomys nivalis* (Martins, 1842). In: Mitchell-Jones AJ, Amori G, Bogdanowicz W et al., eds. The Atlas of European Mammals. London, UK: Academic Press, 1999: 256–7.

[0499] Krapp F. *Microtus nivalis* (Martins, 1842) – Schneemaus. In: Niethammer J, Krapp F, eds. Handbuch der Säugetiere Europas. Band 2/II. Rodentia II. Wiesbaden, Germany: Akademische Verlagsgesellschaft, 1982: 260–283.

[0500] Paspalev GV, Martino KV, Peschev ZCh. Sur certains rongeurs du massiv Vitocha. Annual of Sofia University Faculty of Biology [Godishnik na Sofiiskiya Universitet] 1951; 47: 193–236.

[0501] Kryštufek B, Vohralík V. *Meriones tristrami* (Thomas, 1892). In: Mitchell-Jones AJ, Amori G, Bogdanowicz W et al., eds. The Atlas of European Mammals. London, UK: Academic Press, 1999: 258–9.

[0502] Kryštufek B. *Nannospalax leucodon* (Normann, 1840). In: Mitchell-Jones AJ, Amori G, Bogdanowicz W et al., eds. The Atlas of European Mammals. London, UK: Academic Press, 1999: 261–3.

[0503] Kryštufek B, Griffiths HI. Species richness and rarity in European rodents. Ecography 2002; 25: 120–8.

[0504] Savić I, Soldatović B. Karyotype evolution and taxonomy of the genus *Nannospalax* Palmer, 1903, Mammalia, in Europe. Serbian Academy of Sciences and Arts (Beograd) 1984; 560(59): 1–104.

[0505] Watson GE. Behavioral and ecological notes on *Spalax leucodon.* J Mammal 1961; 42(3): 359–64.

[0506] Paraschi L. *Spalax leucodon* (Nordmann, 1840). In: Karandinos M, Paraschi L, eds. The Red Data Book of Threatened Vertebrates of Greece. Athens, Greece: Hellenic Zoological Society; Hellenic Ornithological Society; Hewlett Packard; WWF, 1992: 279–80.

[0507] Giagia E, Savić I, Soldatović B. Chromosomal forms of the mole rat *Microspalax* from Greece and Turkey. Z Saugetierkd 1982; 47: 231–6.

[0508] Thomas O. On the Spalax of the grecian Archipelago. Annals and Magazine of Natural History London 1917; 118(20): 315–7.

[0509] Hadjissarantos C, Nikolopoulos C, Santas A. Rodentia et Insectivora de Grèce (in Greek). Athens, Greece: Ecole de Hautes Etudes Agronomiques a Athènes, Laboratoire de Zoologie Agricole et de Sériculture, 1962: 1–10.

[0510] Savić I, Soldatović B. Studies on the karyotype and distribution range of the mole rat (*Spalax leucodon* Normann) in Greece. Caryologia 1978; 31(1): 63–75.

[0511] Kummer J. Zum Vorkommen von *Spalax* auf Lemnos. Der Zoologische Garten (Jena) 1982; 52: 46–7.

[0512] Montgomery WI. *Apodemus flavicollis* (Linnaeus, 1758). In: Mitchell-Jones AJ, Amori G, Bogdanowicz W et al., eds. The Atlas of European Mammals. London, UK: Academic Press, 1999: 270–1.

[0513] Spitzenberger F, Bauer K. Mammalia. In: Symeonidis N, Bachmayer F, Zapfe H, eds. Grabungen in der Zwergelefanten-Höhle "Charkadio" auf der Insel Tilos (Dodekanes, Griechenland). Ann Nathist Mus Wien 1973; 77: 136.

[0514] Giagia E, Soldatović B, Savić I, Zimonjić D. Karyotype study of the genus *Apodemus* (Kaup, 1829) populations from the Balkan peninsula. Acta Veterinaria (Beograd) 1985; 35(5–6): 289–98.

[0515] Storch G. *Apodemus mystacinus* (Danford and Alston, 1877). In: Mitchell-Jones AJ, Amori G, Bogdanowicz W et al., eds. The Atlas of European Mammals. London, UK: Academic Press, 1999: 272–3.

[0516] Çolak E, Yiğit N, Çolak R, Sözen M, Özkurt Ş, Kankiliç T. Taxonomic status and distribution of *Apodemus mystacinus* (Danford and Aston, 1877) (Mammalia: Rodentia) in Turkey. Turk J Zool 2004; 28: 285–94.

[0517] Storch G. Die Ausbreitung der Felsenmaus (*Apodemus mystacinus*): Zur Problematik der Inselbesiedlung und Tiergeographie in der Ägäis. Natur und Museum 1977; 107(6): 174–82.

[0518] Niethammer J, Krapp F, eds. Handbuch der Säugetiere Europas. Band I. Rodentia I. Wiesbaden, Germany: Akademische Verlagsgesellschaft, 1978.

[0519] Matsakis J, Tsiourlis GM, Karamaouna M, Legakis A, Paraschi L, Blandin P, Lamotte M. Étude d'un écosystème de maquis à *Juniperus phoenicea* L. (Naxos, Cyclades, Grèce): présentation générale. Bulletin d'Ecologie 1992; 23(1–2): 49–58.

[0520] Papamichael G, Akriotis T, Giampoulakis C, Iliopoulos G, Stasinopoulos G. Range use of rock mice (*Apodemus mystacinus*) in a Mediterranean landscape. In: Abstracts of the 10th International Congress on the Zoogeography and Ecology of Greece and Adjacent Regions [ICZEGAR] (Patras, 26–30 June 2006). Hellenic Zoological Society; Dept of Biology, University of Patras; Hellenic Ornithological Society; Hellenic Ministry of Rural Development and Food, 2006: 91.

[0521] Mayhew DF. Late Pleistocene small mammals from Arnissa (Macedonia, Greece). K Ned Akad van Wet-B 1978; 81: 182–214.

[0522] Filippucci MG, Storch G, Macholán M. Taxonomy of the genus Sylvaemus in western Anatolia – morphological and electrophoretic evidence. Senckenberg Biol 1996; 75: 1–14.

[0523] Pamukoğlu N, Albayrak İ. The rodents of Kastamonu Province (Mammalia, Rodentia). Ankara, Turkey: Communications de la Faculté des Sciences de l'Université d'Ankara, Série C Biologie 1996; 14: 1–22.

[0524] Kryštufek B, Mozetič Francky B. Mount Hermon field mouse *Apodemus iconicus* is a member of the European mammal fauna. Folia Zool 2005; 54(1–2): 69–74.

[0525] Kryštufek B. Identity of four *Apodemus (Sylvaemus)* types rom the eastern Mediterranean and Middle East. Mammalia 2002; 66(1): 43–51.

[0526] Rovatsos MT, Mitsainas GP, Tryfonopoulos GA, Stamatopoulos C, Giagia-Athanasopoulou EB. A chromosomal study on Greek populations of the genus *Apodemus* (Rodentia, Muridae) reveals new data on B chromosome distribution. Acta Theriologica 2008; 53(2): 157–67.

[0527] Niethammer J. *Apodemus sylvaticus* (Linnaeus, 1758) – Waldmaus. In: Niethammer J, Krapp F, eds. Handbuch der Säugetiere Europas. Band 1. Wiesbaden, Germany: Akademische Verlagsgesellschaft, 1978: 337–58.

[0528] Rovatsos MT, Mitsainas GP, Giagia-Athanasopoulou EB, Chondropoulos V, Fraguedakis-Tsolis SE. *Apodemus witherby* (Thomas, 1902). In: Legakis A, Maragou P, eds. The Red Data Book of Endangered Animals of Greece (in Greek with English summary). Athens, Greece: Ministry of the Environment; Hellenic Zoological Society, 2009: 369–70.

[0529] Filippucci MG, Macholán M, Michaux JR. Genetic variation and evolution in the genus *Apodemus* (Muridae: Rodentia). Biol J Linn Soc 2002; 75: 395–419.

[0530] Zagorodnyuk IV. What is *Apodemus sylvaticus arianus* (Blanford, 1881)? Vestnik Zoologii 1996; 30(3): 20.

[0531] Zagorodnyuk IV, Boyeskorov GG, Zykov AE. Variation and taxonomic status of the steppe forms of genus *Sylvaemus "sylvaticus"* (*falzfeini – fulvipectus – hermonensis – arianus*). Vestnik Zoologii 1997; 31(5–6): 37–56.

[0532] Mezhzherin SV. Revision of mice genus *Apodemus* (Rodentia, Muridae) of northern Eurasia. Vestnik Zoologii 1997; 31(4): 29–41.

[0533] Filippucci MG, Simson S, Nevo E. Evolutionary biology of the genus *Apodemus* Kaup, 1829, in Israel. Allozymic and biometric analyses with description of a new species: *Apodemus hermonensis* (Rodentia, Muridae). Boll Zool 1989; 56: 361–7.

[0534] Mezhzherin SV, Zagorodnyuk IV. A new species of mice of the genus *Apodemus* (Rodentia, Muridae). Vestnik Zoologii 1989; 23(4): 55–9.

[0535] Çolak E. A study on *Apodemus iconicus* Heptner, 1948 (Mammalia: Rodentia) in Turkey. Turk J Zool 2003; 27: 61–3.

[0536] Jones JK, Johnson DH. Synopsis of the lagomorphs and rodents of Korea. Lawrence, KS, USA: University of Kansas Publications, Museum of Natural History 1965; 16: 357–407.

[0537] Kowalski K, Hasegawa Y. Quaternary rodents from Japan. Bulletin of the National Science Museum (Tokyo), Series C (Geology and Paleontology) 1976; 2(1): 31–66.

[0538] de Graaf G. The rodents of Southern Africa. Durban, South Africa: Butterworths, 1981.

[0539] Kawamura Y. Quaternary rodent faunas in the Japanese islands (Part 1). Memoirs of the Faculty of Science, Kyoto University, Series of Geology and Mineralogy, 1989; 53: 31–348.

[0540] Dobson M. Patterns of distribution in Japanese land mammals. Mammal Rev 1994; 24(3): 91–111.

[0541] Atkinson IAE. The spread of commensal species of *Rattus* to oceanic islands and their effects on island avifaunas. In: Moors PJ, ed. Conservation of island birds. International Council for Bird Preservation. Technical Publication No 3. 1985: 35–81.

[0542] Amori G, Cristaldi M. *Rattus norvegicus* (Berkenhout, 1769). In: Mitchell-Jones AJ, Amori G, Bogdanowicz W et al., eds. The Atlas of European Mammals. London, UK: Academic Press, 1999: 278–9.

[0543] Pascal M, Lorvelec O, Vigne J-D. Invasions biologiques et extinctions: 11000 ans d'histoire des vertébrés en France. Paris, France: Belin, 2006.

[0544] Vigne J-D. Aproximacions arqueozoològiques de la relació home-animal en els territoris insulars: L'exemple mediterrani [Archaeozoological approaches for the relationships between man and animals on islands: Mediterranean examples; Approches archéozoologiques de la relation homme-animal dans les territoires insulaires: l'exemple Mediterranéen]. Cota Zero 1995; 11: 61–70.

[0545] Cucchi T, Vigne J-D. Origin and diffusion of the house mouse in the Mediterranean. Human Evolution 2006; 21(2): 95–106 [Special Issue on the Anthropochorous Fauna (Ed. by M Masseti)].

[0546] Lesser S. Die tierschutzrechtliche Bewertung der Wasserprivation bei Laboratorien während wissenschaftlicher Untersuchungen auf der Grundlage bisheriger physiologischer und ethologischer Erkenntnisse. Dissertation. Hanover [Hannover], Germany: Institut für Tierschutz und Verhalten, Tierärztliche Hochschule Hannover, 2003.

[0547] Sellami A, Maurel D, Kosa E, Siaud P. Réponses hormonales du mérion, un rongeur désertique, à la privation d'eau prolongée: comparaison avec le rat [Hormonal responses in the shaw's jird, a desert rodent, subjected to water-restriction: comparison with the rat]. Mésogée 2005; 61: 7–15.

[0548] Traveset A, Nogales M, Alcover JA, Delgado JD, López-Darias M, Godoy D, Igual JM, Bover P. A review on the effects of alien rodents in the Balearic (Western Mediterranean Sea) and Canary Islands (Eastern Atlantic Ocean). Biol Invasions 2009; 11(7): 1653–70.

[0549] Russell JC, Clout MN. Rodent incursions on New Zealand islands. In: Parkes J, Statham M, Edwards G, eds. Proceedings of the 13th Australasian Vertebrate Pest Conference (Wellington, 2–6 May 2005). Lincoln, New Zealand: Manaaki Whenua/Landcare Research, 2005: 324–30.

[0550] Meehan A. Rats and mice: their biology and control. East Grinstead, UK: Rentokil Initial UK Limited, 1984.

[0551] Fritts EI. Wildlife and People at Risk: A Plan to Keep Rats Out of Alaska. Juneau, AK, USA: Alaska Department of Fish and Game, 2007.

[0552] Niethammer J. Zur Taxonomie und Ausbreitungsgeschichte der Hausratte (*Rattus rattus*). Zool Anz 1975; 194: 405–15.

[0553] Brooks JE, Rowe FP. Commensal rodent control. Unpublished Document. World Health Organization. WHO/VBC/79.726, 1979.

[0554] Pucek Z, ed. Keys to vertebrates of Poland. Mammals. Warsaw [Warszawa], Poland: Wydawnictwo Naukowe PWN/Polish Scientific Publishers, 1981: 366.

[0555] Cheylan G. Evolution rapide de petites populations insulaires méditerranéennes de *Rattus rattus*. In: Alcover JA, ed. Ecologia de les Illes. Palma de Mallorca, Spain: Monografies de la Societat d'Història Natural de les Balears 6, 1999: 83–103.

[0556] Burton RF. Visit to Lissa and Pelagosa. J Roy Geogr Soc Lond 1879; 49: 151–90.

[0557] Feldman W. Unit character inheritance of color in the black rat, *Mus rattus* L. Genetics 1926; 11: 456–65.

[0558] King CM, ed. The Handbook of New Zealand Mammals. Auckland, New Zealand: Oxford University Press, [1]1990.

[0559] Amori G, Cristaldi M. *Rattus rattus* (Linnaeus, 1758). In: Mitchell-Jones AJ, Amori G, Bogdanowicz W et al., eds. The Atlas of European Mammals. London, UK: Academic Press, 1999: 280–1.

[0560] Austria, Archduke Ludwig Salvator of [Erzherzog Ludwig Salvator von Österreich-Toskana]. Paxos und Antipaxos im Jonischen Meer. Würzburg, Germany; Vienna [Wien], Austria: Leo Woerl, [1]1887; [2]1889. [Internet resource and electronic version: http://www.ludwig-salvator.com/biblio.htm and http://www.ludwigsalvator.com/digi/paxos/index.php (accessed May 5, 2012).]

[0561] Fric J, Christie D, Karris G, Dimalexis A. Rat eradication in the Northern Sporades National Park for the improvement of breeding habitat of Eleonoras's falcon. In: Abstracts of the 10th International Congress on the Zoogeography and Ecology of Greece and Adjacent Regions [ICZEGAR] (Patras, 26–30 June 2006). Hellenic Zoological Society; Dept of Biology, University of Patras; Hellenic Ornithological Society; Hellenic Ministry of Rural Development and Food, 2006: 25.

[0562] Tortonese E. Appunti faunistici relativi all'isola di Rodi. Atti del Museo Civico di Storia Naturale di Trieste 1973; (28-I)12: 271–9.

[0563] Legakis A. Νίσυρος. Το νησί του Πολυβώτη. Επτά Ημέρες/Η καθημερινή, 22 July 2001: 29.

[0564] Geil WE. The isle that is called Patmos. London, UK: Marshall Brothers, 1904.

[0565] Vigne J-D. Zooarchaeology and the biogeographical history of the Mammals of Corsica and Sardinia since the last ice age. Mammal Rev 1992; 22(2): 87–96.

[0566] Tchernov E. Of mice and men. Biological markers for long-term sedentism: a reply. Paléorient 1991; 17(1): 153–60.

[0567] Auffray J-C, Tchernov E, Nevo E. Origine du commensalisme de la souris domestique *Mus musculus domesticus* vis-à-vis de l'homme. CR Acad Sci Paris 1988; 307: 517–22.

[0568] Tchernov E. Biological evidence for human sedentism in Southwest Asia during the Natufian. In: Bar-Yosef O, Valla FR, eds. The Natufian Culture in the Levant. Ann Arbor, MI, USA: International Monographs in Prehistory, Archaeological Series 1, 1991: 315–40.

[0569] Armitage P, West B, Steedman K. New evidence of black rat in Roman London. The London Archaeologist 1984; 4(14): 375–83.

[0570] Masseti M. Current knowledge on the early occurrence of the black rat, *Rattus rattus* L., 1758 (Muridae, Rodentia), on the Italian mainland and islands. In: Atti del 1° Convegno Nazionale di Archeozoologia (Rovigo, 5–7 March 1993). Padusa Quaderni; Collana di Studi Monografici del Centro Polesano di Studi Storici, Archeologici ed Etnografici [CPSSAE] 1995; 1: 348–58.

[0571] Barrett-Hamilton GEH, Hinton MAC. A history of British mammals. London, UK: Gurney and Jackson, 1916.

[0572] Ortalli G. Gli animali nella vita quotidiana dell'Alto Medioevo: termini di un rapporto. In: L'uomo di fronte al mondo animale nell'Alto Medioevo. Atti della XXXI Settimana di studio del Centro Italiano di Studi sull'Alto Medioevo (Spoleto, 7–13 April 1983). Spoleto, Italy: Arti grafiche Panetto & Petrelli, 1985: 1389–443.

[0573] McNamee G. [Aelian's] On the nature of animals [*De natura animalium*]. San Antonio, TX, USA: Trinity University Press, 2011.

[0574] O'Connor TP. Pets and pests in Roman and Medieval Britain. Mammal Rev 1992; 22(2): 107–13.

[0575] Macdonald DW, Barrett P. Mammals of Britain and Europe. London, UK: HarperCollins, 1993.

[0576] Boursot P, Din W, Anand R, Darviche D, Dod B, von Deimling F, Talwar GP, Bonhomme F. Origin and radiation of the house mouse: mitochondrial DNA phylogeny. J Evol Biol 1996; 9: 391–415.

[0577] Marshall Jr JT. Identification and scientific names of Eurasian house mice and their European allies, subgenus *Mus* (Rodentia: Muridae). Springfield, VA, USA: Privately printed, 1998.

[0578] Bonhomme F. Evolutionary relationships in the genus *Mus*. Curr Top Microbiol 1986; 127: 119–24.

[0579] Auffray J-C, Vanlerberghe F, Britton-Davidian J. The house mouse progression in Eurasia: a palaeontological and archaeozoological approach. Biol J Linn Soc 1990; 41: 13–25.

[0580] Frynta D, Slábová M, Váchová H, Volfová R, Munclinger P. Aggression and commensalism in house mouse: a comparative study across Europe and the Near East. Aggr Behav 2005; 31: 283–93.

[0581] Crowcroft P, Rowe FP. Social organization and territorial behaviour in the wild house mouse (*Mus musculus*). Proc Zool Soc Lond 1963; 140(3): 517–31.

[0582] Hurst JL. The function of urine marking in freeliving population of house mice, *Mus domesticus* Rutty. Anim Behav 1987; 35: 1433–42.

[0583] Wolff RJ. Mating behaviour and female choice: their relation to social structure in wild caught House mice (*Mus musculus*) housed in a semi-natural environment. J Zool Lond 1985; 207(1): 43–51.

[0584] Brown PR, Singleton GR. Rate of increase as a function of rainfall for house mouse *Mus domesticus* populations in a cereal-growing region in southern Australia. J Appl Ecol 1999; 36: 484–93.

[0585] Choquenot D, Ruscoe WA. Mouse population eruptions in New Zealand forests: the role of population density and seedfall. J Anim Ecol 2000; 69: 1058–70.

[0586] Krasnov BR, Chochlova JS. Spatial ethological structure of groups. In: Kotenkova E, Bulatova NS, eds. The House Mouse: Origin, distribution, systematics, behavior. Moscow [Moskva], Russia: Nauka, 1994: 188–214.

[0587] Lowe S, Browne M, Boudjelas S, de Poorter M. 100 of the World's Worst Invasive Alien Species. A selection from the Global Invasive Species Database. Auckland, New Zealand: The Invasive Species Specialist Group (ISSG), a specialist group of the Species Survival Commission [SSC] of the World Conservation Union [IUCN]. ISSG Office, School of Geography and Environmental Sciences [SGES], University of Auckland (Tamaki Campus), 2000. Updated and reprinted: November 2004. [Electronic version: www.issg.org/booklet.pdf (accessed May 5, 2012).]

[0588] Copson GR. The diet of the introduced rodents *Mus musculus* and *Rattus rattus* on subantarctic Macquarie Island. Aust Wildlife Res 1986; 13: 441–6.

[0589] Rowe-Rowe DT, Green B, Crafford JE. Estimated impact of feral house mice on sub-Antarctic invertebrates at Marion Island [Indian Ocean]. Polar Biol 1989; 9(7): 457–60.

[0590] Newman DG. Effects of a mouse, *Mus musculus*, eradication programme and habitat change on lizard populations of Mana Island, New Zealand, with special reference to McGregor's skink, *Cyclodina macgregori*. N Z J Zool 1994; 21: 443–56.

[0591] Cole FR, Loope LL, Medeiros AC, Howe CE, Anderson LJ. Food habits of introduced rodents in high-elevation shrubland of Haleakala National Park, Maui, Hawai'i. Pacific Science 2000; 54(4): 313–29.

[0592] Ruscoe WA, Murphy EC. House mouse. In: King CM, ed. The Handbook of New Zealand Mammals. Melbourne, Australia: Oxford University Press, ^{2}2005: 204–21.

[0593] Amori G, Cristaldi M, Federici R. Faunal and cytotaxonomic observations on Aeolian island rodents. Boll Zool 1983; 50: 9–14.

[0594] Macholán M. *Mus macedonicus* (Petrov and Ružić, 1983). In: Mitchell-Jones AJ, Amori G, Bogdanowicz W et al., eds. The Atlas of European Mammals. London, UK: Academic Press, 1999: 284–5.

[0595] Capizzi D, Santini L. Topo selvatico *Mus domesticus* Schwarz et Schwarz, 1943. In: Spagnesi M, Toso S, eds. Iconografia della fauna d'Italia. Ozzano dell'Emilia (Bologna), Italy: Istituto Nazionale per la Fauna Selvatica "Alessandro Ghigi", 1999: 134.

[0596] Marshall JT, Sage RD. Taxonomy of the house mouse. In: Berry RJ, ed. Biology of the House Mouse. Symposium of the Zoological Society of London No 47. London, UK: Academic Press 1981; 47: 15–25.

[0597] Sage RD. Wild mice. In: Foster HL, Small JD, Fox JG, eds. The mouse in biomedical research. Volume 1: History, genetics and wild mice. New York, NY, USA: Academic Press, 1981: 39–90.

[0598] Thaler L, Bonhomme F, Britton-Davidian J, Hamar M. The house mouse complex of species: Sympatric occurrence of biochemical groups Mus 2 and Mus 4 in Romania. Z Saugetierkd 1981; 46: 169–73.

[0599] Bonhomme F, Catalan J, Gerasimov S, Orisini P, Thaler L. Le complexe d'espèces du genre *Mus* en Europe Centrale et Orientale. I. Genétique. Z Saugetierkd 1983; 48: 78–85.

[0600] Kratochvíl J. *Mus abbotti* – eine kleinasiatisch-Balkanische Art (Muridae-Mammalia). Folia Zool 1986; 35(1): 3–20.

[0601] Bonhomme F, Thaler L. L'evolution de la souris. Recherche 1988; 199: 606–16.

[0602] Zima J, Macholán M. Robertsonian fusion 5.12 in a population of *Mus musculus musculus*. Folia Zool 1989; 38(3): 233–8.

[0603] Mitsainas GP, Giagia-Athanasopoulou EB. Studies on the Robertsonian chromosomal variation of *Mus musculus domesticus* (Rodentia, Muridae) in Greece. In: Britton-Davidian J, Searle JB, eds. The genus *Mus* as a model for evolutionary studies. Biol J Linn Soc 2005; 84: 503–13.

[0604] Mitsainas GP, Giagia-Athanasopoulou EB. Possible involvement of whole-arm reciprocal translocations (WARTs) in the evolution of a *Mus musculus domesticus* Robertsonian system from Greece. Atti della Accademia Nazionale dei Lincei; Rendiconti Lincei; Scienze fisiche e naturali 2009; 20(2): 153–62.

[0605] Mitsainas GP, Rovatsos MT, Giagia-Athanasopoulou EB. The HSR and heterochromatin distribution in natural populations of *Mus musculus domesticus* (Rodentia: Muridae) from Greece. Caryologia 2009; 62(1): 53–61.

[0606] Rebuzzini P, Castiglia R, Nergadze SG, Mitsainas G, Munclinger P, Zuccotti M, Capanna E, Redi CA, Garagna S. Quantitative variation of LINE-1 sequences in five species and three subspecies of the subgenus *Mus* and in five Robertsonian races of *Mus musculus domesticus*. Chromosome Res 2009; 17(1): 65–76.

[0607] Bonhomme F, Orth A, Cucchi T, Rajabi-Maham H, Catalan J, Boursot P, Auffray J-C, Britton-Davidian J. Genetic differentiation of the house mouse around the Mediterranean basin: matrilineal footprints of early and late colonization. Proc Biol Sci 2011; 278(1708): 1034–43.

[0608] Berry JG, Scriven PN. The house mouse: a model and motor for evolutionary understanding. Biol J Linn Soc 2005; 84: 335–47.

[0609] Wilson DE, Reeder D-AM. Mammal species of the world. A taxonomic and geographic reference. 3rd Edition, volume 1. Baltimore, MD, USA: The Johns Hopkins University Press, 2005.

[0610] Masseti M. Anthropochorous mammals of the Old World in the West Indies. Mammalia 2011; 75: 113–42.

[0611] Britton-Davidian J. Genetic differentiation in *Mus musculus domesticus* populations from Europe, the Middle East and North Africa: geographic patterns and colonization events. Biol J Linn Soc 1990; 41: 27–45.

[0612] Sage RD, Prager EM, Tichy H, Wilson AC. Mitochondrial DNA variation in house mice, *Mus domesticus* (Rutty). Biol J Linn Soc 1990; 41(1–3): 105–23.

[0613] Cucchi T, Vigne J-D, Auffray J-C. First occurrence of the house mouse (*Mus musculus domesticus* Schwarz and Schwarz, 1943) in the Western Mediterranean: a zooarchaeological revision of subfossil occurrences. In: Britton-Davidian J, Searle JB, eds. The genus *Mus* as a model for evolutionary studies. Biol J Linn Soc 2005; 84: 429–45.

[0614] Chondropoulos BP, Markakis G, Fraguedakis-Tsolis SE. Morphometric and immunological relationships among some Greek *Mus* L. populations (Mammalia, Rodentia, Muridae). Z Saugetierkd 1995; 60: 361–72.

[0615] Giagia-Athanasopoulou EB, Fraguedakis-Tsolis SE, Chondropoulos BP. Chromosomal variations and geographical distribution of some wild mice populations of the genus *Mus* (Mammalia, Rodentia) in Greece. Biol Gallo-Hell 1994; 22: 193–202.

[0616] Fraguedakis-Tsolis SE, Chondropoulos BP, Giagia EB. Contribution to the study of the taxonomy and zoogeography of the wild house mouse, genus *Mus* L. (Mammalia, Rodentia, Muridae) in Greece. I. Serological and immunological study of three populations from southern Greece. Acta Veterinaria (Beograd) 1986; 36(2–3): 95–106.

[0617] Waterhouse GR. [Characters of new species of the genus *Mus*, from the collection of Mr Darwin.] P Zool Soc Lond 1837: 15–21, 27–32.

[0618] Haas G. On the Vertebrate Fauna of the Lower Pleistocene site 'Ubeidiya'. Jerusalem, Israel: The Israel Academy of Sciences and Humanities, 1966: 1–68.

[0619] Jones PE, Johannesdottir F, Günduz I, Richards MB, Searle JB. The expansion of the house mouse into North-western Europe. J Zool 2010; 13: 257–68.

[0620] Guilaine J. De l'Orient à l'Occident: la néolithisation de la Méditerranée – questions ouvertes. In: Pessina A, Muscio G, eds. La Neolitizzazione tra Oriente e Occidente. Udine, Italy: Convegno di Studi. Museo Friulano di Storia Naturale, 2000: 11–21.

[0621] Pulak C, Rogers E. The 1993–1994 Turkish shipwreck surveys. INA Quarterly [Institute of Nautical Archaeology] 1994; 21(4): 17–21. [Electronic version: http://ina.tamu.edu/QUARTER/surv.htm (accessed May 5, 2012)].

[0622] Cucchi T. Le commensalisme murin et les premières sociétés méditerranéennes. Thèse de 3e cycle. Paris, France: Muséum national d'Histoire naturelle, 2005.

[0623] Cucchi T. Uluburun shipwreck stowaway house mouse: molar shape analysis and indirect clues about the vessel's last journey. J Archaeol Sci 2008; 35(1): 2953–59.

[0624] Vigne J-D. Les transferts anciens de mammifères en Europe occidentale: histoires, mécanismes et implications dans les sciences de l'homme et les sciences de la vie. In: Colloques d'histoire des connaissances zoologiques. Volume 5. Bodson L, Libois R, eds. Des animaux introduits par l'homme dans la faune de l'Europe (Journée d'étude – Université de Liège, 20 March 1993). Liège, Belgium: Université de Liège 1994: 15–38.

[0625] Trantalidou K. The contribution of the study of animal bones in the social understanding of Early Iron Age Oropos. In: Mazarakis Ainian A, ed. Oropos and Euboea in the Early Iron Age. Acts of an international round table. Volos, Greece: University of Thessaly Press, 2004: 379–425.

[0626] Kryštufek B, Vohralík V. *Mus macedonicus*. In: The IUCN Red List of Threatened Species. Version 2011.2 [http://www.iucnredlist.org/ (accessed May 5, 2012)].

[0627] Marshall JT. Taxonomy. In: Foster HL, Small JD, Fox JG, eds. The mouse in biomedical research. Volume 1: History, genetics and wild mice. New York, NY, USA: Academic Press, 1981: 17–26.

[0628] Orth A, Auffray J-C, Bonhomme F. Two deeply divergent mitochondrial clades in the wild mouse *Mus macedonicus* reveal multiple glacial refuges south of Caucasus. Heredity 2002; 89: 353–7.

[0629] Macholán M, Vyskočilová M, Bonhomme F, Kryštufek B, Orth A, Vohralík V. Genetic variation and phylogeography of free-living mouse species (genus *Mus*) in the Balkans and the Middle East. Mol Ecol 2007; 16: 4774–88.

[0630] Corbet GB. The mammals of the Palaearctic Region. A taxonomic review. Supplement. London, UK: British Museum (Natural History), 1984.

[0631] Macholán M. *Mus spicilegus* (Petényi, 1882). In: Mitchell-Jones AJ, Amori G, Bogdanowicz W et al., eds. The Atlas of European Mammals. London, UK: Academic Press, 1999: 288–9.
[0632] Macholán M, Vohralík V. Note on the distribution of *Mus spicilegus* (Mammalia: Rodentia) in the south-western Balkans. Acta Societatis Zoologicae Bohemicae 1997; 61: 219–26.
[0633] Sofianidou T, Vohralík V. Notes on the distribution of small mammals (Insectivora, Rodentia) in Epeirus, Greece. Bonn Zool Beitr 1991; 42: 125–35.
[0634] Kryštufek B, Macholán M. Morphological differentiation in *Mus spicilegus* and the taxonomic status of mound-building mice from the Adriatic coast of Yugoslavia. J Zool 1998; 245(2): 185–96.
[0635] Dieterlein F. Zur Kenntnis der Kreta-Stachelmaus, *Acomys (cahirinus) minous* Bate. Z Saugetierkd 1963; 28: 47–57.
[0636] Matthey R. Polymorphisme chromosomique intraspécifique et intraindividuel chez *Acomys minous* Bate (Mammalia – Rodentia – Muridae). Chromosoma 1963; 14: 468–97.
[0637] Spitzenberger F. Die Säugetierfauna Zyperns. II: Chiroptera, Lagomorpha, Carnivora und Artiodactyla. Ann Nathist Mus Wien 1979; 82: 439–65.
[0638] Macholán M, Zima J, červená A, červený J. Karyotype of *Acomys cilicicus* Spitzenberger, 1978 (Rodentia, Muridae). Mammalia 1995; 59(3): 397–402.
[0639] Zima J. *Acomys minous* (Bate, 1906). In: Mitchell-Jones AJ, Amori G, Bogdanowicz W et al., eds. The Atlas of European Mammals. London, UK: Academic Press, 1999: 292–3.
[0640] Barome P-O, Lymberakis P, Monnerot M, Gautun J-C. Molecular phylogeny brings new elements on Mediterranean *Acomys* species (*Rodentia, Muridae*). In: Reig S, ed. Abstracts of the Euro-American Mammal Congress (Santiago de Compostela, 20–24 July 1998). Santiago de Compostela, Spain: Universidad de Santiago de Compostela 1998: 15.
[0641] Tchernov E. Commensal animals and human sedentism in the Middle East. In: Clutton-Brock JC, Grigson C, eds. Animal and Archaeology. British Archaeological Reports International Series 1984; 202: 91–115.
[0642] Kryštufek B. *Glis glis* (Rodentia: Gliridae). Mammalian Species 2010; 42(865): 195–206.
[0643] Lever C. Naturalized animals. London, UK: T & AD Poyser, 1994.
[0644] Yalden D. The history of British mammals. London, UK: Academic Press, 1999.
[0645] Kryštufek B. *Glis glis* (Linnaeus, 1766). In: Mitchell-Jones AJ, Amori G, Bogdanowicz W et al., eds. The Atlas of European Mammals. London, UK: Academic Press, 1999: 294–5.
[0646] Storch G. *Glis glis* (Linnaeus, 1766) – Siebenschläfer. In: Niethammer J, Krapp F, eds. Handbuch der Säugetiere Europas. Band 1/I. Rodentia 1. Wiesbaden, Germany: Akademische Verlagsgesellschaft, 1978: 243–58.
[0647] Kahmann H. Notes sur le statut actuel de quelques mammifères menacés dans la région méditerranéenne. Mammalia 1959; 23(3): 329–31.
[0648] Legakis A. *Glis glis argenteus* (Zimmermann, 1952). In: Karandinos M, Paraschi L, eds. The Red Data Book of Threatened Vertebrates of Greece. Athens, Greece: Hellenic Zoological Society; Hellenic Ornithological Society; Hewlett Packard; WWF, 1992: 278.
[0649] Dimaki M. First record of the edible Dormouse *Glis glis* (Linnaeus, 1766), from the Greek island of Andros. Annales Musei Goulandris 1999; 10: 181–3.
[0650] Kryštufek B, Petkovski S. Annotated checklist of the mammals of the Republic of Macedonia. Bonn Zool Beitr 2003; 51: 229–54.
[0651] Siewert H. Die Siebenschläfer Kretas. Z Saugetierkd 1953; 17: 52–7.
[0652] Carpaneto GM, Cristaldi M. Dormice and man: a review of past and present relations. Hystrix It J Mamm NS 1994; 6(1–2): 303–30.
[0653] Colonnelli G, Carpaneto GM, Cristaldi M. Uso alimentare e allevamento del ghiro (*Myoxus glis*) presso gli antichi romani: materiale e documenti. In: Atti del 2° Convegno Nazionale di Archeozoologia (Asti, 14–16 November 1997). Forlì, Italy: Abaco Edizioni 2000: 315–25.

[0654] Masseti M. Zoologia storica e archeologica dei Felidi italiani. In: Randi E, Ragni B, Bizzarri L, Agostini N, Tedaldi G, eds. Biologia e conservazione dei Felidi in Italia. Atti del convegno (Santa Sofia FC, 7–8 November 2008). Ente Parco Nazionale Foreste Casentinesi: 9–28.

[0655] King A. Mammals. Evidence from wall paintings, sculpture, mosaics, faunal remains, and ancient literary sources. In: Feemster Jashemski W, Meyer FG, eds. The Natural History of Pompeii. Cambridge, UK: Cambridge University Press, 2002: 401–50.

[0656] Morris PA. *Muscardinus avellanarius* (Linnaeus, 1758). In: Mitchell-Jones AJ, Amori G, Bogdanowicz W et al., eds. The Atlas of European Mammals. London, UK: Academic Press, 1999: 296–7.

[0657] Sarà M. I mammiferi delle isole del Mediterraneo. Palermo, Italy: L'Epos società Editrice, 1998.

[0658] Bright P, Morris P. The dormouse. 2nd Edition. London, UK: The Mammal Society, 1992.

[0659] Hosey GR. The Bosporous Land-Bridge and Mammal Distribution in Asia Minor and the Balkans. Säugetierkundliche Mitteilungen 1982; 30: 53–62.

[0660] Por FD. The Levantine Landbridge: Historical and Present Patterns. In: Krupp F, Schneider W, Kinzelbach R, eds. Proceedings of the Symposium on the Fauna and Zoogeography of the Middle East (Mainz, 1985). In: Blume H, Frey W, eds. Tübinger Atlas des Vorderen Orients [TAVO]. Beihefte. Reihe A: Naturwissenschaften. A28. Wiesbaden, Germany: Dr Ludwig Reichert, 1987: 23–8.

[0661] Kryštufek B. *Canis aureus* (Linnaeus, 1758). In: Mitchell-Jones AJ, Amori G, Bogdanowicz W et al., eds. The Atlas of European Mammals. London, UK: Academic Press, 1999: 312–3.

[0662] Perlini R. Gli sciacalli più vicini all'Italia. Le Vie d'Italia 1924; 3: 268–71.

[0663] Kühn W. Die dalmatinischen Schakale. Z Saugetierkd 1935; 10: 144–6.

[0664] Dulić B, Tortić M.Verzeichnis der Säugetiere Jugoslawiens. Säugetierkundliche Mitteilungen 1960; 8(1–2): 1–12.

[0665] Kryštufek B, Tvrtković N. Variability and identity of the jackals (*Canis aureus*) of Dalmatia. Ann Nathist Mus Wien 1990; 91: 7–25.

[0666] Kryštufek B, Murariu D, Kurtonur C. Present distribution of the Golden Jackal *Canis aureus* in the Balkans and adjacient regions. Mammal Rev 1997; 27: 109–14.

[0667] Douma-Petridou E. Systematics and geographical distribution of the families Canidae and Mustelidae in the Peloponnese (in Greek). Unpublished PhD thesis. Patras, Greece: University of Patras, 1977.

[0668] Adamakopoulos P, Adamakopoulos T, Bousbouras D, Giannatos G, Hatzirvassanis V, Ioannidis Y, Papaioannou DH, Sfougaris A. Les grands mammifères de Grèce (carnivores et artiodactyles): situation actuelle, répartition, habitat – les espèces menacées, perspectives de protection. Biol Gallo-Hell 1991; 18(1): 107–26.

[0669] Ioannidis Y, Giannatos G. *Canis aureus* (Linnaeus, 1758). In: Karandinos M, Paraschi L, eds. The Red Data Book of Threatened Vertebrates of Greece. Athens, Greece: Hellenic Zoological Society; Hellenic Ornithological Society; Hewlett Packard; WWF, 1992: 283–4.

[0670] Demeter A, Spassov N. *Canis aureus* – Goldschakal. In: Stubbe M, Krapp F, eds. Handbuch der Säugetiere Europas. Band 5. Raubsäuger-Carnivora (Fissipedia). Teil I: Canidae, Ursidae, Procyonidae, Mustelidae 1. Wiesbaden, Germany: AULA-Verlag, 1993: 107–38.

[0671] Atanassov N. The jackal (*Canis aureus* L.) in Bulgaria. CR Acad Bulg Sci Sofia 1955; 8(4): 61–3.

[0672] Ioannidis Y, Giannatos G. Preliminary survey of the distribution and status of jackal (*Canis aureus* L., 1758) in southern Greece. Biol Gallo-Hell 1991; 18(1): 67–74.

[0673] Giannatos G. Conservation Action Plan for the golden jackal *Canis aureus* L. in Greece. Athens, Greece: WWF, 2004: 47.

[0674] Giannatos G, Marinos Y, Maragou P, Catsadorakis G. The status of the Golden Jackal (*Canis aureus* L.) in Greece. Belg J Zool 2005; 135(2): 145–9.

[0675] Giannatos G, Legakis A. Habitat use of the Golden jackal (*Canis aureus*) in human dominated landscapes in southern Greece. In: Abstracts of the 10th International Congress on the Zoogeography and Ecology of Greece and Adjacent Regions [ICZEGAR] (Patras, 26–30 June 2006). Hellenic Zoological Society; Dept of Biology, University of Patras; Hellenic Ornithological Society; Hellenic Ministry of Rural Development and Food, 2006: 37.

[0676] Frieseman H. Description historique et géographique de l'Archipel, rédigée d'après de nouvelles observations, et particulièrement utile aux Négocians et aux Navigateurs. Neuwied sur le Rhin [Neuwied], Germany: Société Typographique, 1789.

[0677] de Tournefort JP. Relation d'un voyage du Levant, fait par Ordre du Roy. 3 Volumes. Lyon, France: Bruyset éditeur, 1717. [Ed. by S Yerasimos. Paris, France: François Maspero, 1982.]

[0678] Clarke ED. Travels in various countries of Europe, Asia and Africa. First Part. Russia, Tartary and Turkey. 2nd Edition. London, UK: T Cadell and W Davies, 1810.

[0679] Pocock RI. The jackals of SW Asia and SE Europe. P Zool Soc Lond 1938; 108: 37–9.

[0680] Spassov N. The position of jackals in the *Canis* genus and life history of the golden jackal (*Canis aureus* L.) in Bulgaria and on the Balkans. Historia naturalis bulgarica 1989; 1: 44–56.

[0681] Lapini L, Molinari P, Dorigo L, Are G, Beraldo P. Reproduction of the golden jackal (*Canis aureus moreoticus* I. Geoffroy Saint Hilaire, 1835) in Julian Pre-Alps, with new data on its range-expansion in the high-Adriatic hinterland (Mammalia, Carnivora, Canidae). Bollettino del Museo Civico di Storia Naturale di Venezia 2009; 60: 169–86.

[0682] Partsch J. Die Insel Korfu. Petermanns Mitteilungen (Ergänzungsheft) 1887; 88: 1–97.

[0683] Giannatos G. *Canis aureus* (Linnaeus, 1758). In: Legakis A, Maragou P, eds. The Red Data Book of Endangered Animals of Greece (in Greek with English summary). Athens, Greece: Ministry of the Environment; Hellenic Zoological Society, 2009: 371–3.

[0684] Stubbe M. *Vulpes vulpes* (Linnaeus, 1758). In: Mitchell-Jones AJ, Amori G, Bogdanowicz W et al., eds. The Atlas of European Mammals. London, UK: Academic Press, 1999: 318–9.

[0685] Calisti M, Ciampalini B, Lovari S, Lucherini M. Food habits and trophic niche variation of the red fox *Vulpes vulpes* (L., 1758) in a Mediterranean coastal area. Rev Ecol-Terre Vie 1990; 45: 309–20.

[0686] Harris S. An estimation of the number of the foxes (*Vulpes vulpes*) in the city of Bristol, and some possible factors affecting their distribution. J Appl Ecol 1981; 18: 455–65.

[0687] Contesse P, Hegglin D, Gloor S, Bontadina F, Deplazes P. The diet of urban foxes (*Vulpes vulpes*) and the availability of anthropogenic food in the city of Zurich, Switzerland. Mamm Biol 2004; 69(2): 81–95.

[0688] Masseti M. Quaternary biogeography of the Mustelidae family on the Mediterranean islands. Hystrix It J Mamm NS 1995; 7(1–2): 17–34.

[0689] Kemf E, Groombridge B, Abreu A, Wilson A. Marine Turtles in the Wild. A WWF Species Status Report. Gland, Switzerland: WWF World Wide Fund For Nature (formerly World Wildlife Fund), 2000.

[0690] Ondrias JC. The mammals of Greece. In: University of Patras, Museum Goulandris, eds. Mammals and herpetofauna of Greece (in Greek). Athens, Greece: Ελεύθερος Τύπος 1996: 13–158.

[0691] Davis SJM. Khirokitia and its mammal remains. A neolithic Noah's Ark. In: Le Brun A, ed. Fouilles récentes à Khirokitia (Chypre) 1977–1981. Paris, France: Editions Recherches sur les Civilisations, Mémoire 1984; 41: 147–62.

[0692] Wilkens B. Faunal remains from Italian excavations on Crete. In: Reese DS, ed. Pleistocene and Holocene fauna of Crete and its first settlers. Madison, WI, USA: Prehistory Press, 1996: 241–61.

[0693] Higgs ES, Clegg IM, Kinnes IA. Appendix VII. Saliagos animal bones. In: Evans JD, Renfrew AC, eds. Excavations at Saliagos near Antiparos. British School at Athens, Supplementary Volume 1968; 5: 114–7.

[0694] Kotjabopoulou E, Trantalidou K. Faunal analysis of the Skoteini Cave. In: Sampson A, ed. Skoteini, Tharrounia: the Cave, the Settlement and the Cemetery. Athens, Greece: Eleftheroudakis, 1993: 392–434.
[0695] Yannouli E. Non-domestic carnivores in Greek prehistory: a review. In: Kotjabopoulou E, Hamilakis Y, Halstead P, Gamble C, Elefanti P, eds. Zooarchaeology in Greece. Recent advances. British School at Athens Studies 2003; 9: 175–92.
[0696] Clutton-Brock J. The animal bones. In: Hood S, ed. Excavations in Chios 1938–1955. Prehistoric Emporio and Ayio Gala. Volume II. British School at Athens, Supplementary Volume 1982; 16: 678–97.
[0697] Bate DMA. Animal remains, fish etc. In: Lamb W, ed. Excavations at Thermi in Lesbos. Cambridge, UK: Cambridge University Press, 1936.
[0698] Wozencraft WC. Order Carnivora. In: Wilson DE, Reeder D-AM, eds. Mammal species of the world. A taxonomic and geographic reference. 3rd Edition, volume 1. Baltimore, MD, USA: The Johns Hopkins University Press, 2005: 532–628.
[0699] King C. Weasels and stoats. Bromley, Kent, UK: Christopher Helm, 1989.
[0700] Pulliainen E. *Mustela nivalis* (Linnaeus, 1766). In: Mitchell-Jones AJ, Amori G, Bogdanowicz W et al., eds. The Atlas of European Mammals. London, UK: Academic Press, 1999: 334–5.
[0701] King CM. Weasel *Mustela nivalis.* In: Corbet GB, Harris S, eds. The Handbook of British Mammals. Oxford/London, UK; Boston, MA, USA: Blackwell Scientific Publications, 1991: 387–96.
[0702] Sheffield SR, King CM. Mustela nivalis. Mammalian species No. 454, pages 1–10. The American Society of Mammalogists, 1994. [Electronic version: http://www.science.smith.edu/msi/pdf/i0076-3519-454-01-0001.pdf (accessed May 5, 2012.]
[0703] McDonald RA, Webbon C, Harris S. The diet of stoats and weasels *(Mustela nivalis)* in Great Britain. J Zool Lond 2000; 252: 363–71.
[0704] Delattre P. La Belette (*Mustela nivalis* Linnaeus, 1766) et l'hermine (*Mustela erminea* Linnaeus, 1758). Nort-sur-Erdre, France: Société Française pour l'Étude et la Protection des Mammifères, 1987.
[0705] King CM, Moors PJ. On co-existence, foraging strategy and the biogeography of weasels and stoats (*Mustela nivalis* and *M. erminea*) in Britain. Oecologia 1979; 39: 129–50.
[0706] Lomolino MV. Species richness of mammals inhabiting near-shore archipelagos: area, isolation, and immigration filters. J Mammal 1994, 75: 39–49.
[0707] de Marinis AM, Masseti M. The weasel (*Mustela nivalis*) on the Mediterranean islands. Mamm Biol 2003; 68: 181–6.
[0708] de Marinis AM. The weasel *Mustela nivalis* in the Mediterranean region: a morphometrical approach. Vie Milieu 1996; 46: 376.
[0709] Reichstein H. *Mustela nivalis* (Linné, 1766) – Mauswiesel. In: Stubbe M, Krapp F, eds. Handbuch der Säugetiere Europas. Band 5/II. Wiesbaden, Germany: AULA-Verlag, 1993: 571–626.
[0710] Toschi A, ed. Fauna d'Italia. Mammalia. Lagomorpha, Rodentia, Carnivora. Ungulata, Cetacea. Bologna, Italy: Edizioni Calderini, 1965.
[0711] Douma-Petridou E, Ondrias JC. Contribution to the Taxonomy and Geographical Distribution of the Weasel *Mustela nivalis* on the Southern Balkan Peninsula. Säugetierkundliche Mitteilungen 1986; 33(2–3): 235–43.
[0712] Gerakis PA, Koutrakis ET, eds. Greek Wetlands. Athens, Greece: Commercial Bank of Greece, 1996.
[0713] de Marinis AM, Masseti M. Biogeographical relationships between the occurrence of weasel *Mustela nivalis* and some ecological variables on the Mediterranean islands. In: Reig S, ed. Abstracts of the Euro-American Mammal Congress (Santiago de Compostela, 20–24 July 1998). Santiago de Compostela, Spain: Universidad de Santiago de Compostela 1998: 344.

[0714] Zimmermann K. Die Carnivora von Kreta. Z Saugetierkd 1953; 17: 58–65.

[0715] Ragni B, Masseti M, Roussos T, Belardinelli A, Cicconi P. The carnivores of the island of Crete, Greece. Contributions to the Zoogeography and Ecology of the Eastern Mediterranean Region 1999; 1: 117–23.

[0716] Graf U. List of mammals observed (excluding agrimi). In: Nievergelt B, Stocker J, eds. Report. Field Course for Ethologists and Wildlife Biologists (Lefka Ori (White Mountains), Western Crete, 11 September–2 October 1985). Unpublished manuscript. Zurich [Zürich], Switzerland: Department of Ethology and Wildlife Research, Institute of Zoology, University of Zurich-Irchel, 1986: 53–9.

[0717] Forsyth Major CI. Étude sur les noms d'animaux. In: de Stefani C, Forsyth Major CI, Barbey W, eds. Karpathos. Étude géologique, paléontologique et botanique. Lausanne, Switzerland: Georges Bridel, 1895: 63–88.

[0718] Boessneck J. Eine vergleichende Dokumentation subfossiler Wieselfunde aus Anatolien. Säugetierkundliche Mitteilungen 1974; 22(4): 304–13.

[0719] van Zyll de Jong, CG. A morphometric analysis of cranial variation in Holoarctic weasels *(Mustela nivalis)*. Z Saugetierkd 1992; 57: 77–93.

[0720] Lebarbenchon C, Poitevin F, Arnal V, Montgelard C. Phylogeography of the weasel (*Mustela nivalis*) in the western-Palaearctic region: combined effects of glacial events and human movements. Heredity 2010; 105: 449–462.

[0721] Snyder LM, Klippel WE. Dark Age subsistence at the Kastro site, East Crete: exploring subsistence change and continuity during the Late Bronze Age-early Iron Age transition. In: Vaughan SJ, Coulson WDE, eds. Aspects of Palaeodiet in the Aegean. Oxford, UK: Oxbow Monographs, 1999; 96: 65–83.

[0722] Bon M, Masseti M. Wild mammal remains from a Roman pit at Oderzo (Treviso, North-Eastern Italy). Bollettino del Museo Civico di Storia Naturale di Venezia 1995; 44: 153–62.

[0723] Zeuner FE. A history of domesticated animals. London, UK: Hutchinson, 1963.

[0724] Reig S. *Martes foina* (Erxleben, 1777) Garduña. In: Palomo J, Gisbert J, eds. Atlas de los Mamiferos Terrestres de España. Madrid, España: Dirección general de la Conservacíon de la Naturaleza-SECEM-SECEMU, 2002: 270–275.

[0725] Lanza B, Azzaroli ML. I mammiferi delle Alpi Apuane. Lavori della Società Italiana di Biogeografia 1970: 1: 677–714.

[0726] Broekhuizen S. *Martes foina* (Erxleben, 1777). In: Mitchell-Jones AJ, Amori G, Bogdanowicz W et al., eds. The Atlas of European Mammals. London, UK: Academic Press, 1999: 342–3.

[0727] Delibes M, Amores F. The Stone Marten *Martes foina* (Erxleben, 1777) (Mammalia, Carnivora) from Ibiza (Pitiusic, Balearic Islands). Misc Zool 1986; 10: 335–45.

[0728] Masseti M. Presence and distribution of the stone marten, *Martes foina* Erxleben, 1777, on the island of Crete (Greece). Hystrix It J Mamm NS 1995; 7(1–2): 73–8.

[0729] Douma-Petridou E. Contribution to the knowledge of *Martes foina* Erxl. (Mammalia, Carnivora) from Achaia, northern Peloponnesus Greece and rest southern Balkan Peninsula. Mammalia 1984; 48(4): 565–72.

[0730] Stubbe M. *Martes foina* – Steinmarder. In: Stubbe M, Krapp F, eds. Handbuch der Säugetiere Europas. Band 5. Raubsäuger-Carnivora (Fissipedia). Teil I: Canidae, Ursidae, Procyonidae, Mustelidae 1. Wiesbaden, Germany: AULA-Verlag, 1993: 427–80.

[0731] Koller O. Zur Verbreitung von *Martes bunites* Bate. Zool Anz Leipzig 1928; 77: 114.

[0732] Trouessart EL. Faune des Mammifères d'Europe. Berlin, Germany: R Friedländer, 1910.

[0733] Steensma KJ, Reese DS. The mustelids of Crete. In: Reese DS, ed. Pleistocene and Holocene fauna of Crete and its first settlers. Madison, WI, USA: Prehistory Press 1996, 159–66.

[0734] Corbet GB, Ovenden D. The mammals of Britain and Europe. London, UK: Collins, 1980.

[0735] Stubbe M, Krapp F, eds. Handbuch der Säugetiere Europas. Band 5. Raubsäuger-Carnivora (Fissipedia). Teil II: Mustelidae 2, Viverridae, Herpestidae, Felidae. Wiesbaden, Germany: AULA-Verlag, 1993.

[0736] Barrett-Hamilton GEH. Note on the Beech-Marten and Badger of Crete. J Nat Hist Ser 7 1899; 4(23) 383–4.

[0737] Rathgeber T, Jantschke H. Cave explorations on the islands of Kárpathos and Kásos (South Aegean, Greece). In: Papers of the 14th International Congress of Speleology (Athens-Kalamos, 21–28 August 2005); Paper No. 216. Stuttgart, 2005: 1–6. [Electronic version: http://science.naturkundemuseum-bw.de/files/Karp_Athen.pdf (accessed May 5, 2012).]

[0738] Kurtén B. The carnivora of the Palestine caves. Acta Zoologica Fennica 1965; 107: 1–74.

[0739] Anderson E. Quaternary evolution of the genus *Martes* (Carnivora, Mustelidae). Acta Zool Fenn 1970; 130: 1–132.

[0740] Llorente L, Nores C, Gaubert P, Morales A. The neverending story: archaeozoological evidence for the dispersal of the beech marten (*Martes foina* Erxleben, 1777) into Europe. In: Abstracts of the 6th European Congress of Mammalogy [ECM] (Paris, 19–23 July 2011). Paris, France: Université P. et M. Curie; Muséum national d'Histoire naturelle, 2011: 91.

[0741] Caloi L. Fossil Carnivora of Simonelli cave. Quad. Accad. Naz. Lincei 1980, 249: 111–114.

[0742] Klippel WE, Snyder LM. Dark age fauna from Kavousi, Crete. The vertebrates from the 1987 and 1988 excavations. Hesperia 1991; 60: 179–86.

[0743] Gamble CS. Formation processes and the animal bones from the sanctuary at Phylakopi. In: Renfrew C, ed. The archaeology of cult. The sanctuary at Phylakopi. British School at Athens, Supplementary Volume 1985; 18: 479–82.

[0744] Coy J. Faunal analysis. In: Joukowsky MS, ed. Prehistoric Aphrodisias 1. Archaeologia Transatlantica 1986; 3: 180–90.

[0745] Grundbacher B. Nachweis des Baummarders, *Martes martes* in der neolitischen Ufresiedlung von Twann (Kanton Bern, Schweiz) sowie Ammerkungen zur osteometroschen Unterschiedung von *Martes martes* und *M. foina*. Z. Saugetierkunde 1992; 57: 201–210.

[0746] Prigioni C. *Lutra lutra* (Linnaeus, 1978). In: Mitchell-Jones AJ, Amori G, Bogdanowicz W et al., eds. The Atlas of European Mammals. London, UK: Academic Press, 1999: 350–1.

[0747] Raulin V. Description physique de l'Île de Crète. Livre I. Bordeaux, France: T Lafargue, 1859. [Also published as: La Crète en 1845. Description physique de l'Île de Crète. Livre I. Coedited with Association Entre-deux-Mers Régions d'Europe (ERE). Saint-Quentin-de-Baron, France: Éditions de l'Entre-deux-Mers, 2009.]

[0748] Miller GS. Some new European Insectivora and Carnivora. Annals and Magazine of Natural History London 7th Series 1907; 20: 389–98.

[0749] Masseti M, Albarella U, de Grossi Mazzorin J. The crested porcupine, *Hystrix cristata* L, 1758, in Italy. Anthropozoologica 2010; 45(2): 27–42.

[0750] Carli E. Le storie di San Benedetto a Monteoliveto Maggiore. Siena, Italy: Silvana Editoriale/Monte dei Paschi di Siena, 1980.

[0751] Costa OG. Fauna del Regno di Napoli. Quadro delle specie indigene, ed acclimatizzate della classe de' mammiferi. Naples [Napoli], Italy: Stamperia d'Azzolino e Compagno, 1839.

[0752] Willemsen GF. 1992. A revision of Pliocene and Quaternary Lutrinae from Europe. Scripta Geol 1992; 101: 1–115.

[0753] Mason CF, MacDonald SM. Otters. Ecology and Conservation. Cambridge, UK: Cambridge University Press, 1986.

[0754] MacDonald SM. The status of the otter in Europe. In: Reuther C, Röchert R, eds. Proceedings of the 5th International Otter Colloquium (Hankensbüttel/Germany, 24–27 November 1989). Habitat 1991; 6: 1–3.

[0755] Liles G, Jenkins L. A field survey for otters *(Lutra lutra)* in Yugoslavia. Journal of Zoology 1984, London; 203: 282–284.

[0756] Gaethlich, M. Otters in western Greece and Corfu. IUCN Otter Spec Group Bull 1988; 3: 17–23.

[0757] Gaethlich M. Initial study for the conservation and development of Chios islands's important habitats (in Greek). Athens, Greece: Elleniki Etaireia (Hellenic Society for the Protection of the Environment and the Cultural Heritage), 1990.

[0758] Grémillet X. Proposal for the conservation of otters *Lutra lutra* L. on Corfu island (Ionian Sea, Greece). Hystrix It J Mamm NS 1995; 7(1–2): 219–22.
[0759] Sfougaris AI. The Wetlands of Northern Euboea. In: Gerakis PA, Koutrakis ET, eds. Greek Wetlands. Athens, Greece: Commercial Bank of Greece, 1996: 313–5.
[0760] Axiotis M, Troumbis T. Ελπίδες ζωής … από ένα πτώμα. Aiolika Néa, 19 May 1997: 24.
[0761] Galanaki A, Gaethlich M. *Lutra lutra* (Linnaeus, 1758), Βίδρα, Eurasian Otter. In: Legakis A, Maragou P, eds. The Red Data Book of Endangered Animals of Greece (in Greek with English summary). Athens, Greece: Ministry of the Environment; Hellenic Zoological Society, 2009: 380–2.
[0762] de Smet K, Lymberakis P. Eurasian Otters (*Lutra lutra*) in Crete?! IUCN Otter Spec Group Bull 2003; 20(1): 72.
[0763] Clutton-Brock J. Domesticated Animals from Early Times. London, UK: William Heinemann; British Museum (Natural History), 1981.
[0764] Ragni B, Possenti M, Sforzi A, Zavalloni D, Ciani F. The wildcat in Central-Northern Italian Peninsula: a biogeographical dilemma. Biogeographia 1993; 17: 553–66.
[0765] Kumerloeve H. Zur Verbreitung kleinasiatischer Raub- und Huftiere sowie einiger Großnager. Säugetierkundliche Mitteilungen 1967; 15(4): 337–409.
[0766] Huş S. Av Hayvanları ve Avcılık [Game and hunting]. İstanbul Üniversitesi Orman Fakültesi Yayınları, İstanbul Üniversitesi Yayın No: 1971, Orman Fakültesi Yayın No: 202. Istanbul [İstanbul], Turkey: Faculty of Forestry, University of Istanbul, 1974.
[0767] Harrison DL. The mammals of Arabia. Volume 2. Carnivora, Hyracoidea, Artiodactyla. London, UK: Ernest Benn, 1968.
[0768] Gasperetti J, Harrison DL, Büttiker W. The Carnivora of Arabia. Fauna of Saudi Arabia 1985; 7: 397–461.
[0769] Masseti M. Wild cats (Mammalia, Carnivora) of Anatolia. With some observations on the former and present occurrence of leopards in south-eastern Turkey and on the Greek island of Samos. Biogeographia 2000; 21: 607–18.
[0770] Smit CJ, van Wijngaarden A. Threatened Mammals of Europe. Wiesbaden, Germany: Akademische Verlagsgesellschaft, 1981.
[0771] Burgio E. Le attuali conoscenze sui mammiferi quaternari della Sicilia. In: Prima Sicilia. Alle origini della società siciliana. Regione Siciliana, Assessorato dei Beni Culturali, Ambientali e della Pubblica Istruzione. Palermo, Italy: Ediprint, 1997: 54–71.
[0772] Nowell K, Jackson P. Wild Cats: Status Survey and Conservation Action Plan. Gland, Switzerland: IUCN [International Union for Conservation of Nature/Union Internationale pour la Conservation de la Nature et de les Ressources (UICN)], 1996.
[0773] Pocock RI. Notes upon some African species of the genus *Felis*, based upon specimens recently exhibited in the Society's Gardens. P Zool Soc Lond 1907; 77: 656–77.
[0774] Vigne J-D, Guilaine J, Debue K, Haye L, Gérard P. Early taming of the cat in Cyprus. Science 2004; 304: 259.
[0775] Schwaab F, Schwaab M, Vincent G. Unpublished report: Compre rendu de nos observations en Crète, du 6-20 Avril 1988. In Adamakopoulos-Matsoukas P. La distribution du chat sauvage (*Felis silvestris*) en Gréce. Biologia Gallo-hellenica 1991; 18(1): 45–52.
[0776] Adamakopoulos-Matsoukas P. La distribution du chat sauvage (*Felis silvestris*) en Gréce. Biologia Gallo-hellenica 1991; 18(1): 45–52.
[0777] Mylonas S. The wildcat (*Felis silvestris* cf. *cretensis*) on Crete (in Greek). Η Φύσις 1996; 75: 23–5.
[0778] Haltenorth T. Die Wildkatzen der alten Welt. Leipzig, Germany: Akademische Verlagsgesellschaft Geest und Portig, 1953.
[0779] Heptner VG, Sludskii AA. Raubtiere (Feloidea). Die Saugetiere der Sowjetunion Band 3. Vol. 2. Jena, Gustav Fischer, 1980.
[0780] Legakis A. *Felis silvestris cretensis* (Haltenorth, 1953). In: Karandinos M, Paraschi L, eds. The Red Data Book of Threatened Vertebrates of Greece. Athens, Greece: Hellenic Zoological Society; Hellenic Ornithological Society; Hewlett Packard; WWF, 1992: 264–5.

[0781] Snyder LM, Klippel WE. The Cretan badger (*Meles meles*) as a food resource at Late Bronze/ Early Iron Age Kavousi-Kastro. In: Reese DS, ed. Pleistocene and Holocene fauna of Crete and its first settlers. Madison, WI, USA: Prehistory Press, 1996: 283–93.
[0782] Tsoukala E. The animal bones from Smari: a preliminary report. In: Reese DS, ed. Pleistocene and Holocene fauna of Crete and its first settlers. Madison, WI, USA: Prehistory Press, 1996: 273–75.
[0783] Tsoukala E, Chatzi-Vallianou D. Fauna and human diet in the Acropolis of Smari in the geometric and the palaeoanactoric period. 8th International Cretological Congress, Herakleion, 9–14 September 1996. Vol. A_3, 2000: 397–420 (in Greek).
[0784] Malek J. The cat in ancient Egypt. London, UK: British Museum Press,1993.
[0785] Hughes D. Europe as consumer of exotic biodiversity: Greek and Roman times. Landscape Res 2003; 28: 21–31.
[0786] Voultsiadou E, Tatolas A. The fauna of Greece and adjacent areas in the Age of Homer: evidence from the first written documents of Greek literature. J Biogeogr 2005; 32: 1875–82.
[0787] Masseti M. In the gardens of Norman Palermo, Sicily (12th century AD). Anthropozoologica 2009; 44(2): 7–34.
[0788] Francis CM. A field guide to the mammals of south-east Asia. London, UK; Cape Town, South Africa; Sydney, Australia; Auckland, New Zealand: New Holland Publishers, 2008.
[0789] Danford CG, Alston ER. On the Mammals of Asia Minor. Part II. P Zool Soc Lond, 1880: 50–64.
[0790] Zei A. Wildcat under glass. New York, NY, USA: Henry Holt, 1963.
[0791] Kumerloeve H. Zum Stand des Vorkommens von *Panthera pardus tulliana* Valenciennes, 1856 in Kleinasiaen. Der Zoologische Garten NF (Leipzig), 1971; 40: 4–22
[0792] Avcı O. Dilek Yarımadası Milli Parkı [Dilek Peninsula National Park]. Unpublished Manuscript. Orman Fakültesi, İstanbul Üniversitesi. Istanbul [İstanbul], Turkey: Faculty of Forestry, University of Istanbul, 1978.
[0793] Ulrich B, Riffel M. New evidence for the occurrence of the Anatolian leopard *Panthera pardus tulliana* (Valenciennes, 1956) in Western Turkey. Zool Middle East 1993; 8: 5–14.
[0794] Lomolino MV, Riddle BR, Whittaker RJ, Brown JH. Biogeography. Fourth edition. Sunderland, MA, USA Sinauer Associates Inc., 2010.
[0795] Kasparek A, Kasparek M. Reiseführer Natur Türkei. Munich [München], Germany: BLV Verlagsgesellschaft, 1990.
[0796] Valenciennes MA. Sur une nouvelle espèce de Panthère tuée par M Tchihatcheff à Ninfi, village situé à huit lieues Est de Smyrne. CR Hebd Seances Acad Sci Paris 1856; 42: 1035–9.
[0797] Pocock RI. The Panthers and Ounces of Asia. Journal of the Bombay Natural History Society 1930; 34: 64–82 [Part I] and 307–336 [Part II].
[0798] Leyhausen P. The subspecies question. Cat News 1990; 15: 17–8.
[0799] Pocock RI. Note on a Panther from Asia Minor, obtained by JE Whitall. Journal of the Bombay Natural History Society 1935; 37: 947–9.
[0800] Mendelssohn H. Anatolian leopard *Panthera pardus tulliana*. Cat Specialist Group Meeting Report. Cat News 1990; 12: 4.
[0801] Tortonese E. Impressioni di un naturalista in Anatolia. Rivista di Scienze Naturali "Natura" 1948; 39: 49–57.
[0802] Banoğlu NA. Turkey, a sportsman's paradise. Ankara, Turkey: The Press, Broadcasting and Tourism Department, 1958.
[0803] Borovali Ö. Turkey. In: The SCI Record Book of Trophy Animals. Tucson, AZ, USA: Safari Club International, 1986: 528–30.
[0804] Kumerloeve H. Die Säugetiere (Mammalia) der Türkei. Veröffentlichungen der Zoologischen Staatssammlung München 1975; 18: 69–158.
[0805] Mendelssohn H. Felids in Israel. Cat News 1989; 10: 2–4.
[0806] Jackson P. Serious plight of the desert leopards. Cat News 1994; 20: 12.

[0807] Riffel M. Der anatolische Leopard ist nicht ausgestorben. Mitteilungen der Zoologischen Gesellschaft für Arten- und Populationsschutz [ZGAP; Zoological society for the conservation of species and populations] 1990; 6(1): 9–10.
[0808] Ulrich B. Neue Gefahren für die letzen Leoparden der Türkei. Mitteilungen der Zoologischen Gesellschaft für Arten- und Populationsschutz [ZGAP; Zoological society for the conservation of species and populations] 1994; 10(1): 14–5.
[0809] Kasparek M. On a historical occurrence of the lion, *Panthera leo*, in Turkey. Zool Middle East 1986; 1: 9–10.
[0810] Morgan L. A Pride of Leopards: A Unique Aspect of the Hunt Frieze from Tell el-Dab'a. Ägypten und Levante/Egypt and the Levant, 2010: 20, 263–301.
[0811] Masseti M. Representation of birds in Minoan art. International Journal of Osteoarchaeology 1997; 7: 354–63.
[0812] Bökönyi S. Hunting in Arslantepe, Anatolia. In: Frangipane M, Hauptmann H, Liverani M, Matthiae P, Mellink M, eds. Between the rivers and over the mountains. Archaeologica Anatolica et Mesopotamica Alba Palmieri dedicata. Rome [Roma], Italy: Dipartimento di Scienze Storiche Archeologiche e Antropologiche dell'Università di Roma "La Sapienza", 1993: 341–59.
[0813] Spitz F. *Sus scrofa* (Linnaeus, 1758). In: Mitchell-Jones AJ, Amori G, Bogdanowicz W et al., eds. The Atlas of European Mammals. London, UK: Academic Press, 1999: 381–2.
[0814] Jaia G. L'isola di Rodi. Bollettino della Società Geografica Italiana 1912; 5(1): 1081–114.
[0815] Groves C. Current views on taxonomy and zoogeography of the genus Sus. In: Albarella U, Dobney K, Ervynck A, Rowley-Conwey P, eds. Pigs and humans. 10,000 years of interaction. Oxford, UK; New York, NY, USA: Oxford University Press, 2007: 15–29.
[0816] Taibel A. Alcune osservazioni intorno ai maiali dell'Isola di Piscopi. Rivista di Zootecnia 1931; 8: 335–45.
[0817] Peters J, Helmer D, von den Driesch A, Saña Segui M. Early animal husbandry in the northern Levant. Palèorient 2000; 25/2: 27–47.
[0818] Vigne J-D, Zazzo A, Saliège JF, Poplin F, Guilaine J, Simmons A. Pre-Neolithic wild boar management and introduction to Cyprus more than 11,400 years ago. Proc Natl Acad Sci USA, 2009; 106(38): 16135–8.
[0819] Peters J, von den Driesch A, Helmer D. The upper Euphrates–Tigris basin: cradle of agropastoralism? In: Vigne J-D, Peters J, Helmer D, eds, The First Steps of Animal Domestication: New Archaeozoological Approaches. Oxford, Oxbow Books 2005: 99–124.
[0820] Larson G, Albarella U, Dobney K, Rowley-Conwy P, Schibler J, Tresset A, Vigne J-D, Edwards CJ, Schlumbaum A, Dinu A, Bălăçsescu A, Dolman G, Tagliacozzo A, Manaseryan N, Miracle P, van Wijngaarden-Bakker L, Masseti M, Bradley DG, Cooper A. Ancient DNA, pig domestication, and the spread of the Neolithic into Europe. Proc Natl Acad Sci USA 2007; 104(39): 15276–81.
[0821] Haltenorth T. Beitrag zur Kenntnis des Mesopotamischen Damhirsches *Cervus (Dama) mesopotamicus* Brooke, 1875 und zur Stammes- und Verbreitungsgeschichte der Damhirsche allgemein. Säugetierkundliche Mitteilungen 1959; 7: 1–89.
[0822] Heidemann G. Damwild, *Cervus dama* Linné, 1758, in Kleinasien. Bestand und Schutz. Säugetierkundliche Mitteilungen 1976; 24: 124–32.
[0823] Heidemann G. *Cervus dama* (Linnaeus, 1758) – Damhirsch. In: Niethammer J, Krapp F, eds. Handbuch der Säugetiere Europas. Band 2/II. Paarhufer. Wiesbaden, Germany: AULA-Verlag, 1986: 140–58.
[0824] Uerpmann H-P. The major faunal areas of the Middle East during the late Pleistocene and early Holocene. Préhistoire du Levant 1981. Paris. Centre Nationale de la Recherche Scientifique: 99–106.
[0825] Uerpmann H-P. The Ancient Distribution of Ungulate Mammals in the Middle East. Fauna and Archeological Sites in Southwest Asia and Northeast Africa. In: Blume H, Frey W, eds. Tübinger Atlas des Vorderen Orients [TAVO]. Beihefte. Reihe A: Naturwissenschaften. A27. Wiesbaden, Germany: Dr Ludwig Reichert, 1987.

[0826] Chapman D, Chapman N. Fallow deer. Machynlleth, UK: Coch-y-bonddu Books, 1997.
[0827] Arslangündoğdu Z, Kasparek M, Saribaşak H, Kaçar MS, Yöntem O, Şahin MT. Development of the population of the European fallow deer, *Dama dama dama* (Linnaeus, 1758), in Turkey. Zool Middle East 2010; 49: 3–12.
[0828] Masseti M. Island of deer. Deer 2007; 14(3): 36–40.
[0829] Albayrak T, Giannatos G, Kabasakal B. Carnivore and ungulate populations in the Beydaglari Mountains (Antalya, Turkey) – border region between Asia and Europe. Pol J Ecology 2012; 60(2), in press.
[0830] Masseti M. The Postglacial diffusion of the genus *Dama* Frisch, 1775, in the Mediterranean region. Supplemento alle Ricerche di Biologia della Selvaggina 1996; 25: 7–29.
[0831] Chapman NG, Chapman DI. The distribution of fallow deer: a worldwide review. Mammal Rev 1980; 10(2–3): 61–138.
[0832] Dicks TRB. Rhodes. Newton Abbot and London, UK: David and Charles, 1974.
[0833] Wilde WR. Narrative of a voyage to Madeira, Tenerife, and along the shores of the Mediterranean, including Algiers, Egypt, Palestine, Tyre, Rhodes, Telmessus, Cyprus, and Greece. Dublin, William Curry, Jun. and Company, 1840.
[0834] Masseti M, de Marinis AM, Theodoridis N, Papastergiou K. Estimate of numbers of the last Rhodian fallow deer, *Dama dama dama* (L, 1758), Greece: preliminary results and main goals for future research. In: Abstracts of the 6th European Congress of Mammalogy [ECM] (Paris, 19–23 July 2011). Paris, France: Université P. et M. Curie; Muséum national d'Histoire naturelle, 2011: 60.
[0835] Masseti M, Cavallaro A, Vernesi C. The population genetics of the fallow deer of Rhodes. In: Masseti M, ed. Island of deer. Natural history of the fallow deer of Rhodes and of the vertebrates of the Dodecanese (Greece). City of Rhodes: Environment Organization, 2002: 165–8.
[0836] Buchholz H-G. Antike und rezente Hirsche auf Zypern. Tier und Museum 2001, 7, 4: 73–83.
[0837] Migli D, Youlatos D. Habitat utilization and activity patterns of an isolated population of fallow deer (*Dama dama*) on the island of Limnos, north-eastern Aegean. In: Abstracts of the 10th International Congress on the Zoogeography and Ecology of Greece and Adjacent Regions [ICZEGAR] (Patras, 26–30 June 2006). Hellenic Zoological Society; Dept of Biology, University of Patras; Hellenic Ornithological Society; Hellenic Ministry of Rural Development and Food, 2006: 182.
[0838] Migli D, Youlatos D. Effects of high population pressure and human disturbance on social structure, behaviour and habitat utlisation of fallow deer (*Dama dama*) on the island of Lemnos, north-eastern Aegean, Greece. In: Prigioni C, Sforzi A, eds. Abstracts of the 5th European Congress of Mammalogy (Siena, 21–26 September 2007). Hystrix It J Mamm NS 2007; I-2 Suppl: 265.
[0839] Chapman DI, Chapman NG. Fallow deer. Their history, distribution and biology. Lavenham, Suffolk, UK: Terence Dalton, 1975.
[0840] Theodoridis N, Boulgaris K, Papastergiou K. Rhodian fallow deer: Special characteristics, impact of wildfires, livestock grazing and poaching. In: Mantzanas KT, Papanastasis VP, eds. Proceedings of the 6th Panhellenic Rangeland Congress: Range Science and Protected Areas (Leonidio Arcadia/Peloponnesus, 2–4 October 2008). Thessaloniki, Greece: Ministry of Rural Development; Food Directorate General for the Development and Protection of Forests and Natural Environment; Hellenic Pasture and Range Society, 2008: 241–325.
[0841] Bökönyi S. Faunal remains. In: Renfrew C, Gimbutas M, Elster E, eds. Excavations at Sitagroi, a Prehistoric Village in Northeast Greece 1. Monumenta Archaeologica 1986; 13: 63–132.
[0842] Masseti M, Rustioni M. Considerazioni preliminari sulla diffusione di *Dama dama* (Linnaeus, 1758) durante le epoche tardiglaciale e postglaciale nell'Italia mediterranea. Studi per l'Ecologia del Quaternario 1988; 10: 93–119.

[0843] Ueckermann E, Hansen P. Das Damwild. Biologie, Hege und Jagd. Stuttgart, Germany: Kosmos, 2002.

[0844] Jullien R. La faune des vertebrés, a l'exclusion de l'homme, des oiseaux, des rongeurs et des poissons. In: Lambert N, ed. La Grotte Préhistorique de Kitsos (Attique). Mission 1968–1978. Volume 2. Athens, Greece: Éditions ADP/École Française d'Athènes, 1981: 569–90.

[0845] Trantalidou K. Ο Ζωϊκός Κόσμος (The animal world). Αρχαιολογία και Τέχνες 1996; 58: 45–53.

[0846] Darlas A, Psathi E. Le Paléolithique supérieur dans la péninsule du Mani (Péloponnèse, Grèce). In: Darlas A, Mihailović D, eds. The Palaeolithic of the Balkans. British Archaeological Reports International Series 2008; 1819: 51–9.

[0847] Starkovich BM, Stiner MC. Upper Paleolithic animal exploitation at Klissoura Cave 1 in southern Greece: Dietary trends and mammal taphonomy. Eurasian Prehistory 2010; 7 (2): 107–132.

[0848] Yannouli E, Trantalidou K. The fallow deer (*Dama dama* Linnaeus, 1758): Archaeological presence and representation in Greece. In: Archäologie in Eurasien. Volume 6. Benecke N, ed. The Holocene History of the European vertebrate Fauna. Modern Aspects of Research. Papers presented at an international workshop (Berlin, 6–9 April 1998). Rahden/Westfalen, Germany: Marie Leidorf, 1999: 247–81.

[0849] Otte M, Yancinkaya I, Leotard J-M, Kartal M, Bar-Yosef O, Kozlowki J, López Bayón I and Marshack A. The Epi-Palaeolithic of Öküzini cave (SW Anatolia) and its mobiliary art. Antiquity 1995; 69(266): 931–44.

[0850] Atici L. Implications of age structures for Epipalaeolithic hunting strategies in the Western Taurus mountains, southwest Turkey. Anthropozoologica 2009; 44(1): 13–39.

[0851] von den Driesch A. Faunenhistorische Untersuchungen am prähistorischen Tierknochenmaterial vom Sirkeli Höyük, Adana/Türkei. Istanbuler Mitteilungen 1996; 46: 27–39.

[0852] Efstratiou N. Agios Petros. A Neolithic site in the Northern Sporades. British Archaeological Reports International Series 241, 1985.

[0853] Schwartz CA. Agios Petros. The vertebrate and molluscan fauna: final report. In: Efstratiou N, ed. Agios Petros. A Neolithic site in the northern Sporades. Aegean relationships during the Neolithic of the 5th millennium. British Archaeological Reports International Series 1985; 241: 151–60.

[0854] Bökönyi S. Angaben zum frühholozänen Vorkommen des Damhirsches *Cervus* (*Dama dama* Linné, 1758) ausserhalb seines natürlichen Verbreitungsgebietes. Säugetierkundliche Mitteilungen 1971; 19: 206–17.

[0855] Halstead P. Man and other animals in later Greek Prehistory. British School at Athens Annual 1987; 82: 71–83.

[0856] Trantalidou K. The Rhodian fallow deer: game and trophy since prehistoric times. In: Masseti M, ed. Island of deer. Natural history of the fallow deer of Rhodes and of the vertebrates of the Dodecanese (Greece). City of Rhodes: Environment Organization, 2002: 159–64.

[0857] Trantalidou K. Assomatos: un site du Bronze Ancien dans la Région Ialysia dans l'Île de Rhodes. Les vestiges zooarchéologiques (Fouilles 1986, 1989, 1991). Athens Annals of Archaeology 1996–1998 [Archaiologika analekta ex Athēnōn 1996–1998], 2000 [publication year]; 29–31: 113–24.

[0858] Trantalidou K. Animals and Human Diet in the Prehistoric Aegean. In: Hardy DA, Keller J, Galanopoulos VP, Flemming NC, Druitt TH, eds. Thera and the Aegean World III. Volume 2. Earth Sciences. Proceedings of the 3rd International Congress (Santorini/Greece, 3–9 September 1989). London, UK: The Thera Foundation, 1990: 392–405.

[0859] Nobis G. Zur antiken Wild- und Haustierfauna Kretas – nach Studien an Tierresten aus den archäologischen Grabungen Poris bei Iraklion und Eléftherna bei Arkadi. Tier und Museum 1993; 3(4): 109–20.

[0860] Vallianos C. The Museum of Cretan Ethnology. Vori/Crete, Greece: Foundation Museum of Cretan Ethnology; Research Centre of the Cretan Ecosystem, 1989.
[0861] Flerov KK. Musk deer and deer: fauna of the USSR. Volume 2. Mammals (in Russian). Moscow, Russia: Académie des sciences d'URSS, 1952.
[0862] von Raesfeld F, Vorreyer F. Das Rotwild. Naturgeschichte, Hege und Jagd. 8th Edition. Hamburg/Berlin, Germany: Paul Parey, 1978.
[0863] Wagenknecht E. Rotwild. Berlin, Germany: J. Neumann-Neudamm, 1981.
[0864] Geist V. Deer of the world. Their evolution, behaviour, and ecology. Shrewsbury, UK: Swan-Hill Press, 1999.
[0865] Koubek P, Zima J. *Cervus elaphus* Linnaeus, 1758. In: Mitchell-Jones AJ, Amori G, Bogdanowicz W et al., eds. The Atlas of European Mammals. London, UK: Academic Press, 1999: 388–89.
[0866] Grubb P. Order *Artiodactyla*. In: Wilson DE, Reeder D-AM, eds. Mammal species of the world. A taxonomic and geographic reference. 3rd Edition, volume 1. Baltimore, MD, USA: The Johns Hopkins University Press, 2005: 637–722.
[0867] Poirazidis K. Το φυσικόπεριβάλλον της Σιθωνίας. Η Φύσις 1987; 37: 12–4.
[0868] Sfougaris AI. *Cervus elaphus* (Linnaeus, 1758). In: Legakis A, Maragou P, eds. The Red Data Book of Endangered Animals of Greece (in Greek with English summary). Athens, Greece: Ministry of the Environment; Hellenic Zoological Society, 2009: 363–4.
[0869] Papika S. Biology of *Cervus elaphus* (in Greek). In: Amorgianiotis G, ed. Καταγραφή της πανίδας του Εθνικού Δρυμού Πάρνηθας. Athens, Greece: Hellenic Ministry of Agriculture; General Secretariat for Forest and Natural Environment; Parnitha, Greece: Forest Authority, 1997: 125–42.
[0870] Petsis T, Gourgourini P. Greek Nature, the Authentic Field Guide. Athens, Lynx Editions, 2004.
[0871] Turan N. Türkiye'nin av ve yaban hayvanlari memeliler [Wild mammals and game in Turkey]. Ankara, Turkey: Ongun Kardeşler Matbaacılık Sanayii, 1984.
[0872] Poirazidis K, Paraschi L. *Cervus elaphus* (Linnaeus, 1758). In: Karandinos M, Paraschi L, eds. The Red Data Book of Threatened Vertebrates of Greece. Athens, Greece: Hellenic Zoological Society; Hellenic Ornithological Society; Hewlett Packard; WWF, 1992: 270–1.
[0873] Barclay EN. The red deer of the Caucasus. P Zool Soc Lond 1934; 104(4): 789–98.
[0874] Dolan Jr JM. A deer of many lands. A guide to the subspecies of the red deer *Cervus elaphus* L. ZooNooz Newsletter 1988; 62(10): 1–34.
[0875] Mattioli S. Antler morphology in an Italian red deer population. In: Abstracts of the 20th Congress of the International Union of Game Biologists (Gödöllő/Hungary, 21–26 August 1991), 1991: 468–470.
[0876] Trantalidou K, Masseti M. Archaeozoology of the red deer in the southern Balkan peninsula and the Aegean region during antiquity: confronting bones and paintings. In: Vigne J-D, Patou-Mathis M, Lefèvre C, eds. Abstracts of the 11th International Conference of ICAZ [International Council for Archaeozoology] (Paris, 23–28 August 2010). Paris, France: Muséum national d'Histoire naturelle, 2010: 235–6.
[0877] Trantalidou K, Masseti M. Archaeozoology of the red deer in the southern Balkan Peninsula and the Aegean Region during antiquity: confronting bones and paintings. In: Sykes N, Baker K, Carden RF, Madgwick R, eds. Deer and People: Past, Present and Future. Transactions of the international conference (University of Lincoln, UK, 8–11 September 2011). Oxford, UK: Windgather Press/Oxbow Books, in press.
[0878] Masseti M. The mythical swimming deer of Mediterranean tradition. In: Masseti M, ed. Island of deer. Natural history of the fallow deer of Rhodes and of the vertebrates of the Dodecanese (Greece). City of Rhodes: Environment Organization, 2002: 121–4.
[0879] Scholfield AF. On the characteristics of animals [*De natura animalium*], by Aelian. Books I–V. Loeb Classical Library. Cambridge, MA, USA: Harvard University Press; London, UK: William Heinemann, 1958.

[0880] Nobis G, von Lehmann E. Ein Geweihstück vom Rothirsch, *Cervus elaphus* Linné, 1758, aus Kition, Zypern. Säugetierkundliche Mitteilungen 1979; 40(27): 158–60.
[0881] Whitehead GK. The Whitehead Encyclopedia of Deer. Shrewsbury, UK: Swan Hill Press, 1993.
[0882] Danilkin A, Hewison AJM. Behavioural ecology of Siberian and European roe deer. London (UK): Chapman and Hall, 1996.
[0883] Sordinas A. Παλαιολιθική έρευνα στον Ιόνιο χώρο κατά την δεκαετία του '60. Αρχαιολογία και Τέχνες 1996; 62: 57–9.
[0884] Bedwin O. Bedwin O. The animal bones. In: Popham M, ed. The Minoan uinexplored mansion at Knossos. London, British School at Athens supplementary volume no. 17 1984; 307–308.
[0885] Taramelli A. Grotte de Miamu. American Journal of Archaeology 1897; 1: 287–312.
[0886] Coy J. Animals remains. In: Coleman JE, ed. Keos I. Kephala: a Late Neolithic settlement and cemetery. Princeton, NJ, USA: American School of Classical Studies at Athens (ASCSA), 1977: 129–33.
[0887] Trantalidou K. Παρατηρήσεις σε μικρά ανασκαφικά σύνολα: το οστεολογικό υλικό της Καλογερόβρυσης. In: Sampson A, ed. Kaloyerovrisi. A Bronze Age settlement at Phylla, Euboea. Athens, Greece: Ephorate for Palaeoanthropology-Speleology of Southern Greece, 1993: 163–8.
[0888] Gamble C. The Bronze Age Animal Economy from Akrotiri: A Preliminary Analysis. In: Doumas C, ed. Thera and the Aegean World I. Papers presented at the 2nd International Scientific Congress (Santorini/Greece, August 1978). London, UK: Thera and the Aegean World, 1978: 745–53.
[0889] Gamble C. Animal husbandry, population and urbanisation. In: Renfrew C, Wagstaff M, eds. An Island Polity. Cambridge, UK: Cambridge University Press, 1982: 161–71.
[0890] Hazzidakis [Hadzidakis] J. Τύλισος Μινωική [Tylissos minoïkē]. Αρχαιολογική Εφημερίς [Arkhaiologike Ephemeris, Archaiologikē Ephēmeris] 1912; 51: 197–233.
[0891] Detournay B, Poursat J-C, Vandenabeele F. Fouilles exécutées à Malia. Le Quartier Mu II. Paris: Études Crétoises 1980; 26: 157–234.
[0892] Drandakis P, ed. Μεγάλη Ελληνική Εγκυκλοπαίδεια, Αθήναι: Ο Φοίνιξ 1926–1934, 24 τόμοι; Volume 18 [Athens, Greece: O Foinix], 1931.
[0893] Cummins J. The hound and the hawk. The art of medieval hunting. London, Weidenfeld and Nicolson: 1989.
[0894] Macgregor A. Deer on the move: relocation of stock between game parks in the Sixteenth and the Seventeenth centuries. Antropozoologica 1992; 16: 167–179.
[0895] Trombetti Budriesi L. Federico II di Svevia. *De arte venandi cum avibus*. L'arte di cacciare con gli uccelli. Roma-Bari, 2000. Editori Laterza.
[0896] von Lehmann E. Über das Vorkommen des spanischen Rothhirsches, *Cervus elaphus bolivari* Cabrera, 1911, in Nordmarokko. Säugetierk. Mitt. 1969; 2: 137–141.
[0897] Wiles GJ, Buden DW, Worthington DJ. History of introduction, population status, and management of Philippine deer (*Cervus mariannus*) on Micronesia islands. Mammalia 1999; 63(2): 193–215.
[0898] Masseti M. Notes on the diffusion in Sicily of the roe deer *Capreolus capreolus* (L, 1758) (Mammalia, Cervidae). Biogeographia 2011; 30: 141–8 [Special Issue "La Biogeografia della Sicilia"].
[0899] Sfougaris T, Chaparis D. *Capreolus capreolus* Linnaeus, 1758. In: Legakis A. & Maragou P, eds, The red data book of endangered animals of Greece (in Greek with English summary). Athens, Hellenic Zoological Society 2009.
[0900] Masseti M. Note on a Near-Eastern relic population of roe deer, *Capreolus capreolus* (L, 1758) (Mammalia, Artiodactyla). Biogeographia 2000; 21: 619–23.
[0901] Masseti M. Artiodactyls of Syria. Zool Middle East 2004; 33: 139–48.
[0902] Wahby A. Vie et mœurs des *Capra aegagrus* (Pallas) des Monts Taurus (région d'Alaya). Archivio zoologico italiano 1930; 16: 545–9.

[0903] Couturier MA-J. Le bouquetin des Alpes. Grenoble, France: chez l'auteur, 1962.
[0904] Masseti M. The prehistorical diffusion of the Asiatic mouflon, *Ovis gmelini* Blyth, 1841, and of the Bezoar goat, *Capra aegagrus* Erxleben, 1777, in the Mediterranean region beyond their natural distributions. In: Hadjisterkotis E, ed. Proceedings of the 2nd International Symposium on Mediterranean Mouflon. The Mediterranean Mouflon: Management, Genetics and Conservation. Nicosia, Cyprus: Game Fund; Ministry of Interior, 1997: 12–27.
[0905] Pedrotti L, Lovari S. *Capra aegagrus* Erxleben, 1777. In: Mitchell-Jones AJ, Amori G, Bogdanowicz W et al., eds. The Atlas of European Mammals. London, UK: Academic Press, 1999: 410–11.
[0906] Dungel J, Gaisler J. Atlas savců české a Slovenské republiky [Atlas of mammals of the Czech and Slovak Republics]. Prague [Praha], Czech: Academia, 2002.
[0907] Payne S. The origins of domestic sheep and goats: a reconsideration in the light of the fossil evidence. Proceedings of the Prehistoric Society 1968; 34: 368–384.
[0908] Masseti M. La capra selvatica nel Mediterraneo. L'Universo 1981, 61(2): 177–218.
[0909] Masseti M. Decadenza delle faune insulari. In: Masseti M, Osio A, Pratesi F, Rainero E, Trento F eds. Mare nostrum. Firenze 1987, WWF/Studio E. Rainero,: 221–231.
[0910] Helmer D. Le dévelopement de la domestication au Proche-Orient da 9500 à 7500B.P.: les nouvelles donneés d'El Kown et de ras Shamra. Paléorient 1989; 15/1: 111–121.
[0911] Ciani F, Masseti M. Considerazioni sull'origine della popolazione ircina dell'isola di Montecristo, nel Mar Tirreno settentrionale. Elementi per un confronto cronologico-culturale con l'antica diffusione artificiale dell'egagro (*Capra aegagrus* Erleben, 1777) nelle isole del Mediterraneo orientale. Supplemento alle Ricerche di Biologia della Selvaggina 1991; 18: 123–33.
[0912] Schultze-Westrum T. Die Wildziegen der ägäischen Inseln. Säugetierkundliche Mitteilungen 1963; 11: 145–82.
[0913] Azzaroli A. Biogeografia dei mammiferi della Sardegna. Lavori della Società Italiana di Biogeografia 1983; 8: 35–52.
[0914] Masseti M. The wild goats, *Capra aegagrus* Erxleben, 1777, of the Mediterranean Sea and the Eastern Atlantic Ocean islands. Mammal Rev 2009; 39(2): 141–57.
[0915] Kostopoulos DS. Greek bovids through time. Hellenic Journal of Geosciences 2006; 41: 141–52.
[0916] von Lorenz-Liburnau L. Die Wildziegen der griechischen Inseln und ihre Beziehungen zu anderen Ziegenformen. Sarajevo, Bosnia and Herzegovina: Zemaljski muzej Bosne i Hercegovine [National Museum of Bosnia and Herzegovina]: Wissenschaftliche Mitteilungen aus Bosnien und der Herzegovina, 1899; 6, 851–86.
[0917] Hablützel C. Caprinae report. Unpublished manuscript. Zurich [Zürich], Switzerland: Department of Ethology and Wildlife Research, Institute of Zoology, University of Zurich-Irchel, 1989.
[0918] Sfougaris AI. Wild Goats (*Capra aegagrus*) in Greece: distribution and status (Poster). In: Abstracts of the 5th International Congress on the Zoogeography and Ecology of Greece and Adjacent Regions [ICZEGAR] (Iraklion [Herakleion], 16–20 April 1990). Athens, Greece: Hellenic Zoological Society, 1990.
[0919] Couturier MJ. Statut actuel des représentants du genre *Capra* dans le bassin Méditerranéen. In: Animaux et végétaux rares de la région Mediterranéenne. Colloque du Service de Sauvegarde de l'Union Internationale pour la Conservation de la Nature et de les Ressources [IUCN], Septième Réunion Technique/Seventh Technical Meeting. Volume 5. Athens, Greece; Bruxelles, Belgium, 1959–1960: 12–9.
[0920] Reichenow A. Die Wildziege der Insel Joura. Zoologische Jahrbücher 1888; 3: 591–6.
[0921] Ciani F, Masseti M, Palazzo M, Cappuccio A, Matassino D. Genetic variations in the Mediterranean wild goat (*Capra aegagrus* Erxleben, 1777). Contributions to the Zoogeography and Ecology of the Eastern Mediterranean Region 1999; 1: 69–74.
[0922] Husband TP, Brown JH, Davis PB. Management Plan for Sapienza Island. Rhode Island (RI), USA: EDC [Environmental Data Center] Laboratory, University of Rhode Island, *sine data*.

[0923] Jarman MR, Jarman HN. The fauna and economy of Early Neolithic Knossos. British School at Athens Annual 1968; 63: 241–76.
[0924] Grzimek B. Grzimek's Tierleben. Zurich [Zürich], Switzerland: Kindler, 1968.
[0925] Danford CG. Notes on the wild goat *Capra aegagrus*. P Zool Soc Lond, 1875: 458–68.
[0926] Kinch KF. La chèvre sauvage (l'égagre). In: Kinch KF, Kinch H. Fouilles de Vroulia (Rhodes). Berlin, Germany: Reimer, 1914: 265–72.
[0927] de Beaux O. Posizione sistematica degli stambecchi e capre selvatiche viventi (*Capra* Linneo 1758) e loro distribuzione geografica. Atti dell'Accademia di Scienze e Lettere 1955; 12: 1–110.
[0928] Ivrea M. Note on wild goats on the Aegean Islands. P Zool Soc Lond 1899: 599.
[0929] Roussou M, ed. The islands of Messenia (in Greek). Athens, Greece: Road Editions, 2000.
[0930] Sfougaris AI. Nutritional ecology of the Cretan agrimi, European mouflon and feral rabbit and effects of grazing on island ecosystems. PhD thesis. Thessaloniki, Greece: Aristotle University, 1991 (in Greek with English summary).
[0931] Reichenow A. Eine neue Wildziege beherbergt der Zoologische Garten in Berlin. Der Zoologische Garten 1888; 29: 29–30.
[0932] Grant M, Grierson K, Henderson C. The wild goats of Yioura. Sporades Islands Expedition [19]82. Unpublished manuscript. Edinburgh, UK: University of Edinburgh, 1982.
[0933] Pashley R. Travels in Crete. 2 Volumes. Cambridge/London, UK: Murray, 1837.
[0934] Butler A. A wild goat of Crete. Field Natural 1951; 197: 128.
[0935] Nievergelt B. *Capra aegagrus* (Erxleben, 1777) – Bezoarziege. In: Niethammer J, Krapp F, eds. Handbuch der Säugetiere Europas. Band 2/II. Paarhufer. Wiesbaden, Germany: AULA-Verlag, 1986: 367–83.
[0936] Nievergelt B, Stocker J. Field Course for Ethologists and Wildlife Biologists (Lefka Ori (White Mountains), Western Crete, 11 September–2 October 1985). Zurich [Zürich], Switzerland: Department of Ethology and Wildlife Research, Institute of Zoology, University of Zurich-Irchel, 1986.
[0937] Scoula M. The biology and the conservation of the Cretan agrimi (*Capra aegagrus cretica*). Unpublished dissertation. Chania, Greece: Mediterranean Agronomic Institute, 1989.
[0938] Ciani F. Rapporto sulla protezione e conservazione dell'agrimi (*Capra aegagrus cretica*) a Creta. Supplemento alle Ricerche di Biologia della Selvaggina 1991; 18: 135–41.
[0939] Paragamian K. *Capra aegagrus cretica* (Schinz, 1838). In: Karandinos M, Paraschi L, eds. The Red Data Book of Threatened Vertebrates of Greece. Athens, Greece: Hellenic Zoological Society; Hellenic Ornithological Society; Hewlett Packard; WWF, 1992: 271–4.
[0940] Nicholson MC, Husband TP, Brown JH, Logan GT. Implications of behavior for management of the Cretan agrimi. Biol Gallo-Hell 1994; 22: 247–54.
[0941] Masseti M. Taxonomic and behavioural aspects of the representation of mammals in Aegean Bronze Age art. In: Kotjabopoulou E, Hamilakis Y, Halstead P, Gamble C, Elefanti P, eds. Zooarchaeology in Greece. Recent advances. British School at Athens Studies 2003; 9: 267–75.
[0942] Masseti M. Origin and artificial distribution of the wild goat, *Capra aegagrus* Erxleben, 1777, on the islands of the Mediterranean Sea and the Eastern Atlantic Ocean. In: Granados Torres JE, Cano-Manuel León J, Fandos Paris P, Cadenas de Llano Aguilar R, eds. Tendencias actuales en el Estudio y Conservación de los Caprinos Europeos. 2° Congreso Internacional del género Capra en Europa (Granada, 20–23 November 2007). Granada, Spain: Junta de Andalucía; Consejería de Medio Ambiente; SanPrint, 2008: 169–95.
[0943] Husband T. Final report on Cretan agrimi to School for Field Studies. Unpublished manuscript. Boston, MA, USA: School for Field Studies, Northeastern University, 1983.
[0944] von Wettstein O. Was ist *Capra dorcas* Reichenow? Z Saugetierkunde 1955; 20: 132–3.
[0945] Shewring W. Homer. The Odyssey. Oxford and New York 1980, Oxford University Press.
[0946] Sampson A. Excavations at the cave of the Cyclope on Youra, Alonessos. In: Alram-Stern E, ed. Die Ägäische Frühzeit. 2. Serie. Forschungsbericht 1975–1993. Volume 1. Das Neolithikum in Griechenland mit Ausnahme von Kreta und Zypern. Reihe: Veröffentlichungen

der Mykenischen Kommission 16. Vienna [Wien], Austria: Verlag der Österreichischen Akademie der Wissenschaften [Austrian Academy of Sciences Press], 1996: 507–20.

[0947] Sampson A. The Cyclops Cave at Youra, Alonissos. In: Papathanassopoulos GA, ed. Neolithic Culture in Greece. Athens, Greece: Nicholas P Goulandris Foundation; Museum of Cycladic Art, 1996: 58–9.

[0948] Sampson A. The Neolithic and Mesolithic occupation of the cave of Cyclope, Youra, Alonessos, Greece. British School at Athens Annual 1998; 93: 1–22.

[0949] Masseti M, Trantalidou K. Boars and goats in a Mesolithic fishing community: Youra (Northern Sporades, Greece), a case study. In: Albarella U, Dobney, Huntley KJ, Rowley-Cowny P, eds, Abstracts, 9th Conference of the International Council of Archaeozoology. Durham 2002, University of Durham: 84–85.

[0950] Horwitz LK, Bar-Gal GK. The origin and genetic status of insular caprines in the Eastern Mediterranean: a case study of free-ranging goats (*Capra aegagrus cretica*) on Crete. Human Evolution 2006; 21(2): 123–38 [Special Issue on the Anthropochorous Fauna (Ed. by M Masseti)].

[0951] Röhrs M. *Ovis ammon* (Linnaeus, 1758). In: Mitchell-Jones AJ, Amori G, Bogdanowicz W et al., eds. The Atlas of European Mammals. London, UK: Academic Press, 1999: 418–9.

[0952] Kaya MA, Bunch TD, Konuk M. On Konya wild sheep, *Ovis orientalis anatolica*, in the Bozdag protected area. Mammalia 2004; 68(2–3): 229–32.

[0953] Ryder ML. Sheep and man. London, UK: Gerald Duckworth, 1983.

[0954] van Haaften JL. The Mouflons of the Mediterranean Region. In: Kjerner I, Bjurholm P, eds. Proceedings of the 11th International Congress of Game Biologists (Stockholm, 3–7 September 1973). Stockholm, Sweden: Swedish National Environment Protection Board, 1974: 155–8.

[0955] Mursaloglu B. Türkiye'de yaban hayatı [Wildlife in Turkey]. Türkiye Tabiatını Koruma Cemiyeti Yayınları [Turkey Nature Conservation Society Publications], 1970; 16: 1–6.

[0956] Turke F, Schminke S. Verbreitung des Muffelwildes auf dem Europäischen Festland. Wild und Hund 1966; 69: 209–13.

[0957] Uloth W. Zur Geschichte der Verbreitung, Einbürgerung und Verkreuzung des Tyrrhenischen Mufflons, *Ovis ammon musimon* Schreber, 1782, in Europa und Ubersee. Unpublished manuscript: 1970.

[0958] Uloth W. To the history and distribution, introduction and crossbreeding of the Tyrrhenian mouflon in Europe and oversea. Acta Theriologica 1972; 17: 412–413.

[0959] Cassola F. Il muflone di Sardegna: importanza, stato attuale e problemi di conservazione. In: Pedrotti F (ed) S.O.S. Fauna. Animali in pericolo in Italia. Camerino 1976, World Wildlife Fund: 67–107.

[0960] Perco F. Il muflone. Bologna 1977, Edagricole Calderini.

[0961] Razzetti E, Sindaco R. Taxa non confermati o meritevoli di conferma (Unconfirmed taxa or taxa in need of confirmation). In: Sindaco R, Doria G, Razzetti E, Bernini F, eds. Atlante degli anfibi e rettili d'Italia/Atlas of Italian amphibians and reptiles. Florence [Firenze], Italy: Societas Herpetologica Italica; Edizioni Polistampa, 2006: 576–83.

[0962] Corti C, Capula M, Luiselli L, Razzetti E, Sindaco R, eds. Fauna d'Italia. Reptilia (Volume 45). Milan [Milano]/Bologna, Italy: Edizioni Calderini de il Sole 24 Ore, 2011.

[0963] Gurnell J, Wauters L. *Sciurus vulgaris* (Linnaeus, 1758). In: Mitchell-Jones AJ, Amori G, Bogdanowicz W et al., eds. The Atlas of European Mammals. London, UK: Academic Press, 1999: 180–1.

[0964] Niethammer J. *Microtus subterraneus* (de Sélys-Longchamps, 1836) – Kurzohrmaus. In: Niethammer J, Krapp F, eds. Handbuch der Säugetiere Europas. Band 2/I. Rodentia II. Wiesbaden, Germany: Akademische Verlagsgesellschaft, 1982: 397–418.

[0965] Niethammer J. *Microtus subterraneus* (de Sélys-Longchamps, 1836) – Kurzohrmaus. In: Niethammer J, Krapp F, eds. Handbuch der Säugetiere Europas. Band 1. Wiesbaden, Germany: Akademische Verlagsgesellschaft, 1978: 398–418.

[0966] Kryštufek B, Vohralík V. Distribution of the Forest Dormouse *Dryomys nitedula* (Pallas, 1779) (Rodentia, Myoxidae) in Europe. Mammal Rev 1994, 24(4): 161–77.
[0967] Corbet GB, Hill JE. A World List of Mammalian Species. New York, NY, USA; Oxford, UK: Natural History Museum Publications and Oxford University Press, 1991.
[0968] von Lehmann E. Eine Kleinsäugerausbeute von Aspromonte (Kalabrien). Sitzungsberichte der Gesellschaft Naturforschender Freunde zu Berlin NF [Neue Folge] 1964; 4(1): 31–47.
[0969] Paolucci P, Battisti A, de Battisti R. The Forest dormouse (*Dryomys nitedula* Pallas, 1779) in the eastern Alps (Rodentia, Gliridae). Biogeographia 1987; 13: 855–66.
[0970] Cagnin M, Aloise G. Current status of Myoxids (Mammalia, Rodentia) in Calabria (Southern Italy). Hystrix It J Mamm NS 1994; 6(1–2): 169–80.
[0971] Lovari S. Evoluzione recente delle popolazioni di grandi mammiferi della fauna d'Italia. In: XIX Seminario sulla "Evoluzione Biologica e i grandi problemi della Biologia. Faune attuali e faune fossili" (Rome, 26–28 February 1992). Rome [Roma], Italy: Accademia Nazionale dei Lincei. Contributi del Centro Linceo Interdisciplinare "Beniamino Segre" 1993; 86: 21–37.
[0972] van den Brink F. Guida dei mammiferi d'Europa. Milano 1969, Edizioni Labor.
[0973] Honacki JH, Kinman KE, Koeppl JW, eds. Mammal species of the world. A taxonomic and geographic reference. Lawrence, KS, USA: Allen Press/The Association of Systematic Collections, 1982.
[0974] Burton JA, Pearson B. Collins Guide to the Rare Mammals of the World. London, UK: Collins, 1987.
[0975] Gisbert J, Garcia-Perea R, Martinez-Medina FJ. *Hystrix cristata* Linnaeus, 1758 –Puercoespín. In: Palomo LJ, Gisbert J, eds. Atlas de los Mamíferos Terrestres de España. Madrid, Spain: Dirección General de Conservación de la Naturaleza SECEM-SECEMU, 2002: 440–3.
[0976] Woods CA, Kilpatrick CW. Infraorder Hystricognathi. In: Wilson DE, Reeder D-AM, eds. Mammal species of the world. A taxonomic and geographic reference. 3rd Edition, volume 1. Baltimore, MD, USA: The Johns Hopkins University Press, 2005: 1538–600.
[0977] Corbet GB, Jones LA. The specific characters of the crested porcupines, subgenus *Hystrix*. P Zool Soc Lond 1965; 144(2): 285–300.
[0978] de Marinis AM, Masseti M, Sforzi A. Note on the non-flying terrestrial mammals of the Tuscan archipelago, Northern Tyrrhenian sea (Italy). Bollettino del Museo Regionale di Scienze Naturali di Torino 1996; 14(1): 275–81.
[0979] Bashir JT. Biodiversity of Farwa Island. In: Poulakakis N, Vardiniyannis K, eds. Abstracts of the 11th International Congress on the Zoogeography, Ecology and Evolution of the Eastern Mediterranean (Herakleion, 21–25 September 2009). Herakleion, Greece: Hellenic Zoological Society, 2009: 187.
[0980] Bartosiewicz L. A lion's share of attention: archaeozoology and the historical record. Budapest, Hungary: Adadémiai Kiadó. Acta Archaeologica Academiae Scientiarum Hungaricae 2009; DOI: 10.1556/AArch.59.2008.2.28.
[0981] Kominos T, Panagiotopoulou M. Recent tangible evidence of the existence of *Lynx* in Greece. Biol Gallo-Hell 1991; 18: 67–74.
[0982] Kominos T. *Lynx lynx* (Linnaeus, 1758). In: Legakis A, Maragou P, eds. The Red Data Book of Endangered Animals of Greece (in Greek with English summary). Athens, Greece: Ministry of the Environment; Hellenic Zoological Society, 2009: 365–6.
[0983] Aymerich M. Situación del lince boreal en Europa. Quercus 1991; 61: 13–6.
[0984] Ognev SI. Zveri Vostochnoi Evropy i Severnoi Azii. Volume 2. Carnivora (Fissipedia). Moscow, Russia: Glavnauka – Gosudarstvennoe Izdatel'stvo, 1931. [Mammals of eastern Europe and northern Asia, translated from Russian by A Birron and ZS Cole for the National Science Foundation and the Smithsonian Institution. Jerusalem, Israel: Israel Program for Scientific Translations, 1962].
[0985] Albayrak T. Kamera kapan yöntemiyle Antalya Termessos Milli parkı ve çevresinde büyük memelilerin statülerinin belirlenmesi [A camera trap method to determine the status of

large mammals in and around the national park of Antalya Termessos]. Antalya, Turkey: TTKD; Akdeniz Üniversitesi, 2005.

[0986] İlemin Y, Gürkan B. Status and activity patterns of the caracal, *Caracal caracal* (Schreber, 1776), in Datça and Bozburun Peninsulas, Southwestern Turkey. Zool Middle East 2010; 50: 3–10.

[0987] Hood S. The Arts in Prehistoric Greece. Harmondsworth, UK: Penguin Books, 1978.

[0988] Platon N. Zakros. The Discovery of a Lost Palace of Ancient Crete. New York, NY, USA: Charles Scribner's Sons, 1971.

[0989] Evely D, ed. Fresco: a passport into the past. Minoan Crete through the eyes of Mark Cameron. Athens, Greece: British School at Athens; Nicholas P Goulandris Foundation; Museum of Cycladic art, 1999.

[0990] Morgan L. Island Iconography: Thera, Kea, Milos. In: Hardy DA, Doumas CG, Sakellarakis JA, Warren PM, eds. Thera and the Aegean World III. Volume 1. Archaeology. Proceedings of the 3rd International Congress (Santorini/Greece, 3–9 September 1989). London, UK: The Thera Foundation, 1990: 252–66.

[0991] Masseti M, Bruner E. The primates of the western Palaearctic. A biogeographical, historical, ethnozoological and archaeozoological review. JAS Journal of Anthropological Sciences 2009; 87: 33–91.

[0992] Masseti M. Le scimmie azzurre. La fauna etiopica degli affreschi minoici di Santorini (Thera). Mondo archeologico 1980; 51: 32–7.

[0993] Hill WCO. Primates. Comparative anatomy and taxonomy. Volume 6. Catarrhini Cercopithecoidea Cercopithecinae. Edinburgh, UK: Edinburgh University Press, 1966.

[0994] Haltenorth T, Diller H. Säugetiere Afrikas und Madagaskars. Munich [München], Germany: BLV Verlagsgesellschaft, 1977.

[0995] Kingdon J. The Kingdon pocket guide to African mammals. London, UK: A and C Black, 2004.

[0996] Groves CP. Order Primates. In: Wilson DE, Reeder D-AM, eds. Mammal species of the world. A taxonomic and geographic reference. 3rd Edition, volume 1. Baltimore, MD, USA: The Johns Hopkins University Press, 2005: 111–84.

[0997] Battaglia S, ed. Grande Dizionario della Lingua Italiana. II. Torino, Italy: Unione Tipografico-Editrice Torinese, 1962.

[0998] Rocci L. Vocabolario Greco-Italiano. Milan [Milano]/Rome [Roma]/Naples [Napoli]/Città di Castello, Italy: Società editrice Dante Alighieri, Società editrice S. Lapi, 1970.

[0999] Marinatos N. An Offering of Saffron to the Minoan Goddess of Nature: The Role of the Monkey and the Importance of Saffron. In: Linders T, Nordquist GC, eds. Gifts to the Gods. Proceedings of the Uppsala Symposium 1985. Uppsala, Sweden: Acta Universitatis Uppsaliensis (Boreas 15) 1987: 123–32.

[1000] Ferrence SC, Bendersky G. Therapy with saffron and the goddess at Thera. Perspect Biol Med 2004; 47(2): 199–226.

[1001] Marinatos S. Excavations at Thera IV–V, 1970–1971 Seasons (in English). Athens, Greece: The Archaeological Society at Athens Library Series No 179. 2nd Edition, 1999.

[1002] Sapouna-Sakellaraki E. Οι τοιχογραφίες της Θήρας σε σχέση με τη μινωική Κρήτη. In: Πεπραγμένα του Δ΄ Διεθνούς Κρητολογικού Συνεδρίου (Herakleion, 29 August–3 September 1976). Volume A2. Athens, Greece: Ministry of Culture 1981: 479–509.

[1003] Masseti M. Antilopi o gazzelle sugli affreschi di Akrotiri? Archeologia Viva 1984; 7–8: 26–7.

[1004] Säflund G. Girls and gazelles. Reflections on Theran fresco imagery. In: Philia epē eis Geōrgion E. Mylōnan, dia ta 60 etē tou anaskaphikou tou ergou (in Greek) [Festschrift for GE Mylonas]. Vivliothēkē tēs en Athēnais Archaiologikēs Hetaireias, ar 103. Athens, Greece: En Athēnais Archaiologikē Hetaireia, 1986: 185–90.

[1005] Porter R. The Cretan Wild Goat (*Capra aegagrus cretica*) and the Theran "Antelopes". In: Reese DS, ed. Pleistocene and Holocene fauna of Crete and its first settlers. Madison, WI, USA: Prehistory Press, 1996: 295–315.

[1006] Masseti M. Exploitation of fauna in the Near East during ancient historical times: observations on the mammalian species represented in Assyrian art. In: Albore Livadie C, Ortolani F, eds. Variazioni climatico-ambientali e impatto sull'uomo nell'area circum-mediterranea durante l'Olocene. Bari, Italy: Edipuglia; Ravello, Italy: Centro Universitario Europeo per i Beni Culturali, 2003: 375–90.

[1007] Funaioli U. Guida breve dei mammiferi della Somalia. Florence [Firenze], Italy: Istituto Agronomico per l'Oltremare, 1971.

[1008] Leuthold W. African ungulates. Berlin/Heidelberg, Germany; New York, NY, USA: Springer, 1977.

[1009] Godart L, Tzedakis Y. The bull in the Minoan-Mycenean world. In: Athanassopoulou S, Azara P, Nicolau i Martí A, eds. The bull in the Mediterranean world: Myths and cults. [Exhibition catalogue.] Barcelona, Spain: Museu d'Història de la Ciutat; Athens, Greece: Mouseio Benaki/Hellenic Ministry of Culture, 2003: 70–6.

[1010] Walther FR. Some reflections on expressive behaviour in combats and courtship of certain horned ungulates. In: Geist V, Walther FR, eds. The Behaviour of Ungulates and its relation to management. Papers of an International Symposium held at the University of Calgary, Alberta, Canada (2–5 November 1971). IUCN Publications New Series No. 24. Volume 1. Morges, Switzerland: Union Internationale pour la Conservation de la Nature et de ses Ressources [UICN]/International Union for Conservation of Nature and Natural Resources [IUCN], 1974: 1; 56–106.

[1011] Masseti M. Did the study of ethology begin in Crete 4000 years ago? Ethol Ecol Evol 2000; 12: 89–96.

[1012] Radi G. Tracce di un insediamento neolitico nell'isola di Lampedusa. Atti della Società Toscana di Scienze Naturali, Memorie, Serie A, 1972; 79: 197–205.

[1013] Karandinos M, Paraschi L, eds. The Red Data Book of Threatened Vertebrates of Greece. Athens, Greece: Hellenic Zoological Society; Hellenic Ornithological Society; Hewlett Packard; WWF, 1992.

[1014] Mertzanis G, Giannopoulos A, Pulidis C. *Ursus arctus* (Linnaeus, 1758). In: Legakis A, Maragou P, eds. The Red Data Book of Endangered Animals of Greece (in Greek with English summary). Athens, Greece: Ministry of the Environment; Hellenic Zoological Society, 2009: 387–9.

[1015] Boessneck J, von den Driesch A. Knochenabfall von Opfermahlen und Weihgaben aus dem Heraion von Samos (7. Jh. v. Chr.). Munich [München], Germany: Institut für Paläoanatomie, Domestikationsforschung und Geschichte der Tiermedizin, Universität München, 1988.

[1016] Jakubiec Z. *Ursus arctos* (Linnaeus, 1758). In: Mitchell-Jones AJ, Amori G, Bogdanowicz W et al., eds. The Atlas of European Mammals. London, UK: Academic Press, 1999: 322–3.

[1017] Stuart AJ, Lister AM. Extinction chronology of the cave lion *Panthera spelaea*. Quaternary Science Reviews 2011; 30(17–18): 2329–40.

[1018] Antunes A, Troyer JL, Roelke ME, Pecon-Slattery J, Packer C, Winterbach C, Winterbach H, Hemson G, Frank L, Stander P, Siefert L, Driciru M, Funston PJ, Alexander KA, Prager KC, Mills G, Wildt D, Bush M, O'Brien SJ, Johnson WE. The Evolutionary Dynamics of the Lion *Panthera leo* Revealed by Host and Viral Population Genomics. PLoS Genet 2008; 4(11): e1000251. DOI: 10.1371/journal.pgen. 1000251.

[1019] Aulagnier S, Thevenot M. Catalogue des mammifères sauvages du Maroc. Rabat, Morocco: Institut Scientifique, Université Mohammed V; Ministère de l'Education Nationale, Royaume du Maroc, 1986.

[1020] Cuzin F. La régression des grands mammifères terrestres du Maroc méridional au cours du XXe siècle. Naturalia Maroccana 2003; 1(1): 93–9.

[1021] Mouterde P. Nouvelle flore du Liban et de la Syria. Volumes 1–3. Beirut [Beyrouth], Lebanon: Éditions de l'Imprimerie Catholique, 1966.

[1022] Fellows C. An Account of Discoveries in Lycia, being a Journal kept during a second Excursion in Asia Minor (1840). London, UK: Murray, 1841.

[1023] Kinnear NB. The past and present distribution of the lion in south eastern Asia. Journal of the Bombay Natural History Society 1920; 27(1): 33–9. [Electronic version: http://www.biodiversitylibrary.org/page/30358578 (accessed May 5, 2012).]

[1024] Ninov LK. Vergleichende Untersuchungen zur Jagd und zum Jagdwild während des Neolithikums und Äneolithikums in Bulgarien. In: Archäologie in Eurasien. Volume 6. Benecke N, ed. The Holocene History of the European vertebrate Fauna. Modern Aspects of Research. Papers presented at an international workshop (Berlin, 6–9 April 1998). Rahden/Westfalen, Germany: Marie Leidorf, 1999: 323–38.

[1025] Salari L. Mosaico Nilotico di Palestrina: animali reali o fantastici? Archeozoologia di una produzione artistica di età ellenistica. Palestrina, Italy: Circolo Culturale Prenestino "Roberto Simeoni", 2006.

[1026] Alden M. Lions in paradise: lion similes in the Iliad and the lion cubs of Il. 18.318–22. Classical Quarterly 2005; 55(2): 335–42.

[1027] Becker C. Kastanas. Die Tierknochenfunde. Prähistorische Archäologie in Südosteuropa, 5. Berlin, Germany: Wissenschaftsverlag Volker Spiess, 1986.

[1028] Lefèvre F, Mulliez D, Oulhen J, Rousset D, Luce J-M, Moretti J-C, Déroche V, Pétridis P. Delphes. [Rapport sur les travaux de l'École Française d'Athènes.] Bulletin de correspondance hellénique 1992; 116(2): 685–711. [DOI: 10.3406/bch.1992.6888]

[1029] Coy J. The faunal remains from period V. Appendix 2. In: Davis JL, ed. Aya Irini: Period V. Mainz, Germany: Philipp von Zabern, 1986: 109–11.

[1030] Friederich WL. Fossil plants from Weichselian Interstadials, Santorini (Greece) II. In: Doumas CG, ed. Thera and the Aegean World II. Papers and proceedings of the Second International Scientific Congress (Santorini/Greece, August 1978). London, UK: Thera and the Aegean World, 1980: 110–128.

[1031] Meyer AB. Bis wie weit in der historischen Zeit zurück ist der Löwe in Griechenland nachweisbar? Der Zoologische Garten 1903; 44(3): 65–73.

[1032] Guggisberg CAW. Simba, the Life of the Lion. Cape Town, South Africa: Howard Timmins, 1961.

[1033] Bate DMA, Note on a new vole and other remains from the Ghar Dalam Cavern, Malta. Geographical Magazine 1920; 57: 208–211.

[1034] Brunet-Lecomte P, Chaline J, Morphological evolution and phylogenetic relationships of the European round voles (Arvicolidae, Rodentia). Lethaia 1991; 24: 45–53.

[1035] Burgio E, Kotsakis T. Presenza di *Pytimis* (Mammalia, Arvicolidae) nel Pleistocene della Sicilia. Il Naturalista siciliano 1986; 10 (104): 27–34.

[1036] Marcolini F, Arca M, Kotsakis T, Tuveri C. The endemic vole, *Tyrrhenicola* (Arvicolidae, Rodentia) from Monte Tuttavista (Sardinia, Italy): new perspectives for phylogeny and biochronology. In: (JA Alcover and P Bover, eds) Proceedings of the Intenational Symposium "Insular Vertebrate Evolution: the Palaeontological Approach". Monogr Soc Hist Nat Balears 2005; 12: 185–192.

[1037] Marcolini FC, Tuveri C, Arca M, Kotsakis T. *Microtus* (*Tyrrhenicola*) *sondaari* n. sp. (Arvicolidae, Rodentia) from Monte Tuttavista (Sardinia, Italy). Hellenic Journal of Geosciences 2006; 41: 73–82.

[1038] Tvrtković N, Dulić B, Mrakovčić M. Distribution of Insectivora and Rodentia on the north-east Adriatic coast (Yugoslavia). Acta Zoologica Fennica 1985; 170: 201–203.

[1039] Vesmanis I, Hutterer R. Nachweise von *Erinaceus, Cocidura* und *Microtus* für die Insel Elba, Italien. Z Saugetierkd 1980; 45(4): 251–3.

[1040] Contoli L, Aloise G, Filippucci MG. Sulla diversificazione trofica di Barbagianni *Tyto alba* e Civetta *Athene noctua* in rapporto al livello diagnostico delle prede. Avocetta 1988; 12: 21–30.

[1041] Agnelli P. I Mammiferi. In: Giubelli G, ed. Isola d'Elba. Geologia, flora, fauna, storia, arte, ambiente. Ferrara, Italy: Pro.Gra.Ms. 1996, 66–73.

[1042] de Marinis AM, Masseti M. First data on winter feeding of pine marten *Martes martes* in the island of Elba (Northern Tyrrhenian Sea, Italy). Vie Milieu 1996; 46(3–4): 382–3.
[1043] Nappi A. I Micromammiferi d'Italia. Naples [Napoli], Italy: Edizioni Simone, 2001.
[1044] Burgio E, Costanza M, Di Patti C, Mannino G. Attuali conoscenze sulle faune a vertebrati dei siti preistorici della Sicilia occidentale. In: Fiore I, Malerba G, Chilardi S, eds. Atti del 3° Convegno Nazionale di Archeozoologia (Siracusa, 3–5 November 2000). Rome [Roma], Italy: Istituto Poligrafico e Zecca dello Stato, 2005: 145–71.
[1045] Cimò V, Di Patti C, Piscopo G. La Fauna della Grotta del Cavallo (Castellammare del Golfo TP): nota preliminare. In: Malerba G, Visentini P, eds. Atti del 4° Convegno Nazionale di Archeozoologia (Pordenone, 13–15 November 2003). Quaderni del Museo Archeologico del Friuli Occidentale 2005; 6: 183–90.
[1046] Casamento A, Sarà M. L'Arvicola terrestris L. in Sicilia. Atti Soc. it. Sci. nat. Museo civ. Stor. nat. Milano 1995; 134 (1): 8–12.
[1047] Dor M. Observations sur les micromammifères trouvés dans les pelotes de la chouette effraie (*Tyto alba*), en Palestine. Mammalia 1947; 11: 49–54.
[1048] Bromley H, Dimentman C, Por FD. Lake Hula. Reconstruction of the fauna and hydrobiology of a lost lake. Jerusalem, Israel: The Israel Academy of Sciences and Humanities, 1992.
[1049] Broodbank C, Strasser TF. Migant farmers and the Neolithic colonization of Crete. Antiquity 1991; 65(247): 233–45.
[1050] Lax E, Strasser TF. Early Holocene extinctions on Crete: the search for the cause. Journal of Mediterranean Archaeology 1992; 5(2): 203–24.
[1051] Evans JD. Knossos Neolithic, Part II: Summary and Conclusion. British School at Athens Annual 1968; 63: 267–76.
[1052] Vigne J-D, Audoin-Rouzeau F. La colonisation de l'Europe occidentale par le rat noire, contraintes méthodologiques, appel à collaborations. Nouvelles de l'archéologie 1992; 47: 42–4.
[1053] Cristaldi M, Amori G, Cagnin M, De Angelis R, Degrassi F, De Sanctis P, Filippucci MG, Ieradi LA, Pacilli AM, Paggi S, Salucci MP, Tommasi M, Van Axel Castelli I, Interrelazioni tra popolamenti a mammiferi e fattori antropici negli ambienti insuliari eoliani. In Giavelli G. and Moroni A. (eds.): L'approccio interdisciplinare elle ricerche sull'arcipelgo eoliano. Ist. Ecologia, Università di Parma 1987: 51–76.
[1054] Foster JB. Evolution of mammals on islands. Nature 1964; 202: 234–5.
[1055] Werner F. Reptilien der Ägäischen Inseln. Sitzungsber Österr Akad Wiss Mathem Naturwiss Kl Abt 1, 1935; 144: 81–117.
[1056] Werner F. Die Amphibien und Reptilien Griechenlands. Zoologica 1938; 35: 1–117.
[1057] Ondrias JC. Liste des Amphibiens et des Reptiles de Grèce. Biol Gallo-Hell 1968; 1: 111–35.
[1058] Townsend C, Begon M, Harper J. Essentials of Ecology. 2nd Edition. Oxford, UK: Blackwell, 2003.
[1059] Hobbs JJ. Bedouin Life in the Egyptian Wilderness. Cairo, Egypt: The American University in Cairo Press, 1989.
[1060] de Marinis AM, Masseti M. Feeding ecology of the European pine marten *Martes martes* L, 1758: a review. Hystrix It J Mamm NS 1995; 7(1–2): 143–50.
[1061] de Marinis AM, Amato F, Masseti M. Pine marten (*Martes martes*) feeding ecology in continental and insular environments of Europe. In: Abstracts of the 4th International *Martes* Symposium "Martes in carnivore communities" (Lisbon, 20–24 July 2004). Lisbon [Lisboa], Portugal: Universidade de Lisboa, 2004: 45.
[1062] de Planhol X. Le paysage animal. L'homme et la grande faune: une zoogéographie historique. Paris, France: Fayard, 2004: 1127.
[1063] Haslam SM, Sell PD, Wolseley PA, eds. A flora of the Maltese islands. Msida, Malta: Malta University Press, 1977.

[1064] Rackham O. The Flora and Vegetation of Thera and Crete Before and After the Great Eruption. In: Doumas C, ed. Thera and the Aegean World I. Papers presented at the 2nd International Scientific Congress (Santorini/Greece, August 1978). London, UK: Thera and the Aegean World, 1978: 755–64.
[1065] Rackham O. Observations on the Historical Ecology of Santorini. In: Hardy DA, Keller J, Galanopoulos VP, Flemming NC, Druitt TH, eds. Thera and the Aegean World III. Volume 2. Earth Sciences. Proceedings of the 3rd International Congress (Santorini/Greece, 3–9 September 1989). London, UK: The Thera Foundation, 1990: 384–91.
[1066] Turner J. The Vegetation of Greece During Prehistoric Times: The Palynological Evidence. In: Doumas C, ed. Thera and the Aegean World I. Papers presented at the 2nd International Scientific Congress (Santorini/Greece, August 1978). London, UK: Thera and the Aegean World, 1978: 765–73.
[1067] Meikle RD. Flora of Cyprus. Volume 1. London/Kew, UK: The Bentham-Moxon Trust; Royal Botanic Gardens, 1977.
[1068] Turland NJ, Chilton L, Press JR. Flora of the Cretan area: annotated checklist and atlas. London, UK: The Natural History Museum, 1993.
[1069] Ntinou M. Vegetation and human communities in prehistoric Greece. In: Saguntum (Papeles del Laboratorio de Arqueología de Valencia). Extra 5. El paisaje en el Neolítico mediterráneo. Valencia, Spain: Universitat de Valencia, 2002: 91–103.
[1070] Masseti M. The Most Ancient Explorations of the Mediterranean. Proceedings of the California Academy of Sciences, Fourth Series, Volume 59, Supplement I, No. 1: 1–18.
[1071] Perlès C. The early Neolithic in Greece. The first farming communities in Europe. Cambridge, UK; New York, NY, USA: Cambridge University Press, 2001.
[1072] Vigne J-D, Buitenhuis H, Davis S. Les premiers pas de la domestication animale à l'Ouest de l'Euphrate: Chypre et l'Anatolie centrale. Paléorient 1999; 25(2): 49–62.
[1073] Willcox G. Présence des céréales dans le Néolithique précéramique de Shillourokombos à Chypre: résultats de campagne 1999. Paléorient 2001; 26(1): 129–35.
[1074] Masseti M. A possible approach to the "conservation" of the mammalian populations of ancient anthropochorous origin of the Mediterranean islands. Folia Zool 2009; 58(3): 303–8.
[1075] Gippoliti S, Amori G. Anthropochorous mammal taxa and conservation lists. Conservation Biology 2002; 16: 1162–4.
[1076] Gippoliti S, Amori G. Mediterranean Island mammals: are they a priority for biodiversity conservation? Biogeographia 2004; 25: 135–44.
[1077] Bar-Gal G, Smith P, Tchernov E, Greenblatt C, Ducos P, Gardeisen A, Horwits KL. Genetic evidence for the origin of the agrimi goat (*Capra aegagrus* cretica). J Zool 2002; 256: 369–77.
[1078] Hajji GM, Charfi-Cheikrouha F, Lorenzini R, Vigne J-D, Hartl GB, Zachos FE. Phylogeography and founder effect of the endangered Corsican red deer (*Cervus elaphus corsicanus*). Biodiversity Conserv 2007; 17: 659–73.
[1079] Coulentianou J. The Misfits of Skyros. The Athenian, May 1981: 18–22.
[1080] Coulentianou J. The Misfits of Skyros. The Athenian, June 1981: 25–29.
[1081] Constantinides P, Kritikos N, eds. The Skyros pony. Athens, Greece: Gema Publications, 2003.
[1082] Summerhays RS, ed. The Book of the Pony. London, UK: The National Horse Association of Great Britain, 1946.
[1083] Hartley Edwards E. Eyewitness Handbooks. Volume 8: Horses. London, UK: Dorling Kindersley, 1993.
[1084] Azzaroli A. Lezioni di Paleontologia dei Vertebrati. Bologna, Italy: Pitagora Editrice, 1990.
[1085] Delaporte R. L'Ile de Chypre. Grenoble/Paris, France: J Baratier/E Larose, 1913.
[1086] Colombi PG. I cavallini di Rodi. Le Vie d'Italia e del Mondo 1936; 6: 122–4.

[1087] Vezzani V. Rodi e il suo problema zootecnico. Milan [Milano], Italy: Natura 1929; 7(1): 1–13.

[1088] Taibel A. Sul cavallo di Rodi. Boll Zool 1930; 1(1): 127–30.

[1089] Menegatos J. The semi-wild Ainos pony. In: Efthymiatou-Katsouni N, ed. Dedication to the National Park of Ainos. Cephalonia and Ithaca, Greece: Museum of Natural History, 1998: 162–6.

[1090] Machin Goodall D. Horses of the world. New York, NY, USA: Macmillan, 1965.

[1091] Samara Pl. The Midilides (in Greek). Mytilene, Lesbos, Greece, 1979.

[1092] Manetti C. L'Anatolia meridionale. Florence [Firenze], Italy: Istituto Agricolo Coloniale Italiano [IACI], 1921.

[1093] Markou-Hecht P. Τα μικρόσωμα Αλογα του δήμου Αρχαγγέλου της νήσου Ρόδου. [The dwarf horses of the Municipality of Arkhangelos on the island of Rhodes.] Arkhangelos/ Rhodes, Greece: Πολιτιστικού Συλλόγου Αρχαγγέλου Ρόδου "Ο Αίθωνας", 1995.

[1094] Giannone M. Marettimo, capre, ponies e tranquillità. Informatore zootecnico 1992; 14: 37–8.

[1095] Della Maria G. Il cavallo sardo nei documenti dell'alto medioevo. L'Isola 1952; 3(7): 15.

[1096] Della Maria G. La Giara e l'"Equus caballus" in Sardegna. Cagliari, Italy: S.T.E.F. Stabilimento Tipografico, Editoriale Fossataro, 1973.

[1097] Camboni G. Archeologia e ambiente della Giara. L'Unione Sarda 1985, 137.

[1098] Bogliani G. Vita e morte nell'ultimo branco. In: Airone [Monthly Magazine]. Milan [Milano], Italy: Editoriale Giorgio Mondadori, 1992; Number 131: 30–45.

[1099] Fadda A. Caratteri faunistici della Giara di Gesturi. Tesi di Laurea. Unpublished manuscript. Calgiari, Italy: Istituto di Zoologia, Università di Cagliari, 1976.

[1100] Demartis B, Mossa L. La Giara di Gesturi. Sassari, Italy: Carlo Delfino Editore, 1991.

[1101] Cetti F. I quadrupedi di Sardegna. Sassari, Italy: G Piattoli, 1774.

[1102] Cornalia E. Fauna d'Italia. Parte prima. Catalogo descrittivo dei mammiferi osservati fino ad ora in Italia. Naples [Napoli]/Palermo, Italy: Vallardi Editore, 1871–72.

[1103] Pais E. Storia della Sardegna e della Corsica durante il dominio romano. Volume 2. Cagliari, Italy: Edizioni Anastatiche Trois, 1923.

[1104] Lilliu G. Sculture della Sardegna nuragica. Cagliari, Italy: Edizioni La Zattera, 1966.

[1105] Barreca F. La Sardegna fenicia e punica. Sassari, Italy: Ed. Chiarella, 1979.

[1106] Delussu F. Lo stato attuale degli studi sulle faune oloceniche della Sardegna centro-settentrionale. In: Atti del 2° Convegno Nazionale di Archeozoologia (Asti, 14–16 November 1997). Forlì, Italy: Abaco Edizioni 2000: 183–92.

[1107] Fedele F. Antropologia e paleoecologia di Tharros. Ricerche sul tofet (1979) e seconda campagna territoriale nel Sinis. Rivista di Studi Fenici 1980; 8(1): 89–98.

[1108] Manconi F. Equidi in Sardegna tra il II sec. A. C. e il VII sec. D. C. In: Atti del 1° Convegno Nazionale di Archeozoologia (Rovigo, 5–7 March 1993). Padusa Quaderni; Collana di Studi Monografici del Centro Polesano di Studi Storici, Archeologici ed Etnografici [CPSSAE] 1995; 1: 319–29.

[1109] de La Boullaye-Le Gouz F. Les voyages et observations du sieur de La Boullaye Le Gouz. Paris, France: Gervais Closier, 1653 and 1657.

[1110] Firouz L. Conservation of a domestic breed. Biol Conserv 1969; 2(1): 53–4.

[1111] Alderson L. Rare breeds. Aylesbury, Bucks, UK: Shire Publications, 1984.

[1112] Littauer MA. The figured evidence for a small pony in the ancient Near East. Iraq 1971; 33: 24–30.

[1113] Littauer MA. Equids at Persepolis. Antiquity 1979; 3: 218–9.

[1114] Marketou T. Excavations at Trianda (Ialysos) on Rhodes: New Evidence for the Late Bronze Age I Period. Rome [Roma], Italy: Atti della Accademia Nazionale dei Lincei; Classe di Scienze Morali, Storiche e Filologiche; Rendiconti Serie 9, 1998; Volume 9: 37–47.

[1115] Marketou T, Facorellis Y, Maniatis Y. New Late Bronze Age chronology for the Ialysos region, Rhodes. Mediterranean Archaeology and Archaeometry 2001; 1(1): 19–29.

[1116] Dickinson OTPK. The Aegean Bronze Age. Cambridge, UK; New York, NY, USA: Cambridge University Press, 1994.
[1117] Plassara K. Τα ελληνικά ζώα χάνονται. Athens, Greece: Ergo Publishing, 2005.
[1118] SAVE eNews. Agri-Culture and Agri-Biodiversity in modern Greece. In: SAVE eNews 4/2010. Quarterly email service of the European SAVE Foundation (Safeguard for Agricultural Varieties in Europe). Constance [Konstanz], Germany: Edition 4/2010. [Electronic version: http://www.save-foundation.net/english/PDF/news/SAVE_eNews_10_4en.pdf (accessed May 5, 2012).]
[1119] Soysal Mİ, Özder M, Özkan E. Preventing the erosion of farm animal genetic biodiversity in Turkey. In: Proceedings of the 3rd International Symposium on Sustainable Agro-Environmental Systems: New Technologies and Applications (Cairo, 26–29 October 2002). Poster. Edirne, Turkey: Department of Animal Science, Faculty of Agriculture, Trakya University, 2004.
[1120] French MH, Johansson I, Joshi NR, McLaughlin EA, eds. Razas europeas de ganado bovino (2 Volumes). Rome [Roma], Italy: Organización des las Naciones Unidas para la agricultura y la alimentación (FAO: Estudios agropecuarios 67), 1969.
[1121] Bradley DG. Genetic hoofprints: the DNA trail leading back to the origins of today's cattle has taken some surprising turns along the way. Natural History Magazine 2003; 112: 35–41.
[1122] Koutelakis CM. Συμβολή στο τοπωνυμικό της Τήλου. Athens, Greece: Στέγη Γραμμάτων και Τεχνών Δωδεκανήσου, 1982: 94–6; also: Athens, Greece: Ιδιωτική έκδοση, 1983.
[1123] Pinna W, Vacca GM, Lai P. Ethno-demographic surveys of the white donkeys in Asinara. Atti della Società Italiana di Scienze Veterinarie 1990; 44, 1763–7.
[1124] Pinna W, Cappio Borlino A, Vacca GM, Lai P. Morphology of adult white donkeys of Asinara. Bollettino della Società italiana di biologia sperimentale 1993; 10: 595–600.
[1125] Cossu A, Monballiu X, Torre A. L'isola dell'Asinara. Sassari, Italy: Carlo Delfino Editore, 1994.
[1126] Georgiou A. Sheep and goat raising in Greece. Athens, Greece: Paper to Meeting of Experts Organised by FAO and EAAP on Production and Utilisation of Sheep and Goat Milk in the Mediterranean Area, 1960.
[1127] Mason IL. The Sheep Breeds of the Mediterranean. Farnham Royal, England, UK: FAO and Commonwealth Agricultural Bureaux, 1967.
[1128] Brooke CH, Ryder ML. Declining breeds of Mediterranean sheep. Rome [Roma], Italy: FAO Animal Production and Health Paper 8, 1978.
[1129] Bökönyi S. History of domestic mammals in Central and Eastern Europe. Budapest, Hungary: Akadémiai Kiadò, 1974.
[1130] Cantù CM. I bufali in Italia. Dall'alto Medioevo ad oggi. Camera di Commercio, Industria, Artigianato e Agricoltura di Varese. Varese, Italy: Edizioni Lativa, 1989.
[1131] Cantalupo P. Il bufalo nella storia e nell'economia del Salernitano. Annali Cilentani Quaderno 1990; 1: 1–77.
[1132] Dobroruka LJ. Mammals. Prague [Praha], Czech: Aventinum Nakladatelstvi, 1990.
[1133] Lombard M. L'Islam dans sa première grandeur (8e–11e siècle). Paris, France: Flammarion, 1971.
[1134] Epstein H. The origin of the domestic animals of Africa. Volume 1. New York, NY, USA: Africana Publishing Corporation, 1971.
[1135] Holder-Egger O, ed. Vita Willibaldi Episcopi Eichstetensis auctore sanctimoniali Heidenheimensi. In: Monumenta Germaniae Historica, Scriptores, 15,1: Supplementa Tomorum I–XII. Pars III. Vitae aliaeque historiae minores, VII: 86–106. Hanover [Hannover], Germany: Verlag Hahnsche Buchhandlung, 1887. [Electronic version of MGH: http://www.mgh.de/dmgh/ (accessed May 5, 2012).]
[1136] Keller O. Die antike Tierwelt. Volume 1. Säugetiere. Leipzig, Germany: W Engelmann, 1909–1913.
[1137] Kazoglou G. Ο ελληνικός βούβαλος. Διηνός 1999; 10: 28–30.
[1138] Paulopoulos B. Greek wetlands. Drosia, Greece: Philopimen Editions, 2008.

[1139] Vlachos M, ed. Τόπος και εικόνα (in Greek). Athens, Greece: Oikos, 1983.
[1140] Cartwright J. Views in the Ionian Islands. London, UK: R Havell and Sons, 1921.
[1141] Desipris G. Corfu Achilleio, the Island of the Phaeacians. Athens, Greece: Editions Toubi's, 1997.
[1142] Paisidou D. Die Herrenhäuser von Korfu. Thessaloniki, Greece: D Altintzis, 1997.
[1143] Vratzalis A. Νιοχωρίτικα (in Greek). Rhodes, Greece: Ekdosi Typographeiou "Techni", 1990.
[1144] Il Messaggiero di Rodi. Tre bufali a spasso per Rodi. 12 December 1926: 3.
[1145] Deblase A, Martin R. Distributional notes on bats (Chiroptera: Rhinolophidae, Vespertilionidae) from Turkey. Mammalia 1973; 37(4): 598–602.
[1146] Iliopoulou-Georgoudaki I. Taxonomy and geographical distribution of the Chiroptera of Greece (in Greek). Unpublished PhD thesis. Patras, Greece: University of Patras, 1977.
[1147] von Helversen O, Weid R. Die Verbreitung einiger Fledermausarten in Griechenland. Bonn Zool Beitr 1990; 41(1): 9–22.
[1148] Fraguedakis-Tsolis SE, Chondropoulos BP. Mammals. In: Dafis S, Papastergiadou E, Georghiou K, Babalonas D, Georgiadis T, Papageorgiou M, Lazaridou E, Tsiaoussi V, eds. Directive 92/43/EEC. The Greek Habitat Project Natura 2000: an overview. Thessaloniki, Greece: Greek Biotope/Wetland Centre (EKBY), 1996: 514–50.
[1149] von Helversen O. *Eptesicus bottae* (Mammalia, Chiroptera) auf der Insel Rhodos. Bonn Zool Beitr 1998; 48(2): 113–21.
[1150] Georgiakakis P, Galanaki K, Mylonas M. Review of the distribution of bats (Mammalia: Chiroptera) in Crete, Greece. In: Abstracts of the 10th International Congress on the Zoogeography and Ecology of Greece and Adjacent Regions [ICZEGAR] (Patras, 26–30 June 2006). Hellenic Zoological Society; Dept of Biology, University of Patras; Hellenic Ornithological Society; Hellenic Ministry of Rural Development and Food, 2006: 135.
[1151] Legakis A, Maragou P, eds. The Red Data Book of Endangered Animals of Greece (in Greek with English summary). Athens, Greece: Ministry of the Environment; Hellenic Zoological Society, 2009.
[1152] van Laar V, Daan S. On some Chiroptera from Greece. Beaufortia (Series of Miscellaneous Publications. Zoological Museum of the University of Amsterdam), 1964;10: 158–66.
[1153] Kryštufek B. *Rinolophus blasii* (Peters, 1866). In: Mitchell-Jones AJ, Amori G, Bogdanowicz W et al., eds. The Atlas of European Mammals. London, UK: Academic Press, 1999: 90–1.
[1154] Ibañez C. *Rinolophus euryale* (Blasius, 1853). In: Mitchell-Jones AJ, Amori G, Bogdanowicz W et al., eds. The Atlas of European Mammals. London, UK: Academic Press, 1999: 92–3.
[1155] Iliopoulou-Georgoudaki I, Legakis A, Ondrias I, Paraschi L, Weid R. Group II: Chiroptera. In: Karandinos M, Paraschi L, eds. The Red Data Book of Threatened Vertebrates of Greece. Athens, Greece: Hellenic Zoological Society; Hellenic Ornithological Society; Hewlett Packard; WWF, 1992: 303–21.
[1156] Ransome RD. *Rhinolophus ferrumequinum* (Schreber, 1774). In: Mitchell-Jones AJ, Amori G, Bogdanowicz W et al., eds. The Atlas of European Mammals. London, UK: Academic Press, 1999: 94–5.
[1157] Lane AC, Alivizatos H. Additional data on the bat species (Chiroptera: Rhinolophidae and Vespertilionidae) of the island of Thasos, north-eastern Greece. Annales Musei Goulandris 2006; 11: 297–301.
[1158] Schofield HW. *Rhinolophus hipposideros* (Bechstein, 1800). In: Mitchell-Jones AJ, Amori G, Bogdanowicz W et al., eds. The Atlas of European Mammals. London, UK: Academic Press, 1999: 96–7.
[1159] Rodrigues L, Palmeirim JM. *Rhinolophus mehelyi* (Matschie, 1901). In: Mitchell-Jones AJ, Amori G, Bogdanowicz W et al., eds. The Atlas of European Mammals. London, UK: Academic Press, 1999: 98–9.
[1160] Paragamian K, Georgiakakis P, Papadatou E. *Rhinolophus mehelyi* (Matschie, 1901). In: Legakis A, Maragou P, eds. The Red Data Book of Endangered Animals of Greece (in Greek

with English summary). Athens, Greece: Ministry of the Environment; Hellenic Zoological Society, 2009: 396–7.
[1161] Iliopoulou-Georgoudaki JG. Intraspecific and interpopulation morphologic variation in the sharp-eared bat, *Myothis blythii* (Tomes, 1857) (Chiroptera: Vespertilionidae), from Greece. Bonn Zool Beitr 1984; 35(1–3): 15–24.
[1162] Topál G. *Myotis blythii* (Tomes, 1857). In: Mitchell-Jones AJ, Amori G, Bogdanowicz W et al., eds. The Atlas of European Mammals. London, UK: Academic Press, 1999: 102–3.
[1163] Guillén A. *Myotis capaccini* (Bonaparte, 1837). In: Mitchell-Jones AJ, Amori G, Bogdanowicz W et al., eds. The Atlas of European Mammals. London, UK: Academic Press, 1999: 106–7.
[1164] Červený J. *Myotis emarginatus* (E. Geoffroy, 1806). In: Mitchell-Jones AJ, Amori G, Bogdanowicz W et al., eds. The Atlas of European Mammals. London, UK: Academic Press, 1999: 112–3.
[1165] Stutz H-P. *Myotis myotis* (Borkhausen, 1797). In: Mitchell-Jones AJ, Amori G, Bogdanowicz W et al., eds. The Atlas of European Mammals. London, UK: Academic Press, 1999: 114–5.
[1166] Gerell R. *Myotis mystacinus* (Kuhl, 1817). In: Mitchell-Jones AJ, Amori G, Bogdanowicz W et al., eds. The Atlas of European Mammals. London, UK: Academic Press, 1999: 116–7.
[1167] Vernier E, Bogdanowicz W. *Pipistrellus kuhlii* (Kuhl, 1817). In: Mitchell-Jones AJ, Amori G, Bogdanowicz W et al., eds. The Atlas of European Mammals. London, UK: Academic Press, 1999: 98–9.
[1168] Jones G. *Pipistrellus pipistrellus* (Schreber, 1774). In: Mitchell-Jones AJ, Amori G, Bogdanowicz W et al., eds. The Atlas of European Mammals. London, UK: Academic Press, 1999: 126–7.
[1169] von Helversen O. *Eptesicus bottae* (Peter, 1869). In: Mitchell-Jones AJ, Amori G, Bogdanowicz W et al., eds. The Atlas of European Mammals. London, UK: Academic Press, 1999: 138–9.
[1170] Georgiakakis P, Papadatou E, Paragamian K. *Eptesicus bottae* (Peters, 1869). In: Legakis A, Maragou P, eds. The Red Data Book of Endangered Animals of Greece (in Greek with English summary). Athens, Greece: Ministry of the Environment; Hellenic Zoological Society, 2009: 379–80.
[1171] Catto CMC, Hutson AM. *Eptesicus serotinus* (Schreber, 1774). In: Mitchell-Jones AJ, Amori G, Bogdanowicz W et al., eds. The Atlas of European Mammals. London, UK: Academic Press, 1999: 142–3.
[1172] Urbańczyk Z. *Barbastella barbastellus* (Schreber, 1774). In: Mitchell-Jones AJ, Amori G, Bogdanowicz W et al., eds. The Atlas of European Mammals. London, UK: Academic Press, 1999: 146–7.
[1173] Masson D. *Pipistrellus savii* (Bonaparte, 1837). In: Mitchell-Jones AJ, Amori G, Bogdanowicz W et al., eds. The Atlas of European Mammals. London, UK: Academic Press, 1999: 128–129.
[1174] Entwistle AC. *Plecotus auritus* (Linnaeus, 1758). In: Mitchell-Jones AJ, Amori G, Bogdanowicz W et al., eds. The Atlas of European Mammals. London, UK: Academic Press, 1999: 148–9.
[1175] Georgiakakis P, Papadatou E, Paragamian K. *Plecotus auritus* (Linnaeus, 1758). In: Legakis A, Maragou P, eds. The Red Data Book of Endangered Animals of Greece (in Greek with English summary). Athens, Greece: Ministry of the Environment; Hellenic Zoological Society, 2009: 401–2.
[1176] Nader IA, Kock D. *Plecotus austriacus* (Fischer, 1829) new to Saudi Arabia, with remarks on taxonomy and zoogeography (Mammalia: Chiroptera: Vespertilionidae). Fauna of Saudi Arabia 1990; 11: 318–22.
[1177] Bogdanowicz W. *Plecotus austriacus* (J. B. Fisher, 1829). In: Mitchell-Jones AJ, Amori G, Bogdanowicz W et al., eds. The Atlas of European Mammals. London, UK: Academic Press, 1999: 150–1.
[1178] Rodrigues L. *Miniopterus schreibersi* (Kuhl, 1817). In: Mitchell-Jones AJ, Amori G, Bogdanowicz W et al., eds. The Atlas of European Mammals. London, UK: Academic Press, 1999: 154–5.

[1179] Hutson AM. *Tadarida teniotis* (Rafinesque, 1814). In: Mitchell-Jones AJ, Amori G, Bogdanowicz W et al., eds. The Atlas of European Mammals. London, UK: Academic Press, 1999: 157–8.

[1180] Masseti M. Holocene mammals of Libya: A biogeographical, historical and archaeozoological approach. Journal of Arid Environments 2010; 74(7): 794–805 [Special Issue "Palaeoenvironment" in Honour of Aharon Horowitz (Ed. by M Weinstein-Evron, L Scott, L Kolska Horwitz)].

[1181] Benda P, Hulva P, Gaisler J. Systematic status of African populations of *Pipistrellus pipistrellus* complex (Chiroptera: Vespertilionidae), with a description of a new species from Cyrenaica, Libya. Acta Chiropterol 2004; 6(2), 193–217.

[1182] Georgiakakis P, Paragamian K. *Pipistrellus hanaki* (Hulva and Benda, 2004). In: Legakis A, Maragou P, eds. The Red Data Book of Endangered Animals of Greece (in Greek with English summary). Athens, Greece: Ministry of the Environment; Hellenic Zoological Society, 2009: 399–400.

[1183] Georgiakakis P, Papadatou E, Paragamian K. *Plecotus macrobullaris* (Kuzjakin, 1965). In: Legakis A, Maragou P, eds. The Red Data Book of Endangered Animals of Greece (in Greek with English summary). Athens, Greece: Ministry of the Environment; Hellenic Zoological Society, 2009: 400–1.

[1184] Spitzenberger F. The genus *Eptesicus* (Mammalia, Chiroptera) in Southern Anatolia. Folia Zool 1994; 43: 437–54.

[1185] Masseti M. Mammals of the Macaronesian islands (the Azores, Madeira, the Canary and Cape Verde islands): redefinition of the ecological equilibrium. Mammalia 2010; 74: 3–34.

[1186] Marchessaux D. A review of the current knowledge of the cetaceans in the Eastern Mediterranean Sea. Vie Marine 1980; 2: 59–66.

[1187] Valoukas V, Legakis A, Ekonomidis P. Group III: Cetacea. In: Karandinos M, Paraschi L, eds. The Red Data Book of Threatened Vertebrates of Greece. Athens, Greece: Hellenic Zoological Society; Hellenic Ornithological Society; Hewlett Packard; WWF, 1992: 322–6.

[1188] Frantzis A, Verriopoulou A, Papaconstantinou K, Panayotis P, Corsini-Foka M. Greek cetacean strandings networks. United Nations Environment Programme [UNEP] Workshop on the monitoring of cetacean strandings in the Mediterranean. In: Mediterranean Action Plan. Meeting of Experts on the implementation of the Action Plans for marine mammals (monk seal and cetaceans) adopted within MAP (Arta/Greece: 28–31 October 1998). Agenda Item 6.

[1189] Carpentieri P, Corsini M, Marini L. Contribution to the knowledge of the presence and distribution of cetaceans in the Aegean sea. Atti della Società Italiana di Scienze Naturali e del Museo Civico di Storia Naturale di Milano 1999; 140(1): 65–75.

[1190] Pirounakis K, Kaloupi S, Moschonas S, Mourelatis Y, Tselentis L, Voutsinas N, Voutsinas V, Panou A. Cetaceans in the Eastern Ionian Sea: results of an observers' network. Contributions to the Zoogeography and Ecology of the Eastern Mediterranean Region 1999; 1: 429–34.

[1191] Frantzis A, Alexiadou P, Politi E, Gannier A, Corsini-Foka M. Cetacean fauna of the Greek Seas: unexpectedly high species diversity. In: Evans PGH, Pitt-Aiken R, Rogen E, eds. Proceedings of the 15th Annual Conference of the European Cetacean Society (Rome, 6–10 May 2001). European Research on Cetaceans 2001; 15: 421–5.

[1192] Corsini-Foka M, Sioulas A. The marine environment of Rhodes. In: Masseti M, ed. Island of deer. Natural history of the fallow deer of Rhodes and of the vertebrates of the Dodecanese (Greece). City of Rhodes: Environment Organization, 2002: 41–50.

[1193] Frantzis A, Alexiadou P, Paximadis G, Politi E, Gannier A, Corsini-Foka M. Current knowledge of the cetacean fauna of the Greek Seas. J Cetacean Res Manage 2003; 5(3): 220–32.

[1194] Dede A, Öztürk B. Cetacean observations in the Marmara and Aegean Sea in spring season 2005. Rapp Comm Int Mer Medit 2007; 38: 455.

[1195] Güçlüsoy H. The first confirmed report of the harbour porpoise (*Phocoena phocoena*) in the Turkish Aegean Sea. JMBA2 Biodiversity Records [published online], 2007: 1–2.

[1196] Güçlüsoy H, Cirik S. Marine mammal strandings on the Turkish Aegean Sea coasts: fishermen reports. Rapp Comm Int Mer Medit 2007; 38: 492.

[1197] Frantzis A. Initiative for the conservation of cetaceans in Greece. Athens, Greece: MOm, Pelagos Cetacean Research Institute, Tethys Research Institute and WWF Greece, 2009.

[1198] Öztürk AA, Tonay AM, Dede A. Strandings of the beaked whales, Risso's dolphins, and a minke whale on the Turkish coast of the Eastern Mediterranean Sea. J Black Sea/Mediterranean Environment 2011; 17(3): 269–74.

[1199] Paximadis G, Frantzis A. *Stenella coeruleoalba* (Meyen, 1833). In: Legakis A, Maragou P, eds. The Red Data Book of Endangered Animals of Greece (in Greek with English summary). Athens, Greece: Ministry of the Environment; Hellenic Zoological Society, 2009: 405–6.

[1200] Uğurluay H, ed. Türkiye'nin Doğa Rehberi [Turkey nature guide]. Istanbul, Turkey: Mart Matasi, 2005.

[1201] Paximadis G, Frantzis A. *Delphinus delphis* (Linnaeus, 1758). In: Legakis A, Maragou P, eds. The Red Data Book of Endangered Animals of Greece (in Greek with English summary). Athens, Greece: Ministry of the Environment; Hellenic Zoological Society, 2009: 376–8.

[1202] Paximadis G, Frantzis A. *Tursiops truncatus* (Montagu, 1821). In: Legakis A, Maragou P, eds. The Red Data Book of Endangered Animals of Greece (in Greek with English summary). Athens, Greece: Ministry of the Environment; Hellenic Zoological Society, 2009: 407–8.

[1203] Paximadis G, Frantzis A. *Grampus griseus* (Cuvier, 1812). In: Legakis A, Maragou P, eds. The Red Data Book of Endangered Animals of Greece (in Greek with English summary). Athens, Greece: Ministry of the Environment; Hellenic Zoological Society, 2009: 392–3.

[1204] Paximadis G, Frantzis A. *Phocoena phocoena* (Linnaeus, 1758). In: Legakis A, Maragou P, eds. The Red Data Book of Endangered Animals of Greece (in Greek with English summary). Athens, Greece: Ministry of the Environment; Hellenic Zoological Society, 2009: 383–5.

[1205] Paximadis G, Frantzis A. *Physeter macrocephalus* (Linnaeus, 1758). In: Legakis A, Maragou P, eds. The Red Data Book of Endangered Animals of Greece (in Greek with English summary). Athens, Greece: Ministry of the Environment; Hellenic Zoological Society, 2009: 385–7.

[1206] Notarbartolo di Sciara G, Birkun Jr A. Conserving Whales, Dolphins and Porpoises in the Mediterranean and Black Seas: an ACCOBAMS Status Report. Monaco, Monaco: ACCOBAMS [Agreement on the Conservation of Cetaceans in the Black Sea, Mediterranean Sea and contiguous Atlantic area], 2010.

[1207] Verriopoulou A, Tounta E, Dendrinos P. First report of a minke whale (*Balaenoptera acutorostrata* Lacépède, 1804) in Hellenic waters. Aquat Mamm 2001; 27(2): 137–9.

[1208] Sheinin A, Kerem D, Goffman O, Spanier E. Rare occurrences of cetaceans along the Israeli Mediterranean coast. FINS, the Newsletter of ACCOBAMS [Agreement on the Conservation of Cetaceans in the Black Sea, Mediterranean Sea and contiguous Atlantic area] 2004; 1(1): 19.

[1209] Notarbartolo di Sciara G. Stranding of a rare beaked whale in Turkey. FINS, the Newsletter of ACCOBAMS [Agreement on the Conservation of Cetaceans in the Black Sea, Mediterranean Sea and contiguous Atlantic area] 2009; 4(2): 15.

[1210] Jefferson TA, Leatherwood S, Webber MA. FAO Species Identification Guide. Marine Mammals of the World. Rome [Roma], Italy: FAO, 1993.

[1211] Papaconstantinou C, ed. Birds of the Aegean. Athens, Greece: Hellenic Ornithological Society, 1996.

[1212] Reijnders PJH, Verriopoulos G, Brasseur SMJM, eds. Status of Pinnipeds relevant to the European Union. IBN Scientific Contributions 8. Wageningen, Netherlands: Institute for Forestry and Nature Research (IBN-DLO), 1997; 8: 12–26.

[1213] Karamanlidis AA, Androukaki E, Adamantopoulou S, Chatzispyrou A, Johnson WM, Kotomatas S, Papadopoulos A, Paravas V, Paximadis G, Pires R, Tounta E, Dendrinos P.

Assessing accidental entanglement as a threat to the Mediterranean monk seal *Monachus monachus*. Endangered Species Research, 2008; 5: 205–13. [DOI: 10.3354/esr00092].
[1214] Simion G, Popescu M. Guide. Musée du Delta du Danube, Tulcea [Muzeul Delta Dunării, Tulcea]. Tulcea, Romania: Section des Sciences Naturelles; Artis publishers, 1964.
[1215] Shirihai H, Jarrett B. Whales, dolphins and seals. A field guide to the marine mammals of the world. London, UK: A and C Black, 2006.
[1216] Zacharias S. The Blue Cave Kastellorizo. EGEO Exótico [Cultural Periodical. The Dodecanese Islands free guide.] 2007; 2: 10–3.
[1217] Dendrinos P, Karamanlidis A, Paravas V. Mediterranean monk seal. In: Legakis A, Maragou P, eds. The Red Data Book of Endangered Animals of Greece (in Greek with English summary). Athens, Greece: Ministry of the Environment; Hellenic Zoological Society, 2009: 366–9.
[1218] Gibbons B. Greece. Traveller's Nature Guides. Oxford, UK: Oxford University Press, 2003.
[1219] Dendrinos P, Adamantopoulou S. The Mediterranean monk seal, Monachus monachus in the Dodecanese. In: Masseti M, ed. Island of deer. Natural history of the fallow deer of Rhodes and of the vertebrates of the Dodecanese (Greece). City of Rhodes: Environment Organization, 2002: 51–4.
[1220] Johnson WM, Lavigne DM. Monk seals in antiquity. The Mediterranean monk seal (*Monachus monachus*) in ancient history and literature. Mededelingen No. 35. Leiden, Netherlands: The Netherlands Commission for International Nature Protection [Nederlandsche Commissie Voor Internationale Natuurbescherming], 1999.
[1222] Rigas G, Ronald K. Observations on the biology and behaviour of the Mediterranean monk seal. Aquat Mamm 1985; 1: 23–6.
[1222] Fureddu A. La foca monaca nel Dodecaneso. Speleologia Sarda 1974; 11: 21–3.
[1223] Tsimenidis N. A 40-day experience with a seal pup (*Monachus monachus*) at Rhodes Aquarium. Working Paper, 15: 145–6. In: Ronald K, Duguy R, eds. The Mediterranean Monk Seal. Proceedings of the 1st International Conference on the Mediteranean Monk Seal (Rhodes, 2–5 May 1978). United Nations Environment Programme [UNEP] Technical Series, 1. New York, NY, USA: Pergamon Press, 1978.
[1224] Vamvakas C, Tsimenidis N, Kainadas H. Contribution to the knowledge of the distribution of the Monk Seal (*Monachus monachus*) in the Greek seas. Conservation plan by the establishment of marine parks. Working Paper, 16: 147–50. In: Ronald K, Duguy R, eds. The Mediterranean Monk Seal. Proceedings of the 1st International Conference on the Mediteranean Monk Seal (Rhodes, 2–5 May 1978). United Nations Environment Programme [UNEP] Technical Series, 1. New York, NY, USA: Pergamon Press, 1978.
[1225] Marchessaux D, Duguy R. Le phoque moine, *Monachus monachus* (Hermann, 1779), en Grèce. Mammalia 1977; 41(4): 419–39.
[1226] Adamantopoulou S, Androukaki E, Kotomatas S. The distribution of the Mediterranean monk seal in Greece based on an information network. Contributions to the Zoogeography and Ecology of the Eastern Mediterranean Region 1999; 1: 399–404.
[1227] Zachariadis G. Burial of a "domestic" seal of 2000 years ago (in Greek). Τα Νέα, 7 February 2002: 28.
[1228] Platon EN. Εμπορικό Λιμάνι της Ρόδου, «Μώλος των Μύλων» (ανασκαφική έκθεση δημοσιευμένη στο ΑρχαιολογικόΔελτίο, 2000; 54, B2: Χρονικά, 991–4).
[1229] Masseti M. Domestic and Anthropochorous Fauna. Human Evolution 2006; 21(2): 95–106 [Special Issue on the Anthropochorous Fauna (Ed. by M Masseti)].
[1230] Davis SJM, Valla FR. Evidence for domestication of the dog 12,000 years ago in the Natufian of Israel. Nature 1978; 276: 608–10.
[1231] Poulianos A.N. The discovery of the first victim of Thera's Bronze Age eruption. Archaeology 1972; 25: 229–230.

Index

Printed and bound by CPI Group (UK) Ltd, Croydon, CR0 4YY
16/09/2026
14963130-0001